DISCRETE SUBGROUPS OF LIE GROUPS

TATA INSTITUTE OF FUNDAMENTAL RESEARCH
STUDIES IN MATHEMATICS

General Editor: K. G. RAMANATHAN

1. M. Hervé: SEVERAL COMPLEX VARIABLES
2. M. F. Atiyah and others: DIFFERENTIAL ANALYSIS
3. B. Malgrange: IDEALS OF DIFFERENTIABLE FUNCTIONS
4. S. S. Abhyankar and others: ALGEBRAIC GEOMETRY
5. D. Mumford: ABELIAN VARIETIES
6. L. Schwartz: RADON MEASURES ON ARBITRARY TOPOLOGICAL SPACES AND CYLINDRICAL MEASURES
7. W. L. Baily, Jr., and others: DISCRETE SUBGROUPS OF LIE GROUPS AND APPLICATIONS TO MODULI

International Colloquium on Discrete Subgroups of Lie Groups and Applications to Moduli, Bombay, 1973.

DISCRETE SUBGROUPS OF LIE GROUPS AND APPLICATIONS TO MODULI

Papers presented at the Bombay Colloquium, 1973, by

BAILY FREITAG GARLAND GRIFFITHS
HARDER IHARA MOSTOW MUMFORD
RAGHUNATHAN SCHMID VINBERG

Published for the
TATA INSTITUTE OF FUNDAMENTAL RESEARCH, BOMBAY
OXFORD UNIVERSITY PRESS
1975

Oxford University Press

OXFORD LONDON GLASGOW NEW YORK
TORONTO MELBOURNE WELLINGTON CAPE TOWN
DELHI BOMBAY CALCUTTA MADRAS KARACHI LAHORE DACCA
KUALA LUMPUR SINGAPORE JAKARTA HONG KONG TOKYO
NAIROBI DAR ES SALAAM LUSAKA ADDIS ABABA
IBADAN ZARIA ACCRA BEIRUT

Printed by V. B. Gharpure at Tata Press Limited, 414 Veer Savarkar Marg, Bombay 400 025, and Published by C. H. Lewis, Oxford University Press, Apollo Bunder, Bombay 400 001

PRINTED IN INDIA

INTERNATIONAL COLLOQUIUM ON DISCRETE SUBGROUPS OF LIE GROUPS AND APPLICATIONS TO MODULI

BOMBAY, 8-15 JANUARY 1973

REPORT

An International Colloquium on 'Discrete Subgroups of Lie Groups and Applications to Moduli' was held at the Tata Institute of Fundamental Research, Bombay, from 8 to 15 January 1973. The purpose of the Colloquium was to discuss recent developments in some aspects of the following topics: (i) Lattices in Lie groups, (ii) Arithmetic groups, automorphic forms and related number-theoretic questions, (iii) Moduli problems and discrete groups. The Colloquium was a closed meeting of experts and of others specially interested in the subject.

The Colloquium was jointly sponsored by the International Mathematical Union and the Tata Institute of Fundamental Research, and was financially supported by them and the Sir Dorabji Tata Trust.

An Organizing Committee consisting of Professors A. Borel. M. S. Narasimhan, M. S. Raghunathan, K. G. Ramanathan and E. Vesentini was in charge of the scientific programme. Professors A. Borel and E. Vesentini acted as representatives of the International Mathematical Union on the Organizing Committee.

The following mathematicians gave invited addresses at the Colloquium: W. L. Baily, Jr., E. Freitag, H. Garland, P. A. Griffiths, G. Harder, Y. Ihara, G. D. Mostow, D. Mumford, M. S. Raghunathan and W. Schmid.

Professor È. B. Vinberg, who was unable to attend the Colloquium, sent in a paper.

The invited lectures were of fifty minutes' duration. These were followed by discussions. In addition to the programme of invited addresses, there were expository and survey lectures and lectures by some invited speakers giving more details of their work.

The social programme during the Colloquium included a Tea Party on 8 January; a Violin recital (Classical Indian Music) on 9 January; a programme of Western Music on 10 January; a performance of Classical Indian Dances (Bharata Natyam) on 12 January; a Film Show (Pather Panchali) on 13 January; and a dinner at the Institute on 14 January.

CONTENTS

DISCRETE SUBGROUPS OF LIE GROUPS

FOURIER COEFFICIENTS OF EISENSTEIN SERIES ON THE ADELE GROUP

By WALTER L. BAILY, JR.

MUCH OF WHAT I wish to present in this lecture will shortly appear elsewhere [3], so for the published part of this presentation I shall confine myself to a restatement of certain definitions and results, concluding with a few remarks on an area that seems to hold some interest. As in [3], I wish to add here also that many of the actual proofs are to be found in the thesis of L. C. Tsao [8].

Let G be a connected, semi-simple, linear algebraic group defined over $\mathbf{Q}$, which, for simplicity, we assume to be $\mathbf{Q}$-simple (by which we mean G has no proper, connected, normal subgroups defined over $\mathbf{Q}$). We assume G to be simply-connected, which implies in particular that $G_{\mathbf{R}}$ is connected ([2], Ch. 7, §5). Assume that $G_{\mathbf{R}}$ has no compact (connected) simple factors and that if K is a maximal compact subgroup of it, then $X = K\backslash G_{\mathbf{R}}$ has a $G_{\mathbf{R}}$-invariant complex structure, *i.e.*, X is Hermitian symmetric. Then [6] strong approximation holds for G. We assume, finally, that $rk_{\mathbf{Q}}(G)$ (the common dimension of all maximal, $\mathbf{Q}$-split tori of G) is > 0 and that the $\mathbf{Q}$-relative root system ${}_{\mathbf{Q}}\Sigma$ of G is of type C (in the Cartan-Killing classification). Then there exists a totally real algebraic number field k and a connected, almost absolutely simple, simply-connected algebraic group G' defined over k such that $G = \mathscr{R}_{k/\mathbf{Q}}G'$; therefore, if G is written as a direct product $\Pi\, G_i$ of almost absolutely simple factors G_i, then each G_i is defined over a totally real algebraic number field, each G_i is simply-connected, each $G_{i\mathbf{R}}$ is connected and the relative root systems ${}_{\mathbf{R}}\Sigma_i = {}_{\mathbf{R}}\Sigma(G_i)$ are of type C [4].

Letting K_i denote a maximal compact subgroup of $G_{i\mathbf{R}}$, the Hermitian symmetric space $X_i = K_i\backslash G_{i\mathbf{R}}$ is isomorphic to a tube domain since ${}_{\mathbf{R}}\Sigma_i$ is of type C [7], hence $X = \Pi_i\, X_i$ is bi-holomorphically equivalent to a tube domain

$$\mathfrak{T} = \{Z = X + iY \in \mathbf{C}^n \mid Y \in \mathfrak{R}\},$$

where $\mathfrak{R}$ is a certain type of open, convex cone in $\mathbf{R}^n$. Let H be the group of linear affine transformations of $\mathfrak{T}$ of the form $Z \longmapsto AZ+B$, where $B \in \mathbf{R}^n$, and A is a linear transformation of $\mathbf{R}^n$ carrying $\mathfrak{R}$ onto itself, and let $\widetilde{H}$ be its complete pre-image in $G_{\mathbf{R}}$ with respect to the natural homomorphism of $G_{\mathbf{R}}$ into Hol($\mathfrak{T}$), the group of bi-holomorphic automorphisms of $\mathfrak{T}$. Then $\widetilde{H} = P_{\mathbf{R}}$, where P is an $\mathbf{R}$-parabolic subgroup of G, and from our assumption that ${}_{\mathbf{Q}}\Sigma$ is of type C, it follows that we may assume P to be defined over $\mathbf{Q}$ (the reasons for which are somewhat technical, but may all be found in [4]).

Assume $G \subset GL\ (V)$, where V is a finite-dimensional, complex vector space with a $\mathbf{Q}$-structure. Let Λ be a lattice in $V_{\mathbf{R}}$, *i.e.*, a discrete subgroup such that $V_{\mathbf{R}}/\Lambda$ is compact, and suppose that $\Lambda \subset V_{\mathbf{Q}}$. Let $\Gamma = \{ \gamma \in G_{\mathbf{Q}} \mid \gamma. \Lambda = \Lambda\}$, and for each finite prime $\mathfrak{p}$, let $\Lambda_{\mathfrak{p}} = \Lambda \otimes_{\mathbf{Z}} \mathbf{Z}_{\mathfrak{p}}$, $K_{\mathfrak{p}} = \{ \gamma \in G_{\mathbf{Q}_{\mathfrak{p}}} \mid \gamma.\ \Lambda_{\mathfrak{p}} = \Lambda_{\mathfrak{p}}\}$. It may be seen, since strong approximation holds for G, that $K_{\mathfrak{p}}$ is the closure $\Gamma_{\mathfrak{p}}$ of Γ in $G_{\mathbf{Q}_{\mathfrak{p}}}$ (in the ordinary $\mathfrak{p}$-adic topology). Now the adele group G_A of G is defined as $\Pi' G_{\mathbf{Q}_{\mathfrak{p}}}$, where Π' denotes restricted direct product with respect to the family $\{K_{\mathfrak{p}}\}$ of compact sub-groups. Define $K_\infty = K$, $K^* = \prod_{\mathfrak{p}<\infty} K_{\mathfrak{p}}$ (Cartesian product).

For all but a finite number of finite $\mathfrak{p}$, we have $G_{\mathbf{Q}_{\mathfrak{p}}} = K_{\mathfrak{p}}. P_{\mathbf{Q}_{\mathfrak{p}}}$, and by changing the lattice Λ at a finite number of places, we may assume [5] that $G_{\mathbf{Q}_{\mathfrak{p}}} = K_{\mathfrak{p}}. P_{\mathbf{Q}_{\mathfrak{p}}}$ for *all* finite $\mathfrak{p}$. In addition, from the Iwasawa decomposition we have $G_{\mathbf{R}} = K_\infty . P^0_{\mathbf{R}}$, where $P^0_{\mathbf{R}}$ denotes the identity component of $P_{\mathbf{R}}$.

We may write the Lie algebra $\mathfrak{g}_{\mathbf{C}}$ of G as the direct sum of $\mathfrak{k}_{\mathbf{C}}$, the complexification of the Lie algebra $\mathfrak{k}$ of K, and of two Abelian subalgebras $\mathfrak{p}^+$ and $\mathfrak{p}^-$, both normalized by $\mathfrak{k}$, such that $\mathfrak{p}^+$ may be identified with $\mathbf{C}^n \supset \mathfrak{T}$. Let $K_{\mathbf{C}}$ be the analytic subgroup of $G_{\mathbf{C}}$ with Lie algebra $\mathfrak{k}_{\mathbf{C}}$ and let $P^{\pm} = \exp(\mathfrak{p}^{\pm})$; then $K_{\mathbf{C}} . P^+$ is a parabolic subgroup of G which we may take to be the same as P, and $P^+ = U$ is its unipotent radical. Now $\mathfrak{p}^+$ has the structure of a

Jordan algebra over $\mathbf{C}$, supplied with a homogeneous norm form $\mathcal{N}$ such that Ad $K_{\mathbf{C}}$ is contained in the similarity group

$$\mathcal{S} = \{g \in GL(n, \mathbf{C}) = GL(\mathfrak{p}^+) \mid \mathcal{N}(gX) = \nu(g)\, \mathcal{N}(X)\}$$

of $\mathcal{N}$, where $\nu : \mathcal{S} \to \mathbf{C}^\times$ is a rational character [7], defined over $\mathbf{Q}$ if we arrange things such that $K_{\mathbf{C}} = L$ is a $\mathbf{Q}$-Levi subgroup of P. (Note that K and $L_{\mathbf{R}}$ are, respectively, compact and non-compact real forms of $K_{\mathbf{C}}$.) Define ν_∞ as the character on $K_{\mathbf{C}}$ given by $\nu_\infty(k) = \nu(Ad_{\mathfrak{p}+}\, k)$. If $\mathbf{p} \leqslant \infty$ let $|\ \ |_{\mathbf{p}}$ be the "standard" $\mathbf{p}$-adic norm, so that the product formula holds. We define (for $\mathbf{p} \leqslant \infty$) $\chi_{\mathbf{p}}$ on $P_{\mathbf{Q}_{\mathbf{p}}}$ by $\chi_{\mathbf{p}}(ku) = |\,\nu(\mathrm{Ad}_{\mathfrak{p}+}k)|_{\mathbf{p}}$, $k \in L_{\mathbf{Q}_{\mathbf{p}}}$, $u \in U_{\mathbf{Q}_{\mathbf{r}}}$ and χ_A on P_A by $\chi_A((p_{\mathbf{p}})) = \Pi_{\mathbf{p}}\, \chi_{\mathbf{p}}(p_{\mathbf{p}})$, which is well defined since for $(p_{\mathbf{p}}) \in P_A$, we have $\chi_{\mathbf{p}}(p_{\mathbf{p}}) = 1$ for all but a finite number of $\mathbf{p}$. Now ν_∞ is bounded on K and ν takes positive real values on $P^0_{\mathbf{R}}$, hence $\nu_\infty(K \cap P^0_{\mathbf{R}}) = \{1\}$. Moreover, $K_{\mathbf{p}}$ is compact and therefore $\chi_{\mathbf{p}}(K_{\mathbf{p}} \cap P_{\mathbf{Q}_{\mathbf{p}}}) = \{1\}$. Now let m be any positive integer. Define $P^0_A = \{(p_{\mathbf{p}}) \in P_A \mid p_\infty = 1\}$, so that $P_A = P_{\mathbf{R}}.P^0_A$, and put $P^*_A = P^0_R P^0_A$. From our previous discussion it is clear that $G_A = K^*.P^*_A$. Define the function φ_m on G_A by

$$\varphi_m(k^*.p_*) = \nu_\infty(k_\infty)^{-m}\, \chi_A(p_*)^{-m},$$

where $p_* \in P^*_A$, $k^* \in K^*$, $k^* = (k_{\mathbf{p}})$. It follows from the preceding that φ_m is well defined.

By the product formula, $\chi_A(p) = 1$ for $p \in P_{\mathbf{Q}}$. Define

$$\tilde{\mathscr{E}}_m(g) = \sum_{\gamma \in G_{\mathbf{Q}}/P_{\mathbf{Q}}} \varphi_m(g\gamma), \quad g \in G_A.$$

By a criterion of Godement, this converges normally on G_A if m is sufficiently large.

Let $g \in G_{\mathbf{R}}$, $Z \in \mathfrak{T} \subset \mathfrak{p}^+$; we may write [4]

$$\exp(Z).\, g = p^-.\, k(Z, g) \exp(Z.g),$$

where $p^- \in P^-$, $k(Z, g) \in K_{\mathbf{C}}$, $Z.g \in \mathfrak{T}$ (this is, in fact, the definition of the operation of $G_{\mathbf{R}}$ on $\mathfrak{T}$ (loc. cit.)); we then define $\nu(Z, g) = \nu_\infty(k(Z, g))$. $\nu(Z, g)$ is a holomorphic function of Z for fixed g. We have $\nu(Z, g_1 g_2) = \nu(Z, g_1)\, \nu(Z.g_1, g_2)$. Letting E denote the (suitably chosen) identity of the Jordan algebra $\mathfrak{p}^+$, we have $iE \in \mathfrak{T}$,

iE is the unique fixed point of K, and it follows from the definitions that if $(iE).\,g = Z$, $g \in G_{\mathbf{R}}$, and if we put

$$\mathscr{E}_m(Z) = \nu(iE, g)^m \,\tilde{\mathscr{E}}_m(g),$$

then $\mathscr{E}_m$ is a holomorphic function on $\mathfrak{T}$ and we have, in fact,

$$\mathscr{E}_m(Z) = \sum_{\gamma \in G_{\mathbf{Q}}/P_{\mathbf{Q}}} \nu(Z, \gamma)^{-m} |\, \nu(p_\gamma) \,| \nu(p_\gamma)|_A^{-m}, \tag{1}$$

where $p_\gamma \in P_A^0$ is such that $\gamma \in K^*.\,p_\gamma$. $P_{\mathbf{R}}^0$, $|\ |_A$ being the adelic norm. Thus $\mathscr{E}_m$ is a holomorphic automorphic form on $\mathfrak{T}$ with respect to Γ.

If we let $\Theta = \Gamma \cap U$, then $\mathscr{E}_m(Z+\theta) = \mathscr{E}_m(Z)$ for all $\theta \in \Theta$, since $\nu(Z, \theta)^m = 1$ and $\mathscr{E}_m$ is an automorphic form with respect to the system of automorphy factors $\{\nu(Z, \gamma)^m\}$; hence, $\mathscr{E}_m$ has a Fourier expansion

$$\mathscr{E}_m(Z) = \sum_{T \in \Theta'} a_m(T)\, \mathscr{E}((T, Z)),$$

where $\epsilon(\ \ ) = e^{2\pi i(\ \ )}$, $(\ ,\ )$ is an inner product on $\mathfrak{p}^+$ such that $(\mathfrak{p}^+{}_{\mathbf{Q}}, \mathfrak{p}^+{}_{\mathbf{Q}}) \subset \mathbf{Q}$, with respect to which the cone $\mathfrak{K}$ is selfdual, and Θ' is the lattice dual to Θ with respect to $(\ ,\ )$. It follows from the "regular" behaviour of $\mathscr{E}_m$ "at infinity" (Koecher's principle) that $a_m(T) = 0$ unless $T \in \Theta' \cap \mathfrak{K}$. Our main concern is with arithmetic problems related to the Fourier coefficients $a_m(T)$.

One can show that the Fourier coefficients of $\mathscr{E}_m$ are all rational numbers. To do this, we first apply the Bruhat decomposition to group the terms in the series (1). We have $G_{\mathbf{Q}} = \bigcup_{0 \leqslant j \leqslant r} P_{\mathbf{Q}} \iota_j P_{\mathbf{Q}}$, where $r = rk_{\mathbf{Q}}(G)$ and ι_j runs over a certain set of double coset representatives of the relative $\mathbf{Q}$-Weyl group of G with respect to the relative $\mathbf{Q}$-Weyl group of L. In particular, ι_r represents the total involution of ${}_{\mathbf{Q}}W(G) = {}_{\mathbf{Q}}W$, and induction shows that it is sufficient, for proving rationality, to consider the series over the biggest cell $\mathfrak{C}^* = P_{\mathbf{Q}} \iota_r P_{\mathbf{Q}}$:

$$\mathscr{E}_m{}^*(Z) = \sum_{\gamma \in \mathfrak{C}^*/P_{\mathbf{Q}}} \nu(Z, \gamma)^{-m} |\, \nu(p_\gamma) \,|_A^{-m}. \tag{2}$$

On the other hand, writing $\iota = \iota_r$, we can see that ι normalizes L, so that $\mathfrak{C}^*/P_{\mathbf{Q}} \cong U_{\mathbf{Q}}\iota$ (as a set of coset representatives). We obtain

$$\mathscr{E}_m^*(Z) = \Sigma_{\gamma \in U_{\mathbf{Q}^\iota}} \nu(Z, \gamma)^{-m} \mid \nu(p_\gamma) \mid_A^{-m}, \tag{3}$$

where $|\ \ |_A$ is the adelic norm. If $u \in U_{\mathbf{Q_p}}$, then $u\iota \in G_{\mathbf{Q_p}}$, so that we may write $u\iota = k.p$, where $k \in K_{\mathbf{p}}$, $p \in P_{\mathbf{Q_p}}$, and we define $\kappa_{\mathbf{p}}(u) = \mid \nu(p) \mid_{\mathbf{p}}^{-1}$. Now U_k, for any field k, is a vector space over k, and in this sense for $u \in U_{\mathbf{Q_p}}$, αu is defined for any $\alpha \in \mathbf{Q_p}$. Crucial for developments is the

PROPOSITION. *If* $\mu \in \mathbf{Z_p}^\times$ *(the group of* **p***-adic units) and* $u \in U_{\mathbf{Q_p}}$, *then* $\kappa_{\mathbf{p}}(\mu u) = \kappa_{\mathbf{p}}(u)$.

This can be proved in most cases, at least, by appealing to classification; however, the smoothest and most general proof [8] appears to depend on the results of Bruhat and Tits [5,9].

One may show, using the Poisson summation formula along the lines of Siegel [10], that each of the Fourier coefficients $a_m^*(T)$ of the series $E_m^*(Z)$ has an "Euler product expansion":

$$a_m^*(T) = S_\infty(T). \prod_{\mathbf{p}<\infty} S_{\mathbf{p}}(T).$$

Here $S_\infty(T)$ is a fractional power of the Jordan algebra norm $N(T)$ of T multiplied by a product of "gamma" factors independent of T, and $S_{\mathbf{p}}(T)$ is given by the formula

$$S_{\mathbf{p}}(T) = \sum_{u \in U_{\mathbf{Q_p}/\Lambda_{\mathbf{p}}}}{}' \epsilon_{\mathbf{p}}((T, u))\, \kappa_{\mathbf{p}}(u)^m,$$

where (,) is the inner product on the Jordan algebra mentioned previously and ϵ_p is a character of absolute value one on the additive group $\mathbf{Q_p}/\mathbf{Z_p}$ whose restriction to $p^{-1}\mathbf{Z_p}/\mathbf{Z_p}$ is non-trivial. By using the proposition, one may see that each factor $S_{\mathbf{p}}(T)$ is a rational number for each finite **p**; and then by specific calculation of $S_{\mathbf{p}}(T)$ for all but a finite number of "bad" **p** and by using known facts about special values of ζ- and L-functions, one may prove that all $a_m{}^*(T)$ are rational numbers. For further details, we refer to [8,3].

We wish to draw attention to the functions $\kappa_{\mathbf{p}}(u)$. It will be noticed that the definition of $\kappa_{\mathbf{p}}$ depends on the choice of local lattice $\Lambda_\mathfrak{p}$; in other words, the collection of functions $\{\kappa_\mathfrak{p}\}$ depends on the choice of arithmetic group Γ, and now we want to ask if there is any natural choice for Γ.

We have $G \subset GL(V)$, G is simply connected, and $G = R_{k/\mathbf{Q}}\, G'$, where k is a totally real algebraic number field and $G' \subset GL(V')$ is almost absolutely simple and simply-connected. And we have assumed conditions that imply for each Galois injection $\sigma : k \to \overline{\mathbf{Q}}$ (the algebraic closure of $\mathbf{Q}$) that $({}^\sigma G')_\mathbf{R}$ is non-compact, thus strong approximation holds for G. Now Λ is a lattice in $V_\mathbf{R}$ contained in $V_\mathbf{Q}$ and Γ is the stabilizer of Λ. Let $\Gamma_\mathfrak{p}$ be the closure of Γ in $G_{\mathbf{Q}_\mathfrak{p}}$ with respect to the $\mathfrak{p}$-adic topology. It can be shown that $\Gamma_\mathfrak{p} = K_\mathfrak{p}$ is "special" maximal compact [5,9] for all but a finite number of finite $\mathfrak{p}$ (H. Hijikata has communicated a proof of this to me), and thus $G_{\mathbf{Q}_\mathfrak{p}} = K_\mathfrak{p} \cdot P_{\mathbf{Q}_\mathfrak{p}}$ at least for all but a finite number of finite $\mathfrak{p}$. Now by "adjusting" the lattice Λ at a finite number of places, we can assume this to be so for all $\mathfrak{p}$. If $\Gamma_\mathfrak{p}$ is special maximal compact in $G_{\mathbf{Q}_\mathfrak{p}}$ for all finite $\mathfrak{p}$, then we call Γ a "special arithmetic subgroup" of $G_\mathbf{R}$. Let Γ' be another special arithmetic subgroup. Then $\Gamma_\mathfrak{p} = \Gamma'_\mathfrak{p}$ (since $\Lambda_\mathfrak{p} = \Lambda'_\mathfrak{p}$) for all but a finite number of $\mathfrak{p}$. If S is the finite set of finite $\mathfrak{p}$ for which this is *not* true, we know, in *some* cases at least, that for each $\mathfrak{p} \in S$ there exists an outer automorphism $\alpha_\mathfrak{p}$ of $G_{\mathbf{Q}_\mathfrak{p}}$ such that $\alpha_\mathfrak{p}(\Gamma_\mathfrak{p}) = \Gamma'_\mathfrak{p}$. In virtue of strong approximation, we can then (in such cases), up to an automorphism of the $\mathbf{Q}_\mathfrak{p}$-relative, extended Dynkin diagram for a finite number of $\mathfrak{p}$, pass from the one special arithmetic group to the other by an inner automorphism of $G_\mathbf{Q}$. This makes it reasonable to wonder whether this is a general phenomenon, and if so, what smallest class of automorphisms would suffice to identify all special arithmetic groups if G is absolutely simple. Moreover, it suggests something "natural" about the arithmetic structure associated to a special arithmetic group. It would seem inviting to look more closely at such groups to see, for example, if or when they are maximal discrete in $G_\mathbf{R}$ or to see if the normalizers of their images in $\mathrm{Hol}(\mathfrak{X})$ are maximal discrete in the latter. This might be worthwhile in connection with results of [1] giving a criterion for the Satake compactification of $\mathfrak{T}/\Gamma$ to be defined over $\mathbf{Q}$. And of course it would be intriguing to know whether in general there is any natural moduli problem attached to such discontinuous groups.

REFERENCES

1. W. L. BAILY, JR.: Eisenstein series on tube domains, "*Problems in Analysis: A symposium in honor of S. Bochner*", Princeton University Press, Princeton, (1970).
2. W. L. BAILY, JR. : *Introductory lectures on automorphic forms*, Iwanami Shoten Publishers, Tokyo, (1973).
3. W. L. BAILY, JR: On the Fourier coefficients of certain Eisenstein series on the adele group, "*Number Theory, in honor of Y. Akizuki*", Kinokuniya, Tokyo, (1973).
4. W. L. BAILY, JR. and A. BOREL: Compactification of arithmetic quotients of bounded symmetric domains, *Annals of Math.*, 84 (1966), 442-528.
5. F. BRUHAT and J. TITS: Four articles, *C. R. Acad. Sci., Paris*, Série A, v. 263 (1966), 598-601, 766-768, 822-825, 867-869.
6. M. KNESER : Strong approximation, *Proc. Symp. Pure Math. IX*, *Amer. Math. Soc.*, Providence, (1966).
7. A. KORÁNYI, and J. WOLF: Realization of Hermitian symmetric spaces as generalized half-planes, *Annals of Math.* 81(1965), 265-288.
8. L. C. TSAO: *On Fourier coefficients of Eisenstein series*, Thesis, University of Chicago, (1972).
9. F. BRUHAT and J. TITS: Groupes réductifs sur un corps local I., *Publ. Math. Inst. des Hautes Etudes Scien. No.* 41 (1972), 5-251.
10. C. L. SIEGEL: Einführung in die Theorie der Modulfuktionen *n*-ten Grades, *Math. Annalen*, 116 (1939), 617-657.

University of Chicago
Chicago, Illinois, U.S.A.

AUTOMORPHY FACTORS OF HILBERT'S MODULAR GROUP

By EBERHARD FREITAG

INTRODUCTION. Let Γ be a group of analytic automorphisms of a domain $D \subset \mathbf{C}^n$. By an automorphy factor of Γ we understand a family of functions $I(z, \gamma)$, $z \in D$, $\gamma \in \Gamma$ holomorphic on D and without zeros, which satisfy the condition

$$I(z, \gamma'\gamma) = I(z, \gamma)\, I(\gamma z, \gamma').$$

The most-occurring factors are the following ones

1) *The trivial factors*

$$I(z, \gamma) = \frac{h(\gamma z)}{h(z)}.$$

Here h is a holomorphic function on D without zeros.

2) *The powers of the complex functional determinant* (Jacobian).

3) *The abelian characters v of* Γ

$$I(z, \gamma) = v(\gamma).$$

The determination of all automorphy factors belonging to a *discontinuous* group is a difficult problem in general. It is roughly equivalent to the calculation of

Pic D/Γ = group of analytic line bundles on D/Γ.

More precisely, if Γ operates without fixed points, we have

$$\text{Pic } D/\Gamma = \frac{\text{group of automorphy factors}}{\text{subgroup of trivial factors}}.$$

There is a well-known isomorphism

$$\text{Pic } D/\Gamma = H^1(D/\Gamma, \mathcal{O}^*)$$

($\mathcal{O}$ = sheaf of automorphic functions,

$\mathcal{O}^*$ = sheaf of invertible automorphic functions).

By means of the exact sequence

$$0 \to \mathbf{Z} \to \mathcal{O} \xrightarrow{\exp} \mathcal{O}^* \to 0,$$

we reduce the original problem to the calculation of

a) the singular cohomology of D/Γ

b) the analytical cohomology $H^{\bullet}(D/\Gamma, \mathcal{O})$.

This program could be carried out almost completely for the domain

$D = H^n = H \times \ldots \times H$, H the usual upper halt-plane.

Matsushima and Shimura succeeded in calculating those groups in case of a compact quotient by means of the Hodge theory [3]. As for the non-compact quotients D/Γ (Hilbert's modular groups) similar complete results have been found.

a) The singular cohomology was investigated by G. Harder [2].

Let us give a very brief indication of the specific problems arising in the non-compact case.

By "cutting off cusps" of D/Γ one gets a manifold with boundary.

There is a natural mapping from the cohomology of the whole space D/Γ to the (well-known) cohomology of the boundary. In the mentioned paper, Harder determined the image and the kernel of this map. His detailed study of this problem leads into the theory of non-analytic modular forms, especially into the theory of Eisenstein series.

b) The analytical cohomology was determined in [1].

To overcome the discrepancy between the standard compactification of D/Γ and a non-singular model, we had to carry out a thorough investigation of the algebraic nature of the cusps. But it was not necessary to get a concrete resolution of the cusps.

1. The main result. In the following let Γ be a group of simultaneously fractional linear substitutions

$$M(z_1, \ldots, z_n) = \left(\frac{a_1 z_1 + b_1}{c_1 z_1 + d_1}, \ldots, \frac{a_n z_n + b_n}{c_n z_n + d_n}\right),$$

of the half space

$$H^n = \{(z_1, \dots, z_n) \in \mathbf{C}^n \,;\, \operatorname{Im} z_\nu > 0 \text{ for } 1 \leqslant \nu \leqslant n\}.$$

We are only interested in the case in which Γ is commensurable with Hilbert's modular group of a totally real number field. We define the complex power

$$a^b = e^{b \log a}, \quad a \neq 0$$

by the principal branch of the logarithm.

THEOREM 1. *In case of $n \geqslant 3$*[†] *the only automorphy factors of Γ are:*

$$I(z, M) = v(M) \prod_{\nu=1}^{n} (c_\nu z_\nu + d_\nu)^{2r_\nu} \frac{h(Mz)}{h(z)}$$

where

a) $r = (r_1, \dots, r_n)$ *is a vector of rational numbers;*

b) $\{v(M)\}_{M \in \Gamma}$ *is a system of complex numbers of absolute value one;*

c) h *is a holomorphic invertible function on H^n.*

This factorisation of I is unique.

By a system of multipliers of weight $r = (r_1, \dots, r_n)$ we understand a family $\{v(M)\}_{M \in \Gamma}$ of complex numbers of absolute value one, such that

$$I(z, M) = v(M) \prod_{\nu=1}^{n} (c_\nu z_\nu + d_\nu)^{2r_\nu}$$

is a factor of automorphy.

AMENDMENT TO THEOREM 1.

1) *The group of abelian characters of Γ is finite.*[‡]

2) *If $r = (r_1, \dots, r_n)$ is the weight of a multiplier system, the components r_ν have to be rational and their denominators are bounded (by a number which may depend on the group).*

† The method used for the proof is valid also in case of $n < 3$. But one has to carry out some separate investigations because in this case the first cohomology groups are not trivial.

‡ A more general result has been proved by Serre [4].

We now discuss an application of the main theorem.

A meromorphic modular form with respect to Γ is a meromorphic function on H^n satisfying the functional equations:

$$f(Mz) = v(M) \prod_{\nu=1}^{n} (c_\nu z_\nu + d_\nu)^{2r_\nu} f(z) \text{ for } M \in \Gamma.$$

We call $r = (r_1, \ldots, r_n)$ the weight and $v(M)$ the multiplier system of f.

We are interested in the zeros and poles of f which we describe by a divisor (f) as usual.

By a divisor we understand a formal and locally finite sum

$$D = \sum_{Y} n_Y Y, \quad n_Y \in \mathbf{Z}$$

the summation being taken over irreducible closed analytic subvarieties of codimension one.

THEOREM 2. *Let $\mathscr{D}$ be a Γ-invariant divisor on H^n, $n \geqslant 3$. There exists a meromorphic modular form f with the property*

$$\mathscr{D} = (f).$$

PROOF. The space H^n is a topologically trivial Stein-space. Therefore we can find a meromorphic function g on H^n with

$$\mathscr{D} = (g).$$

The function

$$I(z, M) = \frac{g(Mz)}{g(z)}, \quad M \in \Gamma$$

is without poles and zeros because $\mathscr{D}$ is Γ-invariant. We therefore can apply Theorem 1. Put

$$f = \frac{g}{h}.$$

2. Sketch of proof. The group Γ operates in a natural way on the multiplicative group $H^0(D, \mathcal{O}^*)$ of holomorphic invertible functions on $D = H^n$.

The automorphy factors are nothing else but the 1-cocycles with regard to the standard complex and the trivial factors $h(\gamma z)/h(z)$ are the 1-coboundaries, i.e.

$$H^1(\Gamma, H^0(D, \mathcal{O}^*)) = \frac{\text{group of automorphy factors}}{\text{subgroup of trivial factors}}$$

Theorem 1 may thus be formulated as follows

THEOREM 3. *The group $H^1(\Gamma, H^0(D, \mathcal{O}^*))$ is finitely generated and of free rank n.*

We now want to pass on a subgroup $\Gamma_0 \subset \Gamma$ of finite index in order to eliminate the elements of finite order. Let $\Gamma_0 \subset \Gamma$ be a normal subgroup of finite index. Putting

$$A = H^0(D, \mathcal{O}^*)$$

we obtain by means of the Hochschild-Serre sequence

$$0 \to H^1(\Gamma/\Gamma_0, \mathbf{C}^*) \to H^1(\Gamma, A) \to H^1(\Gamma_0, A)^{\Gamma/\Gamma_0} \to H^2(\Gamma/\Gamma_0, \mathbf{C}^*).$$

(Since every holomorphic modular function is constant, we have

$$A^{\Gamma_0} = \mathbf{C}^*.)$$

The groups $H^\nu(\Gamma/\Gamma_0, \mathbf{C}^*)$ are finite. This is proved by means of the sequence

$$0 \to \mathbf{Z} \to \mathbf{C} \to \mathbf{C}^* \to 0.$$

In general, the cohomology groups of a finite group which acts trivially on $\mathbf{Z}$, are finite.

We therefore can assume without loss of generality:

The group Γ is a congruence-subgroup of Hilbert's modular group without torsion.

In the case at hand it is easy to be seen

$$H^1(\Gamma, H^0(D, \mathcal{O}^*)) = H^1(D/\Gamma, \mathcal{O}^*),$$

i.e. there is a one-to-one correspondence between the factor classes and the classes of isomorphic analytical line bundles on $X_0 = D/\Gamma$.

We now treat the group

$$\text{Pic } X_0 = H^1(X_0, \mathcal{O}^*), \ X_0 = D/\Gamma$$

by means of the sequence

$$0 \to \mathbf{Z} \to \mathcal{O} \xrightarrow{\exp} \mathcal{O}^* \to 0.$$

Hereby $\mathcal{O}$ is the sheaf of holomorphic functions on X_0. From the long cohomology sequence results

$$H^1(X_0, \mathcal{O}) \to \text{Pic } X_0 \to H^2(X_0, \mathbf{Z}).$$

We thus have to calculate the groups $H^1(X_0, \mathcal{O})$ and $H^2(X_0, \mathbf{Z})$.

THEOREM 4. *The groups $H^\nu(X_0, \mathcal{O})$ vanish for $1 \leqslant \nu \leqslant n-2$.*

PROOF. Let S be the finite set of cusp classes of Γ and

$$X = X_0 \cup S$$

the standard compactification of $X_0 = D/\Gamma$. There is a long exact sequence, which combines the cohomology with supports in S with the usual cohomology of sheafs

$$H^\nu_S(X, \mathcal{O}) \to H^\nu(X, \mathcal{O}) \to H^\nu(X_0, \mathcal{O}) \to H^{\nu+1}_S(X, \mathcal{O}) \to H^{\nu+1}(X, \mathcal{O}).$$

From my paper [1] the result (Theorem 7.1)

$$H^\nu_S(X, \mathcal{O}) \cong H^\nu(X, \mathcal{O}) \quad \text{for } 1 \leqslant \nu < n$$

is taken.

An analysis of the proof shows that this isomorphism is induced by the natural mapping

$$H^\nu_S(X, \mathcal{O}) \to H^\nu(X, \mathcal{O}).$$

In case of $n \geqslant 3$ we now obtain the exact sequence

$$0 \to \text{Pic } X_0 \to H^1(X_0, \mathbf{Z}).$$

Obviously the free rank of Pic X_0 is not smaller than n because the automorphy factors

$$I_\nu(z, M) = (c_\nu z_\nu + d_\nu)^2 \qquad (1 \leqslant \nu \leqslant n)$$

are independent of each other.

Therefore Theorem 3 has been proved if one knows that $H^2(X_0, \mathbf{Z})$ is of free rank n. That means

THEOREM 5. *In case of $n \geqslant 3$ we have*

$$\dim_{\mathbf{C}} H^2(X_0, \mathbf{C}) = n, \quad X_0 = D/\Gamma.$$

PROOF. We derive the calculation of the 2nd Betti number of X_0 from Harder's investigations on the singular cohomology of $X_0 = D/\Gamma$ [2].

This will be explained briefly in the following.

By cutting off cusps we obtain a bounded manifold X^* which is homotopically equivalent to X_0. (The boundary component at the cusp is given by

$$\prod_{\nu=1}^{n} \operatorname{Im} z_\nu = C; \quad C \gg 0.)$$

In the paper quoted above, Harder gives a decomposition of the singular cohomology of X_0

$$\begin{aligned} H^\bullet(X_0, \mathbf{C}) &= H^\bullet(X^*, \mathbf{C}) \\ &= H^\bullet_{\mathrm{inf}}(X^*, \mathbf{C}) \oplus H^\bullet_{\mathrm{univ}}(X^*, \mathbf{C}) \oplus H^\bullet_{\mathrm{cusp}}(X^*, \mathbf{C}). \end{aligned}$$

This decomposition has the following properties:

(1) The canonical mapping

$$\zeta^* : H^\bullet(X^*, \mathbf{C}) \to H^\bullet(\partial X^*, \mathbf{C})$$

defines an isomorphism of $H_{\mathrm{inf}}(X_0, \mathbf{C})$ onto the image of ζ^*.

(2) $H_{\mathrm{univ}}(X_0, \mathbf{C})$ is a subring, generated by the cohomology-classes attached to the universal harmonic forms

$$\frac{dx_\nu \wedge dy_\nu}{y_\nu^2}, \; 1 \leqslant \nu \leqslant n.$$

(3) The cohomology classes in $H^\bullet_{\mathrm{cusp}}(X_0, \mathbf{C})$ can be represented by harmonic cusp-forms (which are rapidly decreasing at infinity).

The image of ζ^* can be represented by means of the theory of Eisenstein-series. One has

$$H^{\nu}_{\mathrm{inf}}(X_0, \mathbf{C}) = 0 \quad \text{for } 1 \leqslant \nu \leqslant n-1.$$

For the subspace of cusp forms $H^{\bullet}_{\mathrm{cusp}}(X_0, \mathbf{C})$ one has a Hodge decomposition

$$H^{r}_{\mathrm{cusp}} = \bigoplus_{p+q=r} H^{p,q}_{\mathrm{cusp}}.$$

This part of the theory coincides with the investigations of Matsushima and Shimura [3] who treated the case of a compact quotient D/Γ. One has

$$H^{\nu}_{\mathrm{cusp}}(X_0, \mathbf{C}) = 0 \text{ for } \nu \neq n \text{ and therefore}$$

$$H^{2}(D/\Gamma, \mathbf{C}) = H^{2}_{\mathrm{univ}}(X_0, \mathbf{C}) \ (n \geqslant 3).$$

A basis of this vector space is represented by the harmonic forms

$$\frac{dx_\nu \wedge dy_\nu}{y_\nu^2} \quad 1 \leqslant \nu \leqslant n.$$

REMARK. The dimension of $H^{p,q}_{\mathrm{cusp}}$ can be calculated explicitly using the methods of [1]. One obtains an expression in terms of Shimizu's rank polynomials.

3. Line bundles on the standard compactification D/Γ. Every Γ-invariant divisor $\mathscr{D}$ on $H^n (n \geqslant 3)$ can be represented by a modular form of a certain weight $r = (r_1, \ldots, r_n)$ according to Theorem 2.

Finally we investigate the problem in which case the weight r satisfies the condition

$$r_1 = \ldots = r_n.$$

Firstly, $\mathscr{D}$ defines a divisor on $X_0 = H^n/\Gamma$ which can be continued to a divisor on the standard compactification $X = D/\Gamma$ due to a well-known theorem of Remmert.

Then, we call $\mathscr{D}$ a Cartier divisor, if the associated divisorial sheaf on X is a line bundle, i.e. for each point $x \in X$ (even if it is a cusp) exists an open neighbourhood U and a meromorphic function $f: U \to \mathbf{C}$ which represents $\mathscr{D}$:

$$(f) = \mathscr{D}/U.$$

At the regular points, this is automatically the case.

THEOREM 6. *In the case of* $n \geqslant 3$ *the following two conditions are equivalent for a* Γ*-invariant divisor:*

i) *The divisor* $\mathscr{D}$ *is defined by a meromorphic modular form of the type*

$$f(Mz) = v(M) \left[\prod_{\nu=1}^{n} (c_\nu z_\nu + d_\nu)^2 \right]^r f(z) \quad \text{for } M \in \Gamma$$

$$(|v(M)| = 1; r \in \mathbf{Q}).$$

ii) *A suitable multiple* $k\mathscr{D}$, $k \in \mathbf{Z}$, *of* $\mathscr{D}$ *is a Cartier divisor.*

PROOF. The condition ii) is only relevant in the case of cusps. One has to observe that in H^n/Γ only a finite number of quotient singularities occur.

Firstly we show i) $\Rightarrow$ ii).

We may assume that r is integer because we can replace f by a suitable power. In that case v has to be an abelian character.

This character has — owing to amendment 2 of Theorem 1 – finite order. Therefore we may assume $v = 1$.

Let ∞ be a cusp of Γ.

The form f is invariant by the affine substitutions of Γ

$$z \to \epsilon z + \alpha$$

and therefore induces a meromorphic function in a neighbourhood of the cusp, which obviously represents the divisor $\mathscr{D}$.

ii) $\Rightarrow$ i):

We may assume that $\mathscr{D}$ is a Cartier divisor. This means for the cusp: There exists a meromorphic function

$$g\colon U_C \to \mathbf{C};\ U_C = \{z \in H^n;\ N(\operatorname{Im} z) > C\}$$

with the properties

a) g is Γ-invariant,

b) $(g) = \mathscr{D}$ in U_C.

The function $h = \frac{f}{g}$ is holomorphic and without zeros. The transformation law

$$h(\epsilon z + \alpha) = v \begin{bmatrix} \epsilon^{1/2} & \epsilon^{-1/2}\alpha \\ 0 & \epsilon^{-1/2} \end{bmatrix} \prod_{\nu=1}^{n} \epsilon_\nu^{r_\nu} h(z)$$

is valid. By means of a variant of Koecher's principle one sees, that the limit value

$$\lim_{N(\operatorname{Im} z) \to \infty} h(z) = C$$

exists and is finite.

The same argument also holds for the function $\frac{1}{h}$ instead of h.

Therefore C has to be different from zero. It follows that

$$v \begin{bmatrix} \epsilon^{1|2} & \epsilon^{-1/2}\alpha \\ 0 & \epsilon^{-1/2} \end{bmatrix} \prod_{\nu=1}^{n} \epsilon_\nu^{r_\nu} = 1$$

and therefore

$$r_1 = \ldots = r_n.$$

Finally we mention an interesting application without supplying the proof.

Conclusion to Theorem 4. *In case of $n \geqslant 3$ we have*

$$\operatorname{Pic} X = \mathbf{Z},\ X = \overline{H^n/\Gamma}\ (\textit{standard-compactification}).$$

We also have got some information about a generator of Pic X. Choose a natural number e, such that

$$\gamma^e = id \qquad \text{for } \gamma \in \Gamma.$$

Then $\mathscr{K}^e$($\mathscr{K}$ = canonical divisor) defines a line-bundle on X.

This bundle generates a subgroup of finite index in Pic X

$$N = [\operatorname{Pic} X : \{\mathscr{K}^\nu, \nu \equiv 0 \bmod e\}].$$

It is possible to calculate the Chern-class of a generator of Pic X

$$c(\mathscr{K}^{e/N}) = \frac{1}{2\pi i}\frac{e}{N}\sum_{\nu=1}^{n}\frac{dz_\nu \wedge d\bar{z}_\nu}{y_\nu^2}.$$

This defines in fact a cohomology class on X.

We now make use of the fact that Chern-classes are always integral, *i.e.*

$$\int_\gamma c(\mathscr{K}^{e/N}) \in \mathbf{Z},$$

where γ is a two-dimensional cycle on X. Such cycles can be constructed by means of certain specializations. The simplest case is

$$z_1 = \ldots = z_n.$$

If one carries out the integration, one obtains conditions for N.

EXAMPLE. If Γ is the full Hilbert-modular group, then

$$\int_{z_1=\ldots=z_n} c(\mathscr{K}^{e/N}) = \frac{en}{3N} \in \mathbf{Z}.$$

REFERENCES

1. E. FREITAG: Lokale und globale Invarianten der Hilbertschen Modulgruppe, *Invent. math.* 17, (1972) 106-134.
2. G. HARDER: On the cohomology of $Sl\,(2, \mathfrak{O})$ (preprint).
3. Y. MATSUSHIMA and G. SHIMURA: On the cohomology groups attached to certain vector-valued differential forms on the product of the upper half planes. *Ann. of Math.* 78, (1963), 417-449.
4. J. P. SERRE: Le problème des groupes de congruence pour Sl_2. *Ann. of Math.*, 92 (1970), 489-527.

ON THE COHOMOLOGY OF DISCRETE SUBGROUPS OF SEMI-SIMPLE LIE GROUPS

By HOWARD GARLAND

1. **Introduction.** We begin with a classical situation. Thus, let M be a compact Riemannian manifold, let $C^q(M)$ denote the space of C^∞ q-forms on M, and let $(\ ,\ )$ denote the positive-definite inner product on $C^q(M)$, coming from the Riemannian metric ds^2. Let $d: C^q(M) \to C^{q+1}(M)$ denote exterior differentiation, let δ denote the adjoint of d (with respect to $(\ ,\ )$) and let Δ denote the Laplacian $\Delta = d\delta + \delta d$. Let $\mathbf{H}^q = \text{kernel}\ (\Delta : C^q(M) \to C^q(M))$ then if $H^q(M, \mathbf{R})$ denotes the qth cohomology group of M with real coefficients, we have the Hodge-deRham Theorem:

$$\mathbf{H}^q \simeq H^q(M, \mathbf{R}). \tag{1.1}$$

At the same time, there is, after lifting to the orthonormal frame bundle, a well-known decomposition of the Laplacian (see [4])

$$\Delta = \Delta^+ + R.$$

where $(\Delta^+\Phi, \Phi)$, $\Phi \in C^q(M)$, is positive semi-definite, and R is defined in terms of the Riemannian curvature. Thus, if $(R\Phi, \Phi)$ is positive-definite, we have that $H^q(M, \mathbf{R}) = 0$, thanks to (1.1). This idea is due to Bochner. A variant of this idea was applied by Calabi and Weil to the study of the cohomology of certain local coefficient systems, for the purpose of proving a rigidity theorem for discrete subgroups of Lie groups (see [11]). Here, I wish to discuss a method for applying Bochner's idea to the study of the cohomology of discrete subgroups of p-adic groups. In particular, I will introduce a notion of "p-adic curvature", which plays a role in cohomology vanishing theorems for discrete subgroups of p-adic groups, analogous to the role played by R in the argument sketched above. The details will appear in [7].

We mention that for rank 2 p-adic groups, the p-adic curvature coincides with a certain incidence matrix used by Feit, Higman, and

Tits to study finite simple groups (thus they proved the following *rigidity theorem*: A finite B-N pair of rank $\geqslant 3$ is of Lie type!). Before giving a more extensive discussion of the p-adic case, we will begin with the case of discrete subgroups of real semi-simple Lie groups. This will motivate the analogy between the real and p-adic cases.

2. The real case. Let G be a real semi-simple, linear Lie group with no compact factors, let $K \subset G$ be a maximal compact subgroup, and $\Gamma \subset G$ a discrete subgroup such that G/Γ is compact. The space $X = K\backslash G$ is topologically a cell. For simplicity, we assume Γ is torsion-free. Then the action of Γ on X is free and proper, and hence $X \to X/\Gamma$ is a covering. Thus we have an isomorphism

$$H^q(\Gamma, \mathbf{R}) \simeq H^q(X/\Gamma, \mathbf{R}),$$

where $H^q(\Gamma, \mathbf{R})$ denotes the qth Eilenberg-MacLane group of Γ with respect to trivial action on $\mathbf{R}$. On the other hand, the space X is a Riemannian symmetric space, and hence X/Γ is a compact, Riemannian, locally symmetric space. Hence one is tempted to apply Bochner's idea, as described earlier, in order to compute the cohomology groups of X/Γ. However, in this case, the curvature form $(R\Phi, \Phi)$ is negative, and hence Bochner's idea does not apply—at least not at first glance. Matsushima was not discouraged by this, and was inspired, in part by the computation of Calabi and Weil, to develop an ingenious modification of the Bochner idea . He then succeeded in calculating $H^q(X/\Gamma, \mathbf{R})$ in a large number of cases (see Matsushima [9] and Nagano-Kaneyuki [10]).

To give a rough description of Matsushima's results, we let X_u denote the compact dual of X. Now the group G acts on X (to the right). We may identify I^q, the space of G-invariant differential q-forms on X, with a space of q-forms on X/Γ, and we do so whenever convenient. In fact, making this identification, I^q is contained in the space of harmonic q-forms on X/Γ. On the other hand I^q is isomorphic (in a natural way) to the space of all harmonic q-forms on X_u, and hence we have a natural map

$$\varphi\colon H^q(X_u, \mathbf{R}) \to H^q(X/\Gamma, \mathbf{R}).$$

Matsushima proved that in a large number of cases the map φ is an isomorphism. Roughly, his idea was the following : he considered the lift of a harmonic form on X/Γ to G/Γ, expanded the lifted form in terms of a basis of right invariant forms, and proved the derivatives with respect to right invariant vector fields, of the resulting coefficients were zero, i.e., he applied the Bochner idea to the right-invariant derivatives of the coefficients.

It is interesting to pursue matters in the real case when Γ is arithmetic but G/Γ is not necessarily compact. First, one should remark that whereas the map φ is injective when G/Γ is compact (this being a consequence of the Hodge theorem), φ is not in general injective when G/Γ is not compact. However, I proved that in a large number of cases φ is surjective. Thus, if $H^2(X_u, \mathbf{R}) = 0$, we obtain a vanishing theorem for $H^2(X/\Gamma, \mathbf{R})$, and this is how I proved the finiteness theorem for K_2 (see [6]). Then Borel proved that in a certain (substantial) range, φ is injective, and thus he obtained the stability theorems (see [1]).

With a further view to developments in the p-adic case, I will conclude this section with some remarks on the proof of surjectivity. The proof of surjectivity breaks into two parts. The *first part* is an adaptation of Matsushima's argument to prove that square integrable harmonic q-forms on X/Γ are in fact in I^q (for those cases where Matsushima's theorem holds). The *second part* is a square integrability criterion: one shows that for given X and for a certain range of q, every cohomology class in $H^q(X/\Gamma, \mathbf{R})$ has a square summable representative (see [8]).

3. The p-adic case. Let k_v be a non-archimedean completion of an algebraic number field or of a function field over a finite field. Let $\mathbf{G}$ be a simply connected algebraic linear group which is defined and simple over k_v (the simplicity assumption causes little trouble, thanks to a spectral sequence argument of Borel). We let $l = \operatorname{rank}_{k_v}\mathbf{G}$, and assume $l > 0$. Let $G = \mathbf{G}_{k_v}$ be the k_v-rational points of $\mathbf{G}$. Then G inherits a topology from k_v. We let $\Gamma \subset G$ be a discrete subgroup and for simplicity we assume Γ is torsion-free. For the time being, we also assume the coset space $\Gamma\backslash G$ is compact.

Bruhat and Tits have associated a certain simplicial complex $\mathfrak{I}$ with G (see [2] and [3]). Without giving an exact description of the Bruhat-Tits complex, let us mention some of its properties :

(i) $\mathfrak{I}$ is contractible.

(ii) G acts (to the left, say) simplicially on $\mathfrak{I}$.

(iii) The action of G on $\mathfrak{I}$ is proper.

(iv) The action of Γ on $\mathfrak{I}$ is free.

(v) $\Gamma\backslash\mathfrak{I}$ is a finite complex.

(vi) $\dim \mathfrak{I} = l$. (3.1)

We remark that (i) is a theorem of Solomon and Tits, that (iv) is a consequence of our assumption that Γ is torsion-free, that (ii), (iii). (v) and (vi) are immediate consequences of the definition of $\mathfrak{I}$. Also, though $\Gamma\backslash\mathfrak{I}$ may not be simplicial, we assume for simplicity that it is. By (i), (iii), and (iv), we have

$$H^q(\Gamma, \mathbf{R}) \simeq H^q(\Gamma\backslash\mathfrak{I}, \mathbf{R}).$$

Serre was the one who observed this isomorphism and of course he also noted the immediate consequence that

$$H^q(\Gamma, \mathbf{R}) = 0, q > l. \tag{3.2}$$

On the other hand, from an entirely different viewpoint, Kazhdan proved that

$$H^1(\Gamma, \mathbf{R}) = 0, \text{ whenever } l \geqslant 2, \tag{3.3}$$

(see [5]).

From (3.2), (3.3) and his computation of the Euler characteristic of Γ, Serre was led to conjecture that

$$H^q(\Gamma, \mathbf{R}) = 0, 0 < q < l.$$

In [7], we have proved

THEOREM 3.4 *Given l, there exists an integer $N(l)$ such that if the cardinality of the residue class field of k_v is at least $N(l)$, if $\mathbf{G}$ is a simply connected algebraic linear group which is defined and simple over k_v,* *and such that* $\operatorname{rank}_{k_v} \mathbf{G} = l$, *if* $G = \mathbf{G}_{k_v}$, *and if* $\Gamma \subset G$ *is a discrete subgroup such that* $\Gamma \backslash G$ *is compact, then* $H^q(\Gamma, \mathbf{R}) = 0$, $0 < q < l$; *i.e., Serre s conjecture holds for* Γ.

REMARK. In Theorem 3.4 we do not assume Γ is torsion-free. In §4 we shall give an indication of the proof. For the present we mention that our proof bears an analogy to the real case, and that in particular, we introduce a notion of p-adic curvature. Our argument then rests on certain estimates for the eigenvalues of this p-adic curvature. It seems likely that our restriction on the cardinality of the residue class field is unnecessary and only results from the fact that our curvature estimates are not sharp.

4. Square summable cohomology and an indication of proofs. We shall axiomatize our proof. We continue to use the notation of §3. In §3 we mentioned a certain complex $\Gamma \backslash \mathfrak{J}$, which for simplicity, we assumed to be simplicial. Here we drop the assumption that $\Gamma \backslash G$ is compact or correspondingly that $\Gamma \backslash \mathfrak{J}$ is a finite complex.

It turns out that we can describe a portion of our arguments in the following general setting: Let $\mathfrak{S}$ be a locally finite simplicial complex of dimension l and assume we are given a strictly positive valued function λ on the simplices of $\mathfrak{S}$ ($\lambda(\sigma) > 0$, for every simplex σ of $\mathfrak{S}$). We shall refer to λ as a Riemannian metric. Furthermore, let $v(\mathfrak{S})$ denote the vertices of $\mathfrak{S}$, and assume we are given a partition

$$v(\mathfrak{S}) = \Lambda_1 \cup \dots \cup \Lambda_{l+1} \tag{4.1}$$

of $v(\mathfrak{S})$ into $l + 1$ mutually disjoint subsets, $\Lambda_1, \dots, \Lambda_{l+1}$, such that no two vertices in the same Λ_i ever span a one-simplex.

Let $C^q(\mathfrak{S})$ denote the real valued, oriented q-cochains on $\mathfrak{S}$, and let

$$d : C^q(\mathfrak{S}) \to C^{q+1}(\mathfrak{S})$$

denote the simplicial coboundary. Let

$$C(\mathfrak{S}) = \oplus_{q \geqslant 0} C^q(\mathfrak{S}) \text{ (direct sum).}$$

Then if $\Phi, \Psi \in C(\mathfrak{S})$ we set

$$(\Phi, \Psi) = \sum_{\substack{\sigma = \text{simplex} \\ \text{of } \mathfrak{S}}} \lambda(\sigma) \Phi(\sigma) \Psi(\sigma),$$

whenever the sum on the right converges absolutely. Since Φ, Ψ are defined on oriented simplices, we have to say just how one defines the right-hand side. For each geometric simplex σ, pick an oriented representative $\hat{\sigma}$. Then for such σ, set $\Phi(\sigma) = \Phi(\hat{\sigma})$ and $\Psi(\sigma) = \Psi(\hat{\sigma})$. Then the products $\Phi(\sigma)\Psi(\sigma)$ do not depend on the choices we made for oriented representatives, and so the right-hand sum is also independent of these choices.

We let $L^q = L^q(\mathfrak{S})$ consist of all $\Phi \in C^q(\mathfrak{S})$ such that $(\Phi, \Phi) < \infty$. Then $L^q(\mathfrak{S})$ is a subspace of $C^q(\mathfrak{S})$, and is a Hilbert space with respect to the inner product $(\,,)$. When $\mathfrak{S}$ is a finite complex, we have of course that $L^q(\mathfrak{S}) = C^q(\mathfrak{S})$ is a finite dimensional vector space.

We now *assume*

$$d(L^q(\mathfrak{S})) \subset L^{q+1}(\mathfrak{S})$$

$$d\colon L^q(\mathfrak{S}) \to L^{q+1}(\mathfrak{S}) \text{ is bounded.} \tag{4.2}$$

Remark. Borel has pointed out that if $\mathfrak{S} = \Gamma\backslash\mathfrak{J}$, then (4.2) holds.

Let δ be the adjoint of d. Thus

$$(d\Phi, \Psi) = (\Phi, \delta\Psi),\ \Phi \in L^q(\mathfrak{S}), \Psi \in L^{q+1}(\mathfrak{S}). \tag{4.3}$$

We set

$$\Delta^+ = \delta d, \qquad \Delta = \delta d + d\delta,$$

and we let

$$\mathbf{H}^q = (\text{kernel } (\Delta)) \cap L^q(\mathfrak{S}).$$

The elements of $\mathbf{H}^q$ are called harmonic (q-) cocycles. From (4.3) we have that $\Phi \in \mathbf{H}^q$ if and only if $d\Phi = \delta\Phi = 0$. It then follows that we have an orthogonal direct sum decomposition

$$L^q(\mathfrak{S}) = \mathbf{H}^q \oplus \overline{\text{im}}\, d \oplus \overline{\text{im}}\, \delta. \tag{4.4}$$

(Here "____" denotes closure). When $\mathfrak{S}$ is finite, (4.4) implies

$$\mathbf{H}^q = H^q(\mathfrak{S}, \mathbf{R}). \tag{4.5}$$

We have obtained (4.4) and (4.5) from our Riemannian metric λ. We now use the partition (4.1) to develop an analogue of Bochner's idea. Thus, for each $\alpha \in \{1, \ldots, l+1\}$, we define

$$\rho_\alpha(\Phi)(\tau) = \begin{cases} \Phi(\tau), & \tau \text{ has a vertex in } \Lambda_\alpha \\ 0, & \text{otherwise,} \end{cases}$$

where of course $\Phi \in C^q(\mathfrak{S})$ and τ is an oriented q-simplex. We note that $P_\alpha(L^q(\mathfrak{S})) \subset L^q(\mathfrak{S})$ and that $\rho_\alpha : L^q(\mathfrak{S}) \to L^q(\mathfrak{S})$ is a projection; i.e., $\rho_\alpha^2 = \rho_\alpha$ and ρ_α is self adjoint. *From now on*, we regard ρ_α as an operator on $L^q(\mathfrak{S})$. We set

$$d_\alpha = \rho_\alpha \circ d \circ \rho_\alpha$$
$$\delta_\alpha = \rho_\alpha \circ \delta \circ \rho_\alpha, \ \alpha = 1, \ldots, l+1.$$

Then d_α and δ_α are adjoints of one another and

$$d_\alpha = d \circ \rho_\alpha, \ \delta_\alpha = \rho_\alpha \circ \delta. \tag{4.6}$$

We set $\Delta_\alpha^+ = \delta_\alpha d_\alpha, d'_\alpha = d - d_\alpha$. Let $\Phi \in \mathbf{H}^q$; then a simple calculation shows that for all α

$$0 = (\delta d\Phi, \Phi) = (\Delta_\alpha^+\Phi, \Phi) - (d'_\alpha \Phi, d'_\alpha\Phi). \tag{4.7}$$

It turns out that when $\mathfrak{S} = \Gamma\backslash\mathfrak{J}$, the following also holds (for $q > 0$)*

$$(d'_\alpha \Phi, d'_\alpha \Phi) = ((1-\rho_\alpha)\Phi, \Phi). \tag{4.8}$$

On the other hand, the operator Δ_α^+ plays the role of a curvature tensor, and has a nice geometric interpretation. Thus we might ask: If Δ_α^+ is an analogue of a tensor field, then what is Δ_α^+ "at a point"; i.e., at a vertex $v \in \Lambda_\alpha$, say?

To answer this question, let Σ_v be the boundary of the star of v. We define a Riemannian metric λ_v on Σ_v as follows: If σ is a $q-1$ simplex ($q > 0$) in Σ_v, we set

$$\lambda_v(\sigma) = \lambda(v \cdot \sigma).$$

where $v \cdot \sigma$ denotes the join of v and σ. We let d_v denote the simplicial coboundary on Σ_v, and we let δ_v denote the adjoint of d_v, relative to the inner product on cochains defined by λ_v. Then Δ_α^+ on $C^q(\mathfrak{S})$, $q > 0$, may be regarded, in a precise manner, as a field of operators assigned to the vertices of Λ_α; namely, to each vertex v of Λ_α, we

*Borel has observed (see [27]) that for any finite simplicial complex $\mathfrak{S}$, that if each Λ_i is a single point (our assumption that $1 + \dim \mathfrak{S} =$ number of sets in the partition is unnecessary) and if $\lambda(\sigma) =$ number of simplices of maximum dimension, with σ as a face, then (4.8) holds.

assign the operator $\delta_v d_v$ acting on $q-1$ cochains of Σ_v. In order to make a more precise (but slightly different) statement, let $(\ ,\ )_v$ denote the inner product on cochains defined by λ_v. Also, if $\Phi \in C^q(\mathfrak{S})$, $q > 0$, and $v \in \Lambda_\alpha$, we let ${}_v\Phi \in C^{q-1}(\Sigma_v)$ be defined by ${}_v\Phi(\sigma) = \Phi(v \cdot \sigma)$. We then have

$$(\Delta_\alpha^+\Phi, \Phi) = \sum_{v \in \Lambda_\alpha} (\delta_v d_v({}_v\Phi), {}_v\Phi)_v. \tag{4.9}$$

Moreover, if $\Phi \in \mathbf{H}^q$, then $\delta\Phi = 0$. Thus the second formula of (4.6) implies $\delta_\alpha \Phi = 0$. This, it turns out, implies that if $v \in \Lambda_\alpha$ and if Σ_v is $q-1$ connected, then the cochain ${}_v\Phi$ is in the positive eigenspace of $\delta_v d_v$! Thus, if κ_v is the minimal positive eigenvalue of $\delta_v d_v$ on $C^{q-1}(\Sigma_v)$, and if

$$\kappa_q^\alpha = \min_{v \in \Lambda_\alpha} \kappa_v, \ \kappa_q = \min_\alpha \kappa_q^\alpha,$$

then (4.9), (4.8) and (4.7), and our assumption that no two vertices of Λ_α span a one-simplex, yield

$$0 > (\kappa_q + 1)(\rho_\alpha\Phi, \rho_\alpha\Phi) - (\Phi, \Phi). \tag{4.10}$$

Summing these inequalities over all α we obtain

THEOREM 4.11. *Let $q > 0$. Then the inequality*

$$\kappa_q > (l - q)(q + 1)^{-1} \tag{4.12}$$

implies $\mathbf{H}^q = 0$. *In particular, if $\mathfrak{S}$ is finite then* (4.12) *implies* $H^q(\mathfrak{S}, \mathbf{R}) = 0$.

REMARK. Thanks to (4.5), the first assertion of the theorem implies the second.

5. Some further comments on the non-compact case. We continue with the notation of §§3 and 4.

DEFINITION. *We will say Δ^+ is W-elliptic in dimension q, in case there exists a $c > 0$ such that*

$$(\Delta^+ \Phi, \Phi) \geq c(\Phi, \Phi),$$

for all $\Phi \in$ (kernel δ) $\cap\, L^q$.

A direct argument then shows:

PROPOSITION. *If Δ^+ is W-elliptic in dimension q, then $d: Lq \rightarrow L^{q+1}$ is closed and $\mathbf{H}_q = \{0\}$.*

But in fact, estimates of the type described in §4 allow us to conclude that for certain $\Gamma\backslash\mathfrak{J}$ and for certain q, it is true that Δ^+ is W-elliptic in dimension q. It follows that for $\Gamma\backslash\mathfrak{J}$ and for certain q, one can prove that every cocycle in $L_q(\Gamma\backslash\mathfrak{J})$ is cohomologous to zero. In particular, consider the case when $\Gamma \subset G$ is discrete and $\Gamma\backslash G$ has finite invariant volume. Then, since for $\Gamma\backslash\mathfrak{J}$ we have now obtained one analogue of part one of the argument described at the end of §2, we are led to ask about part two. That is, when is it true that every cocycle in $C^q(\Gamma\backslash\mathfrak{J})$ is cohomologous to a square-summable cocycle? Since we are assuming $\Gamma\backslash G$ has finite invariant volume, this question is only interesting in the function field case. The feeling then is that when Γ is arithmetic, the answer is always yes. If so, one could obtain the function field analogues of the results described in §2 for arithmetic groups in the number field case, provided, one could circumvent the problem that in the function field case Γ need not have a torsion-free subgroup of finite index.

REFERENCES

1. A. BOREL: Cohomologie réelle stable de groupes S-arithmétiques classiques. *C.R. Acad. sc. Paris* 274 (1972), 1700-1702.

2. F. BRUHAT and J. TITS: Groupes algébriques, simples sur un corps local, *Proc. Conf. Local Fields*, Springer-Verlag, (1967).

3. F. BRUHAT and J. TITS: "Groupes réductifs sur un corps local", *Publ. Math. I. H. E. S.* 41 (1972), 5-252.

4. S. S. CHERN: On a generalization of Kähler geometry, *Algebraic Geometry and Topology*, A Symposium in honor of S. Lefschetz, Edited by R. H. Fox et al, Princeton Univ. Press.

5. C. DELAROCHE and A. KIRILLOV: Sur les relations entre l'espace dual d'un groupe et la structure de ses sousgroupes fermés (d'après D.A. Kajdan), *Sém, Bourbaki*, 1967/68, exposé 343, Benjamin, New York, (1969).

6. H. GARLAND: A finiteness theorem for K_2 of a number field, *Ann. of Math.*, 94 (1971), 534-548.

7. H. GARLAND: p-adic curvature and the cohomology of discrete subgroups of p-adic groups, *Ann. of Math.*, 97(1973), 375-423.

8. H. GARLAND and W. C. HSIANG: A square integrability criterion for the cohomology of arithmetic groups, *Proc. Nat. Acad. Sci. U. S. A.*, 59 (1968), 354-360.

9. Y. MATSUSHIMA: On Betti numbers of compact, locally symmetric Riemannian manifolds, *Osaka Math. J.*, 14(1962), 1-20.

10. T. NAGANO and S. KANEYUKI: "Quadratic forms related to symmetric Riemannian spaces", *Osaka Math. J.*, 14(1962), 241-252.

11. A. WEIL: On discrete subgroups of Lie groups, II, *Ann. of Math.*, 75(1962), 578-602.

12. A. BOREL: Cohomologie de certains groupes discrets et Laplacien p-adique, Sém. Bourbaki, 1973/74 Exposé No. 437.

RECENT DEVELOPMENTS IN HODGE THEORY: A DISCUSSION OF TECHNIQUES AND RESULTS

By PHILLIP GRIFFITHS and WILFRIED SCHMID

INTRODUCTION. In this paper, we shall review several recent developments in *Hodge theory*, as applied to the study of the cohomology of algebraic varieties. In some sense, we are continuing the report [21] of the first author, in which the then current work in Hodge theory was discussed without proof and a number of open problems were raised. Here we shall be concerned primarily with *methods of proof*, i.e. understanding in as transparent terms as possible the techniques utilized in this recent work in Hodge theory. We shall also present some results, due to the second author [41], which have just now been published, and shall bring up to date the status of the problems raised in [21].

One of the recent developments we shall discuss is Deligne's theory of *mixed Hodge structures* ([12], [13], [14]). In this work, Deligne extends classical Hodge theory first to open, smooth varieties [13], then to complete, singular varieties [14], and finally to general varieties, also in [14]. The heuristic reasoning explaining why such a theory should be possible is given in [12].

Deligne's technique is to use *resolution of singularities* [29], in order to be able in each case to write the cohomology of the variety in question as being derived from the cohomology of Kähler manifolds by homological algebra. Typically this process gives the cohomology of the variety as the abutment of a spectral sequence whose E_1 or E_2 term is the cohomology of a smooth projective variety. Thus the E_1 or E_2 term has a *Hodge structure*, and in order for this structure to survive as a Hodge structure on E_∞, inducing the desired mixed Hodge structure on the cohomology of the variety, it is necessary that the spectral sequence degenerates. Following a discussion of the formalism of Hodge structures and mixed Hodge structures in §1, we have in §2 (a), §4, and §5(d) presented several typical degeneration arguments in as direct a manner as we could.

In §4 we construct the mixed Hodge structure on the cohomology of the simplest singular complete varieties, namely those having only *normal crossings* as singularities. Here the main reason for the various degeneration theorems can be clearly isolated. The result in §4 stops far short of proving the existence of a mixed Hodge structure on the cohomology of a general singular variety [14]. However, it is the method by which one most frequently *calculates* this mixed Hodge structure (cf. [10], for instance), once it is known to exist.

In §5, we have reproved the main result in the open case [13] from a more analytic and less homological point of view. Our main idea is, instead of using the customary de Rham complex of C^∞ forms on a compact Kähler manifold, to utilize a larger complex containing L^1-forms with certain precise types of singularities, and where the *Gysin map* can be given on the form level preserving the Hodge filtration. This complex is discussed in §2(b), where it is pointed out that the introduction of singular forms is necessary in order to have such a Gysin map on the form level. Operating inside this complex allows us to see clearly the differentials in the relevant spectral sequence in the open case, and to conclude the degeneracy result from the principle of two types (§§5(d), (e)).

Section 6 is devoted to some applications of Deligne's theory. First in §6(a), we give his "theorem on the fixed part", which is the main tool in Deligne's study of the moduli of Hodge structures. Then, in §6(b), we give a direct proof of an interesting result from [13], concerning meromorphic differential forms on algebraic varieties; and finally we discuss an application of mixed Hodge structures to *intermediate Jacobians* in §6(c).

The second technique which we shall explore in some depth is the use of *hyperbolic complex analysis*, as it applies to variation of Hodge structure. Hyperbolic complex analysis is the study of the influence of *negative curvature* on holomorphic mappings. The classifying spaces for variation of Hodge structure are negatively curved, relative to the holomorphic maps which might arise in algebraic geometry (cf. [11], [25], and §3(a), (b)), and so it is natural to apply the general philosophy in this case.

Following a discussion of the basic *Ahlfors lemma* and its variants in §7(a), we have given Borel's proof of the quasi-unipotence of the *Picard-Lefschetz transformation* in §7(b); this should illustrate in a simple fashion the power of the method.

Perhaps the most penetrating use of the philosophy of hyperbolic complex analysis occurs in the *Nevanlinna theory* [24], which affords a general mechanism for analyzing the singularities of a holomorphic mapping. Following a preliminary result from Nevanlinna theory in §8(a), we have used this technique to give rather simple, geometric proofs of *Borel's extension theorem* [5] in §8(b), and of the *Riemann extension theorem for variation of Hodge structure* [19] in §8(c).

A final recent development we shall discuss is the work by the second author [41] and joint work by him and Clemens [10], concerning the asymptotic behavior of the Hodge structures on the cohomology groups of an algebraic variety as it acquires singularities. In §9(a), we have used the theorem on *regular singular points* (§3(c)), together with the Ahlfors lemma, to give an alternate proof of the first theorem from [41]. This result, the *nilpotent orbit theorem*, reduces the case of a general degeneration of Hodge structure to the study of a special kind of *nilpotent orbit* in a classifying space for variation of Hodge structure. It seems possible to use Nevanlinna theory in place of the theorem on regular singular points to prove the same result, but we have not pursued this here.

The second main theorem from [41], the *SL_2-orbit theorem*, gives a detailed and somewhat technical description of the nilpotent orbits which can come up when a one-parameter family of Hodge structures degenerates. The proof depends heavily on Lie theory. In §9(b), besides stating the theorem, we describe the observations which originally led to the proof, as well as to the statement, of the theorem.

Some applications of these two theorems will be mentioned in §10; we also summarize joint results of Clemens and the second author about the topology of a degenerating family of projective manifolds, which again are partly based on the two theorems.

We conclude with an appendix, reviewing the current status of the problems and conjectures contained in the report [21] of the first author.

1. Basic definitions. (a) *Hodge structures.* Let $H_{\mathbf{R}}$ be a finite dimensional real vector space, containing a lattice $H_{\mathbf{Z}}$, and let $H = H_{\mathbf{R}} \otimes_{\mathbf{R}} \mathbf{C}$ be its complexification.

(1.1) DEFINITION. *'A Hodge structure of weight m' on H consists of a direct sum decomposition*

$$H = \bigoplus_{p+q=m} H^{p,q}, \text{ with } H^{q,p} = \bar{H}^{p,q}$$

(Barring denotes complex conjugation.)

REMARK. The prototypical example is the decomposition according to Hodge type of the m-th complex cohomology group of a compact Kähler manifold. In this case, $m, p, q \geqslant 0$; however, it will be convenient to admit also negative values for m, p, and q. For example, the *Hodge structure of Tate* $T(1)$ is defined by

$$H_{\mathbf{Z}} = \mathbf{Z},\ H_{\mathbf{R}} = \mathbf{R},\ H = \mathbf{C},\ m = -2, \text{ and } H = H^{-1,-1}.$$

For any two Hodge structures H, H', both of weight m, the direct sum $H \oplus H'$ carries an obvious Hodge structure, also of weight m. Similarly, if H and H' have possibly different weights m and m',

$$H \otimes H',\ \mathrm{Hom}\,(H, H'),\ \Lambda^p H,\ H^*$$

inherit Hodge structures of weights $m + m'$, $m' - m$, pm, and $-m$, respectively: $\lambda \in \mathrm{Hom}(H, H')$ has Hodge type (p, q) if $\lambda\,(H^{r,s}) \subset (H')^{p+r,q+s}$ for all r, s; in particular, this definition applies to $H^* = \mathrm{Hom}(H, \mathbf{C})$, with $\mathbf{C}$ carrying the trivial Hodge structure of weight 0; $H \otimes H'$ can be identified with Hom (H^*, H'), and $\otimes^p H$ induces a Hodge structure on its subspace $\Lambda^p H$.

(1.2) DEFINITION. *A linear map* $\varphi: H \to H'$ *between vector spaces with Hodge structures will be called a morphism (of Hodge structures) if it is defined over* $\mathbf{Q}$, *relative to the lattices* $H_{\mathbf{Z}}$, $H'_{\mathbf{Z}}$, *and if* $\varphi\,(H^{p,q}) \subset (H')^{p,q}$, *for all p, q. More generally,* φ *is a morphism of type* (r, r) *if again it is defined over* $\mathbf{Q}$, *and if it has type* (r, r) *when viewed as an element of* Hom (H, H').

As a trivial, but nevertheless important, observation we note that a morphism of type (r, r) must vanish unless the weights m and m' of H and H' satisfy $m' = m + 2r$.

To each Hodge structure $H = \oplus_{p+q=m} H^{p,q}$ of weight m one associates the *Hodge filtration*

$$H \supset \ldots \supset F^{p-1} \supset F^p \supset F^{p+1} \supset \ldots \supset 0 \text{ with } F^p = \oplus_{i \geqslant p} H^{i,m-i}. \quad (1.3)$$

It may be convenient to visualize the definition by means of the picture below:

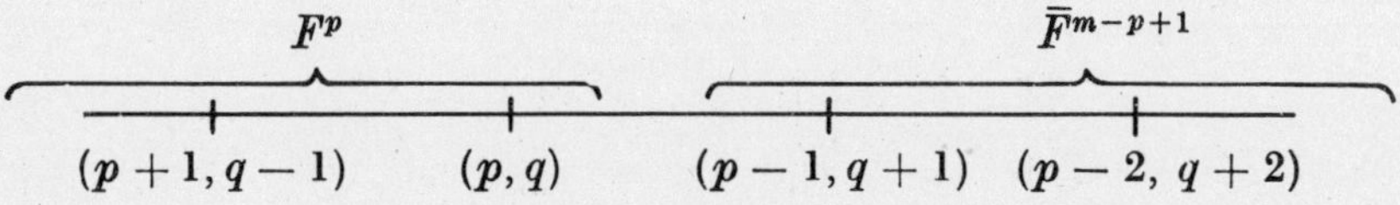

The Hodge filtration determines the Hodge structure completely, since

$$H^{p,q} = F^p \cap \bar{F}^q \quad (1.4)$$

Conversely, a descending filtration $\{F^p\}$ of H arises as the Hodge filtration of some Hodge structure of weight m if and only if

$$F^p \oplus \bar{F}^{m-p+1} \xrightarrow{\approx} H, \quad \text{for all } p. \quad (1.5)$$

Thus one has a 1:1 correspondence between Hodge structures and Hodge filtrations, *i.e.* filtrations satisfying (1.5).

In terms of this latter description, a linear map $\varphi: H \to H'$, which shall be defined over $\mathbf{Q}$, becomes a morphism of type (r, r) exactly when it preserves the Hodge filtration, with a shift by r; in other words, when

$$\varphi(F^p) \subset F'^{p+r}, \quad \text{for all } p. \quad (1.6)$$

Now let φ be a morphism of type (r, r), v a vector in $F'^{p+r} \cap \operatorname{Im} \varphi$. By decomposing a vector in the inverse image of v according to Hodge type, one finds that v lies in the image of F^p. Thus:

> a morphism of Hodge structures of type (r, r) preserves the Hodge filtrations strictly, with a shift by r, in the sense that $\varphi(F^p) = F'^{p+r} \cap \operatorname{Im} \varphi$, for all p. (1.7)

We consider a Hodge structure $H = \bigoplus_{p+q=m} H^{p,q}$ and a bilinear form Q on H, which shall be defined over $\mathbf{Q}$. Also, Q shall be symmetric if m is even, skew if m is odd.

(1.8) DEFINITION. *The Hodge structure is polarized by Q if*

$$Q(H^{p,q}, H^{p',q'}) = 0 \qquad \textit{unless } p = q', q = p',$$

$$(\sqrt{-1})^{p-q}\, Q(v, \bar{v}) > 0 \qquad \textit{for } v \in H^{p,q},\ v \neq 0.$$

Apparently, the polarization form Q must be nondegenerate. The Weil operator $C\colon H \to H$ of the Hodge structure is defined by

$$Cv = (\sqrt{-1})^{p-q}\, v, \qquad \text{for } v \in H^{p,q}. \tag{1.9}$$

In terms of the Hodge filtration and the Weil operator, the two conditions in (1.8) become equivalent to

$$\left.\begin{aligned} Q(F^p, F^{m-p+1}) &= 0 \\ Q(Cv, \bar{v}) &> 0 \quad \text{for } v \neq 0. \end{aligned}\right\} \tag{1.10}$$

The example we have in mind is the Hodge bilinear form on the primitive part of the cohomology of a smooth, projective variety over **C**, as will be discussed below.

It should be mentioned that the operations of tensor product, Hom, exterior product, and duality can also be performed in the context of polarized Hodge structures. For example, if Q and Q' are polarization forms for Hodge structures H and H', then the induced bilinear form on $H \otimes H'$ polarizes the product Hodge structure.

(b) *Mixed Hodge structures*. The symbols H, $H_{\mathbf{R}}$, $H_{\mathbf{Z}}$ shall have the same meaning as in the previous section.

(1.11) DEFINITION. *'A mixed Hodge structure' on H consists of two filtrations,*

$$0 \subset \ldots \subset W_{m-1} \subset W_m \subset W_{m+1} \subset \ldots \subset H,$$

the 'weight filtration' which shall be defined over **Q**, *and*

$$H \supset \ldots \supset F^{p-1} \supset F^p \supset F^{p+1} \supset \ldots \supset 0,$$

the 'Hodge filtration', such that the filtration induced by the latter on $Gr_m(W_*) = W_m/W_{m-1}$ defines a Hodge structure of weight m, for each m (the induced filtration on $Gr_m(W_*)$ is given by

$$F^p(Gr_m(W_*)) = W_m \cap F^p / W_{m-1} \cap F^p).$$

Remark. The notion of a mixed Hodge structure contains that of a Hodge structure of weight m as a special case; as Hodge filtration one takes the Hodge filtration in the old sense, and the weight filtration is defined by $W_m = H$, $W_{m-1} = 0$.

According to the definition of a mixed Hodge structure, only the successive quotients of the weight filtration have direct sum decompositions according to Hodge type. However, the following lemma of Deligne [13] provides a more subtle global decomposition of H. For any pair of integers (p, q), we consider the subspace

$$I^{p,q} = (F^p \cap W_{p+q}) \cap (\bar{F}^q \cap W_{p+q} + \bar{F}^{q-1} \cap W_{p+q-2} + \\ + \bar{F}^{q-2} \cap W_{p+q-3} + \dots).$$

It is certainly not the case that $I^{p,q} = I^{q,p}$, but one does have the congruence $I^{p,q} \equiv I^{q,p} \mod W_{p+q-2}$, as will follow from the proof of lemma (1.12) below. This congruence $I^{p,q} \equiv \overline{I^{q,p}} \mod W_{p+q-2}$ explains why every mixed Hodge structure with a weight filtration of length two splits over **R**, into a sum of two Hodge structures of pure weight. This splitting, of course, may be incompatible with the rational structure. As soon as the weight filtration has length greater than two, a "general" mixed Hodge structure will not split over **R**.

(1.12) **Lemma.** (cf. Lemma 1.2.8 of [13].) *Under the projection* $W_m \to Gr_m(W_*)$, $I^{p,q}$, *with* $p + q = m$, *maps isomorphically onto the Hodge subspace* $Gr_m(W_*)^{p,q}$. *Moreover,*

$$W_m = \oplus_{p+q \leqslant m} I^{p,q},$$

and

$$F^p = \oplus_{i \geqslant p} \oplus_q I^{p,q}.$$

Proof. In view of (1.5), the definition of a mixed Hodge structure amounts to the following:

> given any $v \in W_m$ and integers p, q, with $p + q = m + 1$, one can write $v = v' + \bar{v}'' + u$, such that $v' \in F^p \cap W_m$, $v'' \in F^q \cap W_m$, and $u \in W_{m-1}$; this decomposition is unique modulo W_{m-1}. (*)

In order to prove the first assertion of the lemma, we fix m, p, q, subject to $m = p + q$, and $\alpha \in Gr_m(W_*)^{p,q}$. Then α can be represented by some $v_0 \in F^p \cap W_m$, and also by some $\bar{u}_0 \in \bar{F}^q \cap W_m$. Both are unique upto W_{m-1}, and $v_0 = \bar{u}_0 + w_0$, for some $w_0 \in W_{m-1}$. By induction on k, starting with $k = 0$, we shall find vectors

$$v_k \in F^p \cap W_m, \qquad w_k \in W_{m-1-k}$$

$$u_k \in F^q \cap W_m + F^{q-1} \cap W_{m-2} + F^{q-2} \cap W_{m-3} + \dots + F^{q+1-k} \cap W_{m-k}$$

which will be unique up to W_{m-k}, such that v_k represents α, and $v_k = \bar{u}_k + w_k$. For $k = 0$, this has been done ($F^{q+1} \subset F^q$!). If v_k, u_k, w_k have been picked, we apply (*) to w_k: we write $w_k = w_k' + \bar{w}_k'' + w_{k+1}$, with $w_k' \in F^p \cap W_{m-1-k}$, $w_k'' \in F^{p-k} \cap W_{m-1-k}$, $w_{k+1} \in W_{m-2-k}$, uniquely modulo W_{m-2-k}. The vectors w_{k+1}, $v_{k+1} = v_k - w_k'$, $u_{k+1} = u_k + w_k''$ then have the desired properties. For large enough k, $W_{m-1-k} = 0$; hence α has a unique representative in $I^{p,q}$. We may deduce that

$$W_m = W_{m-1} \oplus (\oplus_{p+q=m} I^{p,q}),$$

and thus $W_m = \oplus_{p+q \leqslant m} I^{p,q}$. As for the last statement of the lemma, the sum of the $I^{p,q}$ is now known to be direct. Also, one containment is obvious. We consider some $v \in F^p$, and we let m be the least integer for which $v \in W_m$. The image of v in $Gr_m(W_*)$ has Hodge components of type $(i, m-i)$, with $i \geqslant p$, because $v \in F^p \cap W_m$. Subtracting off components in the spaces $I^{i,m-i}$, with $i \geqslant p$, we can push v into W_{m-1}. Continuing with descending induction on m, we find that $v \in \oplus_{i \geqslant p} \oplus_q I^{i,q}$, as was to be shown.

A *morphism* between two mixed Hodge structures $\{H, W_m, F^p\}$, $\{H', W_m', F'^p\}$ is a rationally defined linear map $\varphi: H \to H'$, such that $\varphi(W_m) \subset W_m'$ and $\varphi(F^p) \subset F'^p$. More generally, a rationally defined linear map $\varphi: H \to H'$ will be called a morphism of mixed Hodge structures of type (r, r) if $\varphi(W_m) \subset W'_{m+2r}$, $\varphi(F^p) \subset F'^{p+r}$, for all p and m. In this case, the induced mapping

$$\varphi: Gr_m(W_*) \longrightarrow Gr_{m+2r}(W_*')$$

becomes a morphism of type (r, r) relative to the two Hodge structures of weights m and $m + 2r$, respectively.

(1.13) Lemma. *A morphism of type (r, r) between mixed Hodge structures is strict with respect to both the weight and Hodge filtrations, with the appropriate shift in indices. More precisely,* $\varphi(W_m) = W'_{m+2r} \cap \operatorname{Im}\varphi$, $\varphi(F^p) = F'^{p+r} \cap \operatorname{Im}\varphi$.

Proof. The definition of the subspaces $I^{p,q}$ immediately gives the containments $\varphi(I^{p,q}) \subset I'^{p+r,q+r}$. Now let $v \in W'_{m+2r} \cap \operatorname{Im}\varphi$, so that $v = \varphi(u)$ for some $u \in H$. According to (1.12),

$$u = \sum_{p,q} u^{p,q}, \text{ with } u^{p,q} \in I^{p,q}.$$

Then $\varphi(u^{p,q}) \in I'^{p+r,q+r}$, and $v = \sum_{p,q} \varphi(u^{p,q}) \in W'_{m+2r}$. Again appealing to (1.12), we deduce that $\varphi(u^{p,q}) = 0$, unless $p + q \leqslant m$. Hence

$$v = \varphi\left(\sum_{p+q\leqslant m} u^{p,q}\right) \in \varphi(W_m).$$

The case of the Hodge filtration is treated similarly.

(1.14) Lemma. *Let $\varphi: H \to H'$ be a morphism of mixed Hodge structures of type (r, r). Then the induced Hodge and weight filtrations put mixed Hodge structure both on the kernel and the cokernel.*

Proof. As for the kernel, given $v \in \ker\varphi \cap W_m$ and any integer p, we must exhibit vectors

$$v' \in \ker\varphi \cap W_m \cap F^p,\quad v'' \in \ker\varphi \cap W_m \cap F^{m-p+1},\quad w \in \ker\varphi \cap W_{m-1},$$

such that $v = v' + \bar{v}'' + u$, and these must be uniquely determined modulo $\ker\varphi \cap W_{m-1}$. The uniqueness already follows from the corresponding statement about H. Also, there do exist $u' \in W_m \cap F^p$, $u'' \in W_m \cap F^{n-p+1}$, such that $v \equiv u' + \bar{u}'' \bmod W_{m-1}$. Since $\varphi(v) = 0$, we conclude that $\varphi(u'), \varphi(u'') \in W'_{m+2r-1}$. By appealing to (1.12) and decomposing u' into its components in the subspaces $I^{s,t} \subset H$, we can find $u'_1 \in W_{m-1} \cap F^p$, so that $\varphi(u') = \varphi(u'_1)$. Similarly, $\varphi(u'') = \varphi(u''_1)$ for some $u''_1 \in W_{m-1} \cap F^{m-p+1}$. The vectors $v' = u' - u'_1$, $v'' = u'' - u''_1$, $w = v - v' - \bar{v}''$ have the desired properties. In order to prove the assertion about the cokernel, one only has to check one nontrivial fact: if $u \in W'_m \cap F'^p$, $v \in W'_m \cap F'^{m-p+1}$, and if $u + \bar{v} \in W'_{m-1} + \operatorname{Im}\varphi$, then $u, v \in W'_{m-1} + \operatorname{Im}\varphi$. Using (1.12), this can be done, in a manner similar to the argument above. Details are left to the reader.

(1.15) COROLLARY. *Let (H^*, d) be a finite dimensional complex with a mixed Hodge structure, and such that the differential d is a morphism of mixed Hodge structures of type (r, r), for some r. Then the induced filtrations on the cohomology determine a mixed Hodge structure.*

As a final remark, whose verification is left to the reader, we want to add the

(1.16) OBSERVATION: Let $0 \to H' \to H \to H'' \to 0$ be an exact sequence of vector spaces. If two filtrations $\{W_l\}$ and $\{F^p\}$ for H induce mixed Hodge structures on both H' and H'', then they determine a mixed Hodge structure on H itself.

2. Classical Hodge theory. (a) *The cohomology of a Kähler manifold.* Let V be a compact, complex manifold of dimension n, and $A^*(V)$ the de Rham complex of C^∞ forms on v. The decomposition into type

$$A^*(V) = \oplus_{p,q} A^{p,q}(V)$$

reflects the complex structure on V, and *via* de Rham's theorem has implications in the cohomology $H^*(V, \mathbf{C})$. However, not very much is known about this unless V is Kähler, or at least nearly Kähler. In this case, there are two main sources for the many profound implications which the complex structure plus the Kähler metric have in the coholomogy, and we shall briefly discuss these.

Suppose that $ds_V^2 = \Sigma_{i,j}\, g_{ij}\, dz_i\, dz_j$ is a Kähler metric with fundamental (1,1)-form $\omega = \frac{\sqrt{-1}}{2} \Sigma_{i,j}\, g_{ij}\, dz_i \wedge d\bar{z}_j$. The operators

$$L: A^k(V) \to A^{k+2}(V)$$
$$\Lambda: A^k(V) \to A^{k-2}(V)$$

are defined by $L(\varphi) = \omega \wedge \varphi$ and $\Lambda =$ adjoint of $L = \pm *L*$, where $* : A^k(V) \to A^{2n-k}(V)$ is the duality or "star" operator.

Letting

$$P: A^k(V) \to A^k(V)$$

be given by $P(\varphi) = (k - n)\,\varphi$, the commutation relations

$$\left.\begin{aligned} [L, \Lambda] &= P \\ [P, L] &= 2L \\ [P, \Lambda] &= -2\Lambda \end{aligned}\right\} \tag{2.1}$$

exactly say that we have a Lie algebra homomorphism

$$\rho : \mathfrak{sl}(2) \to \mathrm{End}\,(A^*(V)),$$

given by

$$\begin{aligned} \rho(E_+) &= L \\ \rho(E_-) &= \Lambda \\ \rho(H) &= P, \end{aligned}$$

where $E_+ = \begin{pmatrix} 0 & 1 \\ 0 & 0 \end{pmatrix}$, $E_- = \begin{pmatrix} 0 & 0 \\ 1 & 0 \end{pmatrix}$, and $H = \begin{pmatrix} 1 & 0 \\ 0 & -1 \end{pmatrix}$ are the usual basis elements for $\mathfrak{sl}(2)$. The first main source for the structure on $H^*(V)$ arises from the commutation relation

$$[\rho, \Delta] = 0 \tag{2.2}$$

where $\Delta = dd^* + d^*d$ is the Laplacian associated to ds_v^2.[1] Letting $\mathscr{H}^*(V) = \{\varphi \in A^*(V) : \Delta\varphi = 0\}$ be the *harmonic forms*, the *Hodge theorem* [44]

$$\mathscr{H}^*(V) \xrightarrow{\approx} H^*_{DR}(V)$$

together with (2.2) tells us that ρ induces a representation

$$\rho_* : \mathfrak{sl}(2) \to \mathrm{End}\,(H^*(V)) \tag{2.4}$$

on the cohomology level. Applying the standard facts about representations of $\mathfrak{sl}(2)$ to ρ_*, one obtains first the so-called *Hard Lefschetz theorem*

$$L^k : H^{n-k}(V) \xrightarrow{\approx} H^{n+k}(V), \tag{2.4}$$

and secondly the *Lefschetz decomposition*

[1] Here we are adopting the viewpoint of Chern [7] (see also [46]), where the proofs of our statements can be found. Alternate sources are [45] or [47].

$$H^l(V) = \oplus_{0 \leqslant k \leqslant [l/2]} L^k P^{l-2k}(V), \tag{2.5}$$

where

$$P^{n-k}(V) = \ker \{H^{n-k}(V) \xrightarrow{L^{k+1}} H^{n+k+2}(V)\} \tag{2.6}$$

is the *primitive part* of $(n-k)$th cohomology group.

We shall briefly discuss an application of (2.4) and (2.5) to prove degeneration of a spectral sequence; the argument is due to Blanchard and Deligne.

Let X be a Kähler manifold (possibly non-compact), S a complex manifold, and

$$f\colon X \to S$$

a smooth, proper holomorphic mapping.[2] The *Theorem of Leray* [17] gives a spectral sequence $\{E_r\}$ with

$$E_2^{p,q} = H^p(S, R^q_{f_*}(\mathbf{C}))$$
$$E_\infty \Rightarrow H^*(X)$$

where the *direct image sheaf* $R^*_{f_*}(\mathbf{C})$ comes from the presheaf

$$U \to H^*(f^{-1}(U), \mathbf{C}).$$

The theorem asserts that $E_2 = E_\infty$.

To prove this, we remark that the Kähler metric on X induces operators L, Λ on the direct image sheaves $R^*_{f_*}(\mathbf{C})$ which commute with the differentials in the spectral sequence. In particular, the hard Lefschetz Theorem (2.4) and Lefschetz decomposition (2.5) become

$$L^k\colon R^{n-k}_{f_*}(\mathbf{C}) \xrightarrow{\approx} R^{n+k}_{f_*}(\mathbf{C})$$
$$R^l_{f_*}(\mathbf{C}) = \oplus_k L^k P^{l-2k}_{f_*}(\mathbf{C}),$$

where $P^{l-2k}_{f_*} = \ker\{L^{k+1}\colon R^{n-k}_{f_*}(\mathbf{C}) \to R^{n+k+2}_{f_*}\}$. We shall check that $d_2 = 0$, the proof that the higher $d_r = 0$ being the same. Using the Lefschetz decomposition, it will suffice to show that $d_2 = 0$ on $P^{n-k}_{f_*}(\mathbf{C})$. Now in the diagram

[2] $f\colon X \to S$ is a differential fibre bundle whose fibres are compact Kähler manifolds; cf. § 3 for further discussion.

$$\begin{array}{ccc} H^p(S, P^{n-k}_{f_*}(\mathbf{C})) & \xrightarrow{L^{k+1}} & H^p(S, R^{n+k+2}_{f_*}(\mathbf{C})) \\ \downarrow d_2 & & \downarrow d_2 \\ H^{p+2}(S, R^{n-k-1}_{f_*}(\mathbf{C})) & \xhookrightarrow{L^{k+1}} & H^{p+2}(S, R^{n+k+1}_{f_*}(\mathbf{C})), \end{array}$$

the bottom row is injective by Hard Lefschetz and the top row is zero by the definition of primitivity. Thus $d_2 = 0$.

The second main source for the structure on $H^*(V)$ is the relation

$$\Delta_d = 2\Delta_{\bar\partial}\text{[3]} \tag{2.7}$$

between the Laplacians for d and $\bar\partial$. It follows from (2.7) that

$$[\Delta, \pi_{p,q}] = 0 \tag{2.8}$$

where $\pi_{p,q}$: $A^*(V) \to A^{p,q}(V)$ is the projection onto the space of (p, q)-forms. Using (2.8) and the isomorphism

$$\mathscr{H}^*(V) \cong H^*(V, \mathbf{C}),$$

we obtain the *Hodge decomposition*

$$H^m(V, \mathbf{C}) = \bigoplus_{p+q=m} H^{p,q}(V),$$
$$H^{p,q}(V) = \overline{H^{p,q}(V)}$$

where $H^{p,q}(V) = \{\varphi \in A^{p,q}\colon d\varphi = 0\}/\{dA^* \cap A^{p,q}\}$.

In particular, $H^m(V, \mathbf{C})$ has a Hodge structure of weight m. Note that the Lefschetz decomposition is topological, whereas the Hodge decomposition reflects the complex structure (or the *moduli*) of V.

Let us assume for the moment that the Kähler metric ds^2_V is induced by a projective embedding of V. In this case, the Kähler operator L, on the cohomology level, is defined over $\mathbf{Q}$. Since the fundamental form ω has Hodge type $(1, 1)$, L turns out to be a morphism of Hodge structures of type $(1, 1)$. From this, one can deduce that the Hodge structure of $H^m(V, \mathbf{C})$ restricts to a Hodge structure on the subspace $P^m(V, \mathbf{C})$. The Hodge bilinear form

$$Q\colon P^m(V, \mathbf{C}) \times P^m(V, \mathbf{C}) \to \mathbf{C}$$

[3] This identity is *equivalent* to the metric being Kählerian.

is defined by

$$Q([\varphi],[\psi]) = (-1)^{\frac{m(m-1)}{2}} \int_V \omega^{n-m} \wedge \varphi \wedge \psi,$$

if $\varphi, \psi \in A^m(V)$ represent $[\varphi], [\psi] \in P^m(V)$. According to the Hodge-Riemann bilinear relations [45],

$$Q(P^m(V) \cap H^{p,q}(V), \quad P^m(V) \cap H^{p',q'}(V)) = 0$$

unless $p = q'$, $q = p'$, and

$$(\sqrt{-1})^{p-q}\, Q(c, \bar{c}) > 0 \text{ if } c \in P^m(V) \cap H^{p,q}(V),\ c \neq 0.$$

Hence:

the Hodge bilinear form Q polarizes the Hodge structure on the primitive part of the cohomology groups (2.9)

(cf. §1(a)).

There are two applications of (2.7) we want to mention. Define the *Hodge filtration* on the de Rham complex by

$$F^pA^*(V) = \bigoplus_{i \geq p} A^{i,*}(V).$$

(2.10) LEMMA. *The exterior derivative d is strict with respect to the Hodge filtration on $A^*(V)$. In other words, if $\varphi \in F^pA^*(V)$ and $\varphi = d\eta$ for some $\eta \in A^*(V)$, then η can be chosen to lie in $F^pA^*(V)$.*

PROOF. Write $\varphi = \varphi_p + \varphi'$ where $\varphi' \in F^{p+1}A^*(V)$. Then $d\varphi = 0 \Rightarrow \bar{\partial}\varphi_p = 0$, and $\varphi = d\eta \Rightarrow \varphi_p = \partial\eta' + \bar{\partial}\eta''$ for some η', η''. Since $\Delta_\partial = \Delta_{\bar{\partial}}$ by (2.7), the harmonic space for $\bar{\partial}$ is orthogonal to $\partial A^*(V)$, as well as to $\bar{\partial}A^*(V)$. Thus tho $\bar{\partial}$-harmonic part of φ_p is zero, and so $\varphi_p = \bar{\partial}\psi_p$ where $\psi_p \in F^pA^*(V)$. Then $\varphi - d\psi \in F^{p+1}A^*(V)$, and we may continue inductively.

Using the general mechanism of the spectral sequence of a filtered complex, the Hodge filtration on the de Rham complex gives rise to the Hodge – de Rham sepectral squence $\{E_r\}$ with

$$E_1 = H^*_{\bar{\partial}}(A^*(V)),$$

$$E_\infty \Rightarrow H^*_{DR}(V).$$

Lemma 2.10 is equivalent to the degeneration assertion

$$E_1 = E_\infty, \tag{2.11}$$

and implies the *Dolbeault isomorphism*

$$H^{p,q}(V) \simeq H^q(V, \Omega^p). \tag{2.12}$$

It also implies that the filtration on $H^*(V, \mathbf{C})$ induced by the filtration $F^pA^*(V)$ on the C^∞ forms is just the usual Hodge filtration.

The second application of (2.7) which we want to mention is the following

(2.13) LEMMA. *If $\varphi \in A^{p,q}(V)$ is an exact form, then we have both*

$\varphi = \partial\eta'$ *for some* $\eta' \in A^{p-1,q}$, *with* $\bar\partial\eta' = 0$; *and*

$\varphi = \bar\partial\eta''$ *for some* $\eta'' \in A^{p,q-1}$ *with* $\partial\eta'' = 0$.

PROOF. The ∂-cohomology class of φ is zero, and thus $\varphi = \partial\eta'$ where $\eta' = \partial^* G_\partial \varphi$, and G_∂ is the *Green's operator* for ∂ ($G_\partial = \Delta_\partial^{-1}$ on the orthogonal complement of the harmonic space [44]). Now η' has type $(p-1, q)$; and $\bar\partial\eta' = 0$, since $[\partial^*, \bar\partial] = 0 = [G_\partial, \bar\partial]$.

The use of Lemma 2.13 comes up in the *principle of two types*: If $[\varphi] \in H^m(V, \mathbf{C})$ can be represented by $\varphi' \in A^{p',q'}(V)$, and also by $\varphi'' \in A^{p'',q''}(V)$ with $p' \neq p''$, then $[\varphi] = 0$. In practice, we may have a "secondary" cohomological construction which involves writing a cocycle as a coboundary, doing some manipulation, and then arriving at a cohomology class. This class may turn out to be zero, using (2.13) and the principle of two types. It is this heuristic reasoning which underlies the degeneration arguments for the various spectral sequences discussed in §§4, 5 below.

(b) *Some comments about the Gysin mapping.* Let V be a compact Kähler manifold, and $D \subset V$ a smooth divisor. Applying *Poincaré duality* to the homology mapping

$$H_p(D) \xrightarrow{i} H_p(V)$$

induced by the inclusion $D \subset V$, one obtains the *Gysin map*

$$H^q(D) \xrightarrow{\gamma} H^{q+2}(V). \tag{2.14}$$

Since both the Poincaré duality isomorphisms and i are morphisms of Hodge structures (of appropriate types), γ is also a morphism, of type (1,1). We shall give a method for computing γ on the form level; as it turns out, this cannot be done in the complex of C^∞ forms, if one wants to preserve the Hodge filtration. The computation will be useful in §5. In fact, the proof of the degeneration of the spectral sequence used in putting a mixed Hodge structure on the cohomology of an open variety will follow from an obvious extension of our computation of γ on the form level.

(i) Definition of Gysin mapping. Let $[D]$ be the holomorphic line bundle associated to D, $\sigma \in \Gamma(V, \mathcal{O}[D])$ a holomorphic section with $(\sigma) = D$ and $|\sigma|$ the length function with respect to a fibre metric for $[D] \to V$. Define

$$\left.\begin{aligned} \eta &= \frac{1}{2\pi\sqrt{-1}}\,\partial \log |\sigma|^2 \\ \omega = \bar\partial\eta &= \frac{\sqrt{-1}}{2\pi}\,\partial\bar\partial \log |\sigma|^2; \end{aligned}\right\} \tag{2.15}$$

ω is a $C^\infty(1,1)$-form on V, which represents the dual cohomology class $c_1([D])$ (cf. §0 of [24]). If D is locally given by $f = 0$, then

$$\eta = \frac{1}{2\pi\sqrt{-1}}\frac{df}{f} + \theta$$

where θ is a $C^\infty(1,0)$ form.

Definition. *$A^*(\log\langle D\rangle)$ is the sub-complex of the de Rham complex $A^*(V - D)$ generated by $A^*(V)$ and η.*[4]

A form $\varphi \in A^*(\log\langle D\rangle)$ may be (non-uniquely) written as

$$\varphi = \alpha \wedge \eta + \beta, \tag{2.16}$$

where $\alpha, \beta \in A^*(V)$. The restriction $\alpha|_D$ is not ambiguous, however. Hence we may define $R\colon A^*(\log\langle D\rangle) \to A^{*-1}(D)$ by

[4] $A^*(\log\langle D\rangle)$ is a special case of the C^∞ *log complex* associated to a divisor with normal crossings, which is discussed in §5(a).

$$R(\varphi) = \alpha\,|_D,^{(5)} \tag{2.17}$$

and let $W^* \subset A^*(\log \langle D \rangle)$ be the kernel of R. There is an obvious inclusion

$$A^*(V) \overset{i}{\hookrightarrow} W^*,$$

and we shall prove shortly the

(2.18) PROPOSITION. *The inclusion i induces an isomorphism on d and $\bar{\partial}$ cohomology.*

Assuming this, the Gysin map on the form level is given as follows: For $\alpha \in A^{p,q}(D)$, Choose $\tilde{\alpha} \in A^{p,q}(V)$ with $\tilde{\alpha}|_D = \alpha$, and set

$$\gamma(\alpha) = d(\tilde{\alpha} \wedge \eta) = d\tilde{\alpha} \wedge \eta \pm \tilde{\alpha} \wedge \omega. \tag{2.19}$$

If α is a closed form on D, then $\gamma(\alpha)$ is a closed form in W^* and defines a class

$$\gamma(\alpha) \in H^*(W^*) \cong H^*_{DR}(V),$$

using (2.18). We claim that this prescription, up to a factor of ± 1, represents the Gysin map (2.14).

PROOF. Given a closed form α on D and a closed form ψ on V, we must show that

$$\int_V \gamma(\alpha) \wedge \psi = \pm \int_D \alpha \wedge \psi.$$

Let T_ϵ be a solid tube of radius ϵ around D. By (2.19) and Stokes' theorem

$$\int_V \gamma(\alpha) \wedge \psi = -\lim_{\epsilon \to 0} \int_{\partial T_\epsilon} \tilde{\alpha} \wedge \eta \wedge \psi = \pm \int_D \alpha \wedge \psi,$$

since $\lim_{\epsilon \to 0} \dfrac{1}{2\pi} \displaystyle\int_0^{2\pi} f(\epsilon\, e^{i\theta})\, d\theta = f(0)$ for any C^∞ function f.

(ii) COMMENTS. (A) The forms in $A^*(\log \langle D \rangle)$ are *integrable* on V, in the sense that

$$\left| \int_V \varphi \wedge \psi \right| < \infty$$

[5] R is the *Poincaré residue operator* discussed in §5(b).

for $\varphi \in A^*(\log \langle D \rangle)$ and any $\psi \in A^*(V)$, and thus they define *currents* on V(cf. §2 in [18]). Now η satisfies the equation of currents

$$d\eta = \omega - \{D\}$$

where $\{D\}$ is the current defined by integration over D, whereas the forms $\varphi \in W^*$ satisfy

$$\begin{pmatrix} d\varphi \text{ in the} \\ \text{sense of currents} \end{pmatrix} = \begin{pmatrix} d\varphi \text{ in the} \\ \text{sense of forms} \end{pmatrix}.$$

This is basically the reason why (2.18) holds.

(B) The Hodge filtration F^p on $A^*(V)$ extends to a filtration

$$F^p\, W^* = \oplus_{i \geqslant p}\, W^{i,*}$$

on the bigraded complex W^*. Since

$$H^*_{\bar\partial}(A^*(V)) \cong H^*_{\bar\partial}(W^*)$$

is an isomorphism by (2.18), it follows from the discussion in 2(a) that the spectral sequence associated to F^pW^* degenerates at E_1, and that the induced filtration on $H^*(W^*) \simeq H^*(V, \mathbf{C})$ is the usual Hodge filtration. Referring to (2.19), we see that

$$F^pA^*(D) \xrightarrow{\ \gamma\ } F^{p+1}\, W^*,$$

which again shows: *The Gysin mapping* (2.14) *is a morphism of Hodge structures of type* (1,1).

(C) Apropos the comment just made, we can see the necessity for going outside the class of C^∞ forms in order to give γ on the form level. If we think of D as a C^∞ manifold, then the extension $\tilde\alpha$ of α may be taken to be closed in a tubular neighborhood of D. Then $d(\eta \wedge \tilde\alpha) = \omega \wedge \tilde\alpha - \eta \wedge d\tilde\alpha$ is C^∞ on V. However, if α lies in the pth level of the Hodge filtration, then in general we cannot find $\tilde\alpha$ which is closed near D and is also in the pth level; the primary obstruction to doing this is a class in

$$H^*(D, \Omega_D^{p-1}[D])$$

which may not be zero. The complex W^* is probably the smallest one in which γ is defined.

(iii) PROOF OF (2.18). First observe that the definitions of $A^*(\log\langle D\rangle)$ and W^* *localize*: that is to say, there are obviously defined complexes of sheaves $\mathscr{A}^*$, $\mathscr{A}^*(\log\langle D\rangle)$, and $\mathscr{W}^*$ on V such that

$$A^*(V) = \Gamma(V, \mathscr{A}^*)$$
$$A^*(\log\langle D\rangle) = \Gamma(V, \mathscr{A}^*(\log\langle D\rangle))$$
$$W^* = \Gamma(V, \mathscr{W}^*).$$

The usual sheaf-theoretic proof of de Rham's theorem will apply if we can prove the Poincaré lemma:[6]

(2.20) LEMMA. *The sheaf sequences on* V

$$0 \longrightarrow \mathbf{C} \longrightarrow \mathscr{W}^0 \xrightarrow{d} \mathscr{W}^1 \xrightarrow{d} \mathscr{W}^2 \longrightarrow \cdots$$
$$0 \longrightarrow \Omega_V^p \longrightarrow \mathscr{W}^p \xrightarrow{\bar\partial} \mathscr{W}^{p,1} \xrightarrow{\bar\partial} \mathscr{W}^{p,2} \longrightarrow$$

are exact.

PROOF. The problem is local around a point $p \in D$, where we choose holomorphic coordinates $(z, w) = (z, w_1, \ldots, w_{n-1})$ on V such that D is given by $z = 0$. Sections of $\mathscr{W}^*$ may be written as (cf. (2.16))

$$\varphi = \alpha \wedge \frac{dz}{z} + \beta$$

where α, β are C^∞ forms, and where (cf. (2.17))

$$\alpha|_{z=0} \equiv 0, \text{ and}$$
$$\beta \text{ does not involve } dz.$$

Suppose that $d\varphi = 0$ and $\deg\varphi > 0$. Write

$$\beta = \gamma \wedge d\bar z + \delta$$

where δ involves only dw and $d\bar w$. Then $d\varphi = 0 \Rightarrow d_w\delta = 0$ (d_w = exterior derivative with respect to the w's), and so $\delta = d_w\theta$ by the usual Poincaré lemma with C^∞ dependence on parameters [16]. Now

$$\varphi - d\theta = \alpha' \wedge \frac{dz}{z} + \beta' \wedge d\bar z.$$

[6] The sheaves $\mathscr{A}^*, \mathscr{A}^*(\log\langle D\rangle), \mathscr{W}^*$ all satisfy $H^q(V,\cdot) = 0$ for $q > 0$.

where β' does not involve dz. Again, $d\varphi = 0 \Rightarrow d_w\beta' = 0$ and so $\beta' = d_w\theta' \equiv \psi \wedge \dfrac{dz}{z}$ (mod exact forms). Write $\psi = \psi' \wedge d\bar{z} + \psi''$, where ψ'' involves only dw, $d\bar{w}$. Then $d_w\psi'' = 0$ and $\psi'' |_{z=0} \equiv 0$. We may write $\psi' = d_w\eta$, with $\eta |_{z=0} \equiv 0$ [16], and then subtracting $d\left(\eta \wedge \dfrac{dz}{z}\right)$ gives

$$\varphi = \tau \wedge d\bar{z} \wedge \frac{dz}{z} \qquad \text{(mod exact forms).}$$

Once more $d_w\tau = 0$ and so $\tau = d_w\omega$, so that

$$\varphi = \rho\, d\bar{z} \wedge \frac{dz}{z}, \quad d_w\rho = 0.$$

Now $\rho = \rho(z, \bar{z})$, and by the $\bar{\partial}$-Poincaré lemma [39]

$$\rho\, d\bar{z} = \bar{\partial}\xi, \ \xi(0) = 0,$$

so that subtracting $d\left(\xi \wedge \dfrac{dz}{z}\right)$ gives finally that φ is exact.

The proof of the $\bar{\partial}$-Poincaré lemma in the present context is done in the same way, using [39].

REMARK. The ∂-Poincaré lemma is *false* in $\mathscr{W}^*$; forms $f(\bar{z})\dfrac{dz}{z}$, with $f(\bar{z}) = \sum_{n=1}^{\infty} a_n\bar{z}^n$, are ∂-closed but not ∂-exact.

3. Variation of Hodge structure. On a compact Kähler manifold, the Hodge decompositions of the complex cohomology groups reflect the complex structure of the manifold. Since a Hodge structure is a much simpler object than a global complex structure, by passing to the Hodge decompositions, one obtains a simplified model of the complex structure of the manifold. In some sense, this process is analogous to looking at the topology of a space in terms of its homology. The study of variation of Hodge structure was begun in [18, 19]. We shall recall the constructions which are relevant for this paper. One can approach the subject from several points of view. Each has its advantages, and so we shall discuss and relate them in the three parts of this section. One more general comment: For technical reasons, which will become apparent below, it is necessary to consider the *polarized* Hodge structures on the primitive parts of the cohomoloy, rather than the Hodge structures on the full cohomology. Since the former completely determine the latter, no information is lost by doing so.

(a) *The Hodge bundles.* Throughout this section, X and S will denote connected complex manifolds, and $\pi: X \to S$ a holmorphic proper mapping with connected fibres, which is everywhere of maximal rank. Moreover, X is assumed to be embedded in some projective space, but not necessarily as a closed submanifold. Each fibre $V_s = \pi^{-1}(s)$, $s \in S$, then becomes a projective manifold. We shall refer to this geometric situation as a *family of polarized algebraic manifolds.*[7] In practice, such families usually arise as follows: let $\bar{X}$ and $\bar{S}$ be projective varieties and $\pi: \bar{X} \to \bar{S}$ a proper algebraic mapping, whose generic fibre is smooth. If we set S equal to the subset of the regular set of $\bar{S}$ over which π has smooth fibres, and $X = \pi^{-1}(S)$, we obtain a family of polarized algebraic manifolds.

Disregarding the complex structures, one may think of $\pi: X \to S$ as a C^∞ fibre bundle. For each integer m between 0 and $2n$ ($n = \dim_{\mathbf{C}} V_s$), the direct image sheaf $R^m_{\pi_*}(\mathbf{C})$ is the sheaf of flat sections of a flat complex vector bundle $\mathbf{H}^m \to S$. The fibre of $\mathbf{H}^m$ over $s \in S$ has a natural identification with $H^m(V_s, \mathbf{C})$. According to harmonic theory with variable coefficients [33], the dimensions of the Hodge subspaces $H^{p,q}(V_s)$, with $p + q = m$, depend upper semicontinuously on s. Since their sum, being a topological invariant, remains constant, so does each of the summands. Again appealing to the results of [33] one now finds that the Hodge subspaces $H^{p,q}(V_s)$ are the fibres of a C^∞-subbundle $\mathbf{H}^{p,q} \subset \mathbf{H}^m$. As a preliminary definition, which will soon be changed slightly, we set $\mathbf{F}^p = \oplus_{i \geqslant p} \mathbf{H}^{i,m-i}$. Let $\mathbf{T}^* \to S$ be the holomorphic cotangent bundle, and

$$\nabla : \mathcal{O}(\mathbf{H}^m) \to \mathcal{O}(\mathbf{H}^m \otimes \mathbf{T}^*),$$

[7] By the polarization of the fibres V_s, we mean the datum of the cohomology class of a projective embedding. Instead of assuming that the total space X lies in some $\mathbf{P}^N$, we only need a polarization for each fibre, which is constant with respect to s, in the sense that the polarizations form a global section of the direct image sheaf $R^2_{\pi_*}(\mathbf{Z})$ on S.

the flat connection of $\mathbf{H}^m$. The following result of the first author provides the starting point of the study of variation of Hodge structures.

(3.1) THEOREM [18]. *Each* $\mathbf{F}^p$ *is a holomorphic subbundle of* $\mathbf{H}^m$. *Furthermore,*

$$\nabla\, \mathcal{O}(\mathbf{F}^p) \subset \mathcal{O}(\mathbf{F}^{p-1} \otimes \mathbf{T}^*).$$

One can paraphrase the second statement roughly by saying that infinitesimally the subspaces $H^{p,q}(V_s)$ get shifted by a change in indices of at most one. When it is restated in terms of period matrices, as we shall do below, it looks like an infinitesimal period relation. For families of algebraic curves, this condition is vacuous. However, in the general case, it becomes a crucial ingredient of virtually all arguments about variation of Hodge structure.

The Kähler operator $L: H^m(F_s) \to H^{m+2}(V_s)$ is defined solely in terms of topological quantities. It therefore extends to a flat bundle map $L: \mathbf{H}^m \to \mathbf{H}^{m+2}$. Let $\mathbf{P}^m$ be the kernel of L^{n-m+1}, acting on $\mathbf{H}^m$. Then $\mathbf{P}^m$ becomes a flat subbundle of $\mathbf{H}^m$, whose fibres correspond to the subspaces $P^m(V_s) \subset H^m(V_s)$. It is the complexification of a flat real subbundle $\mathbf{P}^m_{\mathbf{R}}$, and $\mathbf{P}^m_{\mathbf{R}}$ in turn contains a flat lattice bundle $\mathbf{P}^m_{\mathbf{Z}}$. In terms of a local flat trivialization, $P^m(V_s) \cap H^{p,q}(V_s)$ is the intersection of a fixed vector space with a family of continuously varying subspaces. Hence the dimension depends semicontinuously on s. The sum of these dimensions, with $p+q=m$, equals the dimension of $P^m(V_s)$, which is constant. We may conclude that $\mathbf{P}^m \cap \mathbf{H}^{p,q}$ has constant fibre dimension, and is therefore a C^∞-subbundle of $\mathbf{P}^m$. Changing notation, we now set

$$\mathbf{F}^p = \oplus_{i \geqslant p} \mathbf{P}^m \cap \mathbf{H}^{i,m-i}.$$

From (3.1), one immediately deduces the two analogous statements

$$\left.\begin{array}{c} \mathbf{F}^p \text{ is a holomorphic subbundle of } \mathbf{P}^m, \text{ and} \\ \nabla \mathcal{O}(\mathbf{F}^p) \subset \mathcal{O}(\mathbf{F}^{p-1} \otimes \mathbf{T}^*). \end{array}\right\} \quad (3.2)$$

Finally, since the Hodge bilinear form Q does not depend on the complex structures of the fibres, we may view it as a flat bilinear form on the bundle $\mathbf{P}^m$.

For some applications, it is convenient to consider collections of vector bundles with the various properties mentioned above, even if the situation does not arise directly from a family of algebraic manifolds. We gather the ingredients in the form of a definition. Let S be a complex manifold. By a *variation of Hodge structure, with base S, of weight m*, we shall mean a collection of the following data:

(i) a flat complex vector bundle $\mathbf{H} \to S$, containing a flat, real subbundle $\mathbf{H}_{\mathbf{R}}$, so that $\mathbf{H}$ is the complexification of $\mathbf{H}_{\mathbf{R}}$, together with a flat bundle of lattices $\mathbf{H}_{\mathbf{Z}} \subset \mathbf{H}_{\mathbf{R}}$;

(ii) a flat bilinear form Q: $\mathbf{H} \times \mathbf{H} \to \mathbf{C}$, with $Q(f, e) = (-1)^m Q(e, f)$, which is rational with respect to $\mathbf{H}_{\mathbf{Z}}$;

(iii) a descending filtration of $\mathbf{H}$ by a family of holomorphic subbundles $\mathbf{H} \supset \ldots \supset \mathbf{F}^{p-1} \supset \mathbf{F}^p \supset \mathbf{F}^{p+1} \supset \ldots \supset 0$, so that $\nabla \mathcal{O}(\mathbf{F}^p) \subset \mathcal{O}(\mathbf{F}^{p-1} \otimes \mathbf{T}^*)$;

these data have to satisfy the conditions that at each $s \in S$, the fibres of the $\{\mathbf{F}^p\}$ at s define a Hodge structure of weight m on the fibre of $\mathbf{H}$, and this Hodge structure is to be polarized by Q.

The bilinear form Q determines indefinite Hermitian metrics on the bundles $\{\mathbf{F}^p\}$. It is thus possible to apply the methods of Hermitian differential geometry, as was done by the first author in [19]. We shall take up these matters again in §10.

(b) *Classifying spaces and the period mapping.* Not surprisingly, the bundles $\{\mathbf{F}^p\}$ of a variation of Hodge structure can be realized as the pullbacks of certain universal bundles over a classifying space. This classifying space parametrizes the polarized Hodge structure on a fixed vector space. In order to recall the construction, which was given in [18], we consider a finite dimensional complex vector space H, with a real form $H_{\mathbf{R}} \subset H$ and a lattice $H_{\mathbf{Z}} \subset H_{\mathbf{R}}$. We also fix an integer m and a rationally defined bilinear form Q on H, which shall be symmetric if m is even, and skew if m is odd. Next, we let $\{h^{p,q}\}$ be a collection of nonnegative integers, corresponding to pairs of indices (p, q) with $p+q=m$, such that $h^{q,p}=h^{p,q}$ and $\Sigma\, h^{p,q} = \dim H$. By $\check{D}$, we denote the set of decreasing filtrations

$$H \supset \ldots \supset F^{p-1} \supset F^p \supset F^{p+1} \supset \ldots \supset 0$$

which satisfy the two conditions

$$\left.\begin{array}{ll} \text{a)} & \dim F^p = \sum_{i \geqslant p} h^{i,m-i}, \\ \text{b)} & Q(F^p, F^{m-p+1}) = 0. \end{array}\right\} \tag{3.3}$$

In a natural way, $\check{D}$ lies as a subvariety in a product of Grassmann varieties. By elementary arguments in linear algebra one finds that the algebraic group

$$\begin{aligned} G_{\mathbf{C}} &= \text{orthogonal group of } Q \\ &= \{T \in Gl(H) \mid Q(Tu, Tv) = Q(u, v) \text{ for all } u, v \in H\} \end{aligned} \tag{3.4}$$

operates transitively on $\check{D}$. In particular, $\check{D}$ cannot have any singularities; it is a projective manifold. The subset D of all those points in $\check{D}$ which correspond to filtrations $\{F^p\}$ with the property

$$(\sqrt{-1})^{2p-m}\, Q(v, \bar{v}) > 0 \text{ if } v \in F^p \cap \overline{F^{m-p}},\ v \neq 0, \tag{3.5}$$

is open in the Hausdorff topology of $\check{D}$. Hence D inherits the structure of a complex manifold from D. Any filtration $\{F^p\}$ belonging to a point D automatically satisfies (1.5), and therefore determines a Hodge structure of weight m on H, which is polarized with respect to Q. In other words, D parametrizes exactly the Hodge structures of weight m on H for which Q is a polarization, and such that $\dim H^{p,q} = h^{p,q}$. We call D a *classifying space* for polarized Hodge structures, and $\check{D}$ its *dual space*.

Almost by definition, the trivial vector bundle $\mathbf{H} = D \times H$ over $\check{D}$ is filtered by decreasing family of holomorphic subbundles

$$\mathbf{H} \supset \ldots \supset \mathbf{F}^{p-1} \supset \mathbf{F}^p \supset \mathbf{F}^{p+1} \supset \ldots \supset 0$$

whose fibres over any point of $\check{D}$ constitute the filtration of H corresponding to the point in question. Let ∇ be the trivial flat connection on $\mathbf{H}$, and x a point of $\check{D}$. We shall say that a tangent vector X at x is *horizontal* if

$$\nabla_X\, \mathcal{O}(\mathbf{F}^p)_x \subset \mathcal{O}(\mathbf{F}^{p-1})_x, \text{ for all } p. \tag{3.6}$$

The transitive action of $G_{\mathbf{C}}$ on D lifts to the family of bundles $\{\mathbf{F}^p\}$ and the tangent bundle. This action maps horizontal tangent vectors again to horizontal tangent vectors. In particular, the spaces of horizontal tangent vectors at the various points of $\check{D}$ have constant dimension, and they fit together, to form a $G_{\mathbf{C}}$-invariant, holomorphic subbundle of the holomorphic tangent bundle $\mathbf{T} \to \check{D}$. We shall call it the *horizontal tangent subbundle*, $\mathbf{T}_h$. A holomorphic mapping f of a complex manifold S into $\check{D}$, or into the open submanifold $D \subset \check{D}$, is said to be *horizontal* if the induced mapping f_* between the tangent spaces takes values in the horizontal tangent subbundle.

Let $G_{\mathbf{R}} \subset G_{\mathbf{C}}$ be the subgroup of real points,

$$G_{\mathbf{R}} = \{T \in G_{\mathbf{C}} \mid TH_{\mathbf{R}} \subset H_{\mathbf{R}}\}. \tag{3.7}$$

The action of $G_{\mathbf{R}}$ preserves $D \subset \check{D}$. By arguments in linear algebra (cf. [18]), one can show that $G_{\mathbf{R}}$ acts transitively on D. Thus D has the structure of a homogeneous space, and this is the key to understanding all of the more subtle properties of D. In order to realize D as a quotient space of $G_{\mathbf{R}}$, we fix a *base point*, or *origin*, $o \in D$. It corresponds to a filtration $\{F_0^p\}$ of H, the *reference Hodge filtration*, which in turn determines the *reference Hodge structure* $\{H_0^{p,q}\}$. The automorphism group $G_{\mathbf{C}}$ of $\check{D}$ operates with isotropy group

$$B_{\mathbf{C}} = \{T \in G_{\mathbf{C}} \mid TF_0^p \subset F_0^p \text{ for all } p\} \tag{3.8}$$

at o; $B_{\mathbf{C}}$ is a parabolic subgroup of $G_{\mathbf{C}}$, and one has the identification $\check{D} \cong G_{\mathbf{C}}/B_{\mathbf{C}}$. We denote the group of real points in $B_{\mathbf{C}}$ by V, *i.e.*

$$V = B_{\mathbf{C}} \cap G_{\mathbf{R}}. \tag{3.9}$$

Then V is the isotropy subgroup of $G_{\mathbf{R}}$ at o, and $D \simeq G_{\mathbf{R}}/V$. Under these identifications, the inclusion $D \subset \check{D}$ corresponds to

$$D \simeq G_{\mathbf{R}}/V = G_{\mathbf{R}}/G_{\mathbf{R}} \cap B_{\mathbf{C}} \hookrightarrow G_{\mathbf{C}}/B_{\mathbf{C}} \simeq \check{D}. \tag{3.10}$$

Let C_0 be the Weil operator of the reference Hodge structure, so that $C_0 v = (\sqrt{-1})^{p-q} v$ if $v \in H_0^{p,q}$. Since V commutes with complex conjugation, it fixes not only the filtration $\{F_0^p\}$, but also the reference Hodge structure, and therefore also the positive definite hermitian form

$$(u, v) = Q(C_0 u, \bar{v}), \quad u, v \in H.$$

Hence:

$$V \text{ is a compact subgroup of } G_{\mathbf{R}}. \tag{3.11}$$

As an arithmetic subgroup of $G_{\mathbf{R}}$,

$$\Gamma = G_{\mathbf{Z}} = \{T \in G_{\mathbf{R}} \mid TH_{\mathbf{Z}} \subset H_{\mathbf{Z}}\}$$

is discrete in $G_{\mathbf{R}}$. Coupled with (3.11) and the identification $D \simeq G_{\mathbf{R}}/V$, this shows:

$$\Gamma \text{ operates properly discontinuously on } D. \tag{3.12}$$

In particular, the quotient $\Gamma\backslash D$ has the structure of a normal analytic space. If we had considered arbitrary Hodge structures, rather than polarized Hodge structures only, the analogous statements would be false.

Before coming back to the properties of the classifying space D, we recall the definition of period mapping. Let $(\mathbf{H}, \mathbf{F}^p)$ be a variation of Hodge structure, of weight m, with base space S— for example, the variation of Hodge structure corresponding to the mth primitive cohomology groups of the fibres of a family of polarized algebraic manifolds $\pi : X \to S$. The pullback of the flat vector bundle $\mathbf{H}$ to the universal covering $\tilde{S}$ of S is canonically trivial. Thus it makes sense to talk of *the* fibre H of this pullback. The flat bundle $\mathbf{H} \to S$ is then associated to the principal bundle $\pi_1(S) \to \tilde{S} \to S$ by a representation $\varphi\colon \pi_1(S) \to Gl(H)$. The flat subbundle $\mathbf{H}_{\mathbf{R}} \subset \mathbf{H}$, the flat lattice bundle $\mathbf{H}_{\mathbf{Z}} \subset \mathbf{H}_{\mathbf{R}}$, and the flat pairing $Q : \mathbf{H} \times \mathbf{H} \to \mathbf{C}$ correspond to, respectively, a real form $H_{\mathbf{R}} \subset H$, a lattice $H_{\mathbf{Z}} \subset H_{\mathbf{R}}$, and a bilinear form Q on H. All of these objects are preserved by the representation φ, so that φ takes values in $\Gamma = G_{\mathbf{Z}}$. The bundles $\mathbf{F}^p \to S$ pull back to holomorphic subbundles of the trivial bundle $H \times \tilde{S}$. At each point of $\tilde{S}$, the fibres of these pullbacks determine a filtration of H, which corresponds to a Hodge structure of weight m on H, with polarization Q. For these Hodge structures, the dimensions $h^{p,q} = \dim H^{p,q}$ are constant. We now consider the classifying space for Hodge structures D which corresponds to the collection of Hodge numbers $\{h^{p,q}\}$. For each $s \in \tilde{S}$, the Hodge structure determined by s corresponds to a definite point in D. This gives a mapping

$F: \tilde{S} \to D$. As a direct consequence of the definition of the complex structure of D, F is holomorphic. Also the condition (iii) for a variation of Hodge structure ensures that F is a horizontal mapping. Next, if the points $s, s' \in \tilde{S}$ are related by an element γ of the fundamental group of S, and if $T = \varphi(\gamma)$,[8] then the Hodge structures corresponding to s and s' are related by T: i.e. $F(s') = TF(s)$. In particular, when F is composed with the projection $D \to \Gamma\backslash D$, the resulting mapping becomes $\pi_1(S)$-invariant. Thus we obtain a mapping $f: S \to \Gamma\backslash D$, which is the *period mapping* of the variation of Hodge structure. As follows from the construction,

the period mapping is holomorphic, locally liftable,
and it has horizontal local liftings. (3.13)

By "locally liftable" we mean that f, restricted to any sufficiently small open set in S, factors through the projection $D \to \Gamma\backslash D$.

This process, which associates to a variation of Hodge structure the period mapping, can almost be reversed. Let $f: S \to \Gamma\backslash D$ be a mapping of the connected complex manifold S into $\Gamma\backslash D$, with all of the properties mentioned in (3.13). Then there exists a holomorphic horizontal map $F: \tilde{S} \to D$, which makes the diagram

$$\begin{array}{ccc} \tilde{S} & \xrightarrow{F} & D \\ \downarrow & & \downarrow \\ S & \xrightarrow{f} & \Gamma\backslash D \end{array}$$

commutative. Moreover, for each $\gamma \in \pi_1(S)$, one can choose an element $T_\gamma \in \Gamma$, so that

$$F(\gamma s) = T_\gamma F(s), \quad \text{for all } s \in \tilde{S}.$$

Since Γ does not operate fixed point-free, T_γ may not be uniquely determined by γ[9], in which case $\gamma \longmapsto T_\gamma$ need not be a representation

[8] For example, if the variation of Hodge structure arises from the mth primitive cohomology groups of the fibres of a family of polarized algebraic manifolds $\pi: X \to \Delta^*$ parametrized by the punctured disc Δ^*, and if $\gamma \in \pi_1(\Delta^*)$ is the canonical generator, $T = \varphi(\gamma)$ represents the action of γ on the mth primitive cohomology group of a typical fibre V_s. This element T is usually called the *Picard-Lefschetz transformation.*

[9] This cannot happen if f is "sufficiently general".

of $\pi_1(S)$. However, if there does exist a homomorphism $\varphi: \pi_1(S) \to \Gamma$, with $\varphi(\gamma) = T_\gamma$ for suitable choices of T_γ, then φ will determine a flat bundle $\mathbf{H} \to S$. All the other ingredients of a variation of Hodge structure can now also be reconstructed; details are left to the reader.

We briefly mention the classical situation, which has motivated the study of variation of Hodge structure. Let $\pi: X \to S$ be a family of principally polarized, g-dimensional abelian varieties, or of nonsingular algebraic curves of genus g. The classifying space for the Hodge structures on the first cohomology groups is then the Siegel upper half plane H_g, the discrete group Γ is the Siegel moduler group, and the period mapping $f: S \to \Gamma\backslash H_g$ associates to each fibre of the family the usual invariant in the quotient $\Gamma\backslash H_g$.

The Siegel upper half plane, as is well known, has a realization as a bounded symmetric domain. The classifying space for the Hodge structures on the cohomology of algebraic $K3$ surfaces also has this property; it is a hermitian symmetric domain of type IV. In general however, D may be very far from being a bounded domain. In fact, D will usually not have any nonconstant holomorphic functions. On the other hand, the classifying spaces behave somewhat like bounded domains, as far as horizontal mappings into them are concerned. The important feature is the existence of a metric which is negatively curved in the horizontal directions. How this affects mappings into D will be taken up in §7. Here we shall only give a precise statement about the metric in question, to which we can refer later.

(3.14) PROPOSITION. *Let D be a classifying space for Hodge structures. Then there exists a $G_{\mathbf{R}}$-invariant hermitian metric on D, whose holomorphic sectional curvatures in all horizontal tangent directions are negative and uniformly bounded away from zero.*

A general discussion of the manifolds which can arise as classifying spaces for Hodge structures is contained in [25]. The proposition is proven in §9 of that paper. Deligne has given a short, self-contained proof in [11]. Incidentally, in order to have results like (3.14), one is again forced to look at *polarized* Hodge structures.

proof in [11]. Incidentally, in order to have results like (3.14), one is again forced to look at *polarized* Hodge structures.

In some applications of the theory of variation of Hodge structure, one is confronted with technical problems of the following type: If $\mathbf{E} \to \check{D}$ is a homogeneous, holomorphic vector bundle[10], and if the restriction of $\mathbf{E}$ to D carries a $G_{\mathbf{R}}$-invariant metric, how does the metric behave as one approaches the boundary of D? In order to illustrate the kind of arguments which are made possible by the homogeneous structure of D, we shall look at this question in particular; the answer will also be of use elsewhere in this paper.

Some preliminary remarks are needed. We recall the identification $\check{D} \simeq G_{\mathbf{C}}/B_{\mathbf{C}}$. A holomorphic representation $\tau : B_{\mathbf{C}} \to Gl(E)$ associates a vector bundle $\mathbf{E}$ to the principal bundle $B_{\mathbf{C}} \to G_{\mathbf{C}} \to D$. Its total space can be identified with $G_{\mathbf{C}} \times E/\sim$, with the equivalent relation $\sim$ defined by $(gb^{-1}, be) \sim (g, e)$ if $b \in B_{\mathbf{C}}$. The action of $G_{\mathbf{C}}$ on the first factor of $G_{\mathbf{C}} \times E$ then induces an action on $\mathbf{E}$ and turns $\mathbf{E}$ into a homogeneous vector bundle. Conversely, every homogeneous vector bundle arises in this fashion. The restriction of $\mathbf{E}$ to $D \simeq G_{\mathbf{R}}/V$ may be thought of as the vector bundle associated to the principal bundle $V \to G_{\mathbf{R}} \to D$ by the representation $\tau|_V$. Because of the compactness of V, one can choose a V-invariant inner product on the vector space E. When E is identified with the fibre of $\mathbf{E}$ over the origin, by translating the inner product via $G_{\mathbf{R}}$, one obtains a $G_{\mathbf{R}}$-invariant metric on $\mathbf{E} \to D$. It should be pointed out that any two $G_{\mathbf{R}}$-invariant metrics will be mutually bounded.

We are interested in comparing a $G_{\mathbf{R}}$-invariant metric to a global Hermitian metric of $\mathbf{E}$ over $\check{D}$, near the boundary of D. Since $\check{D}$ is compact, the choice of a global metric will not matter. However, there exist metrics with which one can calculate particularly easily, because they are derived from the homogeneous structure of $\check{D}$. We shall proceed to describe them. As before, C_0 shall denote the Weil operator of the reference Hodge structure. Then

$$(u, v) = Q(C_0 u, \bar{v}), \quad u, v \in H, \tag{3.15}$$

[10] i.e. a vector bundle to which the action of G on $\check{D}$ lifts.

defines a positive definite inner product on H. Let $M \subset G_{\mathbf{C}}$ be the intersection of $G_{\mathbf{C}}$ with the unitary group of this inner product; it is a compact subgroup of $G_{\mathbf{C}}$. One can check directly that

$$\dim_{\mathbf{R}} M = \dim_{\mathbf{C}} G_{\mathbf{C}} = \dim_{\mathbf{R}} G_{\mathbf{R}}. \tag{3.16}$$

Next, we claim that

$$M \cap B_{\mathbf{C}} = V; \tag{3.17}$$

in fact, every $g \in M \cap B_{\mathbf{C}}$ leaves both the reference Hodge filtration and the inner product (3.15) invariant, and must therefore also keep the Hodge subspaces of the reference Hodge filtration fixed. In particular, g commutes with C_0. Thus g preserves the Hermitian form $Q(u,\bar{v})$, as well as the bilinear form Q. This is possible only if $g \in G_{\mathbf{R}}$, so that

$$M \cap B_{\mathbf{C}} \subset G_{\mathbf{R}} \cap B_{\mathbf{C}} = V.$$

The reverse containment is clear, and (3.17) is proven. The M-orbit of the origin in $\check{D}$ can now be indentified with M/V; because of (3.16), it has the same dimension as $D \simeq G_{\mathbf{R}}/V$ and must be open in $\check{D}$. On the other hand, the compactness of M forces the orbit to be closed. Thus

$$\begin{gathered} M \text{ operates transitively on } \check{D}, \text{ with isotopy group } V \\ \text{at the origin, so that } \check{D} \simeq M/V. \end{gathered} \tag{3.18}$$

Just as a V-invariant inner product on E gives rise to a $G_{\mathbf{R}}$-invariant Hermitian metric for $\mathbf{E} \to D$, such an inner product can be translated around by M, to give an M-invariant Hermitian metric for $\mathbf{E}$ over all of $\check{D}$.

We now consider two Hermitian metrics h_1, h_2 for $\mathbf{E}$, of which the first is $G_{\mathbf{R}}$-invariant and defined over D, and the second M-invariant and defined over all of $\check{D}$. We also assume that the two metrics coincide on the fibre over the origin. Let x be a point of D and e a vector in the fibre of $\mathbf{E}$ over x. We can write x as the g-translate of the origin, for some $g \in G_{\mathbf{R}}$, and also as the m-translate of the origin, for some $m \in M$. Since $B_{\mathbf{C}}$ is the isotropy subgroup of $G_{\mathbf{C}}$ at the origin, $g = mb$, with $b \in B_{\mathbf{C}}$. In order to compute the h_1-length of e, we may translate e by g^{-1} to the origin and compute the length there. Similarly, the h_2-length of e is the length of its translate by m^{-1} at the origin. It follows that h_1 and h_2 at x are mutually bounded by, respectively, the operator norm of $\tau(b)$ and the operator norm of $\tau(b^{-1})$, relative to the V-invariant inner product on E which corresponds to the two metrics. The matrix entries of $\tau(b)$ and $\tau(b^{-1})$ are rational functions of those of b, when b is viewed as an element of the matrix group $G_{\mathbf{C}}$. Because of the compactness of M, the matrix entries of $b = m^{-1}g$ are bounded by a constant multiple of the largest matrix entry of g. For any $g \in G_{\mathbf{R}}$, we let $\|g\|$ denote the operator norm of g on H, relative to some inner product on H. According to what has been argued above, the metrics h_1 and h_2 on the fibre of $\mathbf{E}$ over x must be mutually bounded by a constant multiple of a suitable power of $\|g\|$. Clearly this remains correct if we replace h_1 by any $G_{\mathbf{R}}$-invariant metric and h_2 by any global Hermitian metric for $\mathbf{E}$ over all of $\check{D}$. We have proven:

(3.19) LEMMA. *Let* $\mathbf{E} \to \check{D}$ *be a homogeneous holomorphic vector bundle,* h_1 *a* $G_{\mathbf{R}}$*-invariant metric for the restriction of* $\mathbf{E}$ *to* D*, and* h_2 *a global Hermitian metric for* $\mathbf{E}$ *over* $\check{D}$*. There exist constants* C, N *with the following property: if* $x \in D$ *is the* g*-translate of the origin, with* $g \in G_{\mathbf{R}}$*, then each of the two metrics* h_1, h_2 *at* x *is bounded by* $C\|g\|^N$ *times the other.*

In order to make the lemma useful, one has to know how $\|g\|$ grows as the point x approaches the boundary of D. For this purpose, we recall some standard facts from the theory of symmetric spaces.[11] Let $G_{\mathbf{R}}$ be a semisimple matrix group, and $K \subset G_{\mathbf{R}}$ a maximal compact subgroup. The quotient $G_{\mathbf{R}}/K$ then carries a $G_{\mathbf{R}}$-invariant, Riemannian metric ds^2, which is essentially unique. In the Lie algebra of $\mathfrak{g}_0$ of $G_{\mathbf{R}}$, the subalgebra $\mathfrak{k}_0$ corresponding to K has a unique Ad K-invariant complement $\mathfrak{p}_0$. We now choose a maximal abelian subspace $\mathfrak{a}_0$ in $\mathfrak{p}_0$, and we denote the subgroup $\exp \mathfrak{a}_0$ of $G_{\mathbf{R}}$ by A. All elements of A act semisimply, under any finite dimensional representation of $G_{\mathbf{R}}$. One then has the (non-unique) decomposition

[11]Helgason's book [27] is a good reference.

$$G_{\mathbf{R}} = K\,A\,K. \tag{3.20}$$

Moreover, with respect to a suitable Euclidean metric on $\mathfrak{a}_0$,[12]

$X \longmapsto \exp XK$ is a locally and globally isometric, totally geodesic embedding of A in $G_{\mathbf{R}}/K$. (3.21)

In our situation, as the "orthogonal group" of a nondegenerate, symmetric or skew symmetric bilinear form, $G_{\mathbf{R}}$ will certainly be semisimple. Because of (3.11), there exists a maximal compact subgroup K of $G_{\mathbf{R}}$ which contains V. Relative to any two $G_{\mathbf{R}}$-invariant metrics, the projection

$$D \simeq G_{\mathbf{R}}/V \to G_{\mathbf{R}}/K \tag{3.22}$$

is bounded. Let $x \in D$ be the g-translate of the origin, with $g \in G_{\mathbf{R}}$. We write $g = k_1\, a\, k_2$, $k_1, k_2 \in K$, $a \in A$. According to (3.21) and the boundedness of the projection (3.22), the Euclidean norm of $\log a$ is bounded by some multiple of the distance $\rho_D(x, o)$. The abelian group A can be simultaneously diagonalized, because all of its elements operate semisimply. Hence the operator norm of a cannot exceed some multiple of a suitable power of $\exp \rho_D(x, o)$. Since $g = k_1\, a\, k_2$, and since K is compact, this gives the same kind of estimate also for the operator norm of g. Thus:

(3.23) LEMMA. *There exist positive constants B, M, with the following property: if $x \in D$ is the g-translate of the origin, with $g \in G_{\mathbf{R}}$, then*

$$\|g\| \leqslant B \exp M\rho_D(x, o).$$

With this lemma, the comparison of the metrics in (3.19) can now be rephrased in a more intrinsic manner.

(c) *Period matrices.* In his book [30] on harmonic integrals, Hodge phrased his results on the cohomology of Kähler manifolds in the language of *period matrices.* For some questions, such as computations of specific examples, it is useful to be able to think in this way, and so we shall give a brief "dictionary", relating the preceding discussion to the language of period matrices. This

[12]If the Riemannian structure ds^2 is the one corresponding to the Killing form, the restriction of the Killing form to $\mathfrak{a}_0$ will be the "suitable Euclidean metric".

description of Hodge structures will be used in §§8, 9 below. We conclude this section with a proof of the theorem of the regularity of the connection on the Hodge bundles.

Let us consider a polarized Hodge structure $\{H^{p,q}\}$ of weight m on the vector space H, with polarization form Q. Once and for all, we assume that $H^{p,q} = 0$ unless $p, q \geqslant 0$. Also to simplify the discussion, we shall limit ourselves mainly to the case when $m = 2$, with only some parenthetical remarks about the general case. Under these hypotheses, the Hodge filtration has length 2:

$$H = F^0 \supset F^1 \supset F^2 \supset 0. \tag{3.24}$$

For $0 \leqslant k \leqslant 2$, F^k and F^{3-k} are perpendicular with respect to Q. On the other hand, Q is nondegenerate, and F^k and F^{3-k} have complementary dimensions, so that $F^{3-k} = F^{k\perp}$. As an immediate consequence, we see that F^2 already determines the remaining subspaces.[13] We now let the polarized Hodge structure vary, keeping the polarization form Q and the Hodge number $r = \dim H^{2,0} = \dim H^{0,2}$ and $s = \dim H^{1,1}$ fixed. The points of the dual space $\check{D}$ of the classifying space D then correspond exactly to the subspaces

$$F^2 \subset H, \text{ with } \dim F^2 = r,\ Q(F^2, F^2) = 0. \tag{3.25}$$

Such a subspace can be completed to a filtration (3.24) by setting $F^1 = F^{2\perp}$. The subspace belongs to a point of D when the appropriate positivity conditions are satisfied. In our special case, they can be compressed into the single condition

$$-Q(v, \bar{v}) > 0 \quad \text{if } v \in F^2,\ v \neq 0. \tag{3.26}$$

Indeed, by conjugation, the condition on $H^{0,2}$ follows from (3.26), and the condition on $H^{1,1} = (H^{2,0} \otimes H^{0,2})^\perp$ is automatic, since Q has exactly s positive eigenvalues.

In order to represent the points of D and $\check{D}$ by period matrices, we pick a basis $\{e_1, \dots, e_{2r+s}\}$ of the lattice $H_{\mathbf{Z}}$, and we denote the dual basis by $\{\lambda^1, \dots, \lambda^{2r+s}\}$. Relative to the basis $\{e_i\}$, the bilinear form Q is specified by a $(2r+s)\times(2r+s)$ matrix, which we shall also call Q. Given a subspace F^2 as in (3.25), we choose a basis $\{v_1, \dots, v_r\}$, and

[13] In general, it suffices to know F^k, for $[m/2] + 1 < k < m$; if $k < [m/2]$, F^k is then determined by $F^k = (F^{m-k+1})^\perp$.

we let Ω be the $(2r+s)\times r$ matrix whose (i, j)-entry is $\langle \lambda^i, v_j \rangle$; this is the period matrix of the Hodge structure in question.[14] It clearly determines the Hodge structure completely. Every nonsingular $(2r+s)\times r$ matrix Ω which satisfies the first of the two bilinear relations

$$\left.\begin{aligned} {}^t\Omega Q\,\Omega &= 0 \\ -{}^t\Omega Q\,\bar{\Omega} &> 0 \end{aligned}\right\} \tag{3.27}$$

corresponds to a point of D. If it satisfies the second relation as well, then it is the period matrix of an actual Hodge structure. Two such matrices Ω and Ω' belong to the same point of $\check{D}$, or of D, exactly when $\Omega' = \Omega A$, for some invertible $r\times r$ matrix A. This equivalence relation reflects the freedom of choice of a basis of F^2. In the preceding discussion, the basis $\{e_i\}$ of $H_{\mathbf{Z}}$ has been kept fixed; changing it has the effect of pre-multiplying the period matrices by the transpose of the change-of-base matrix.

The set of all nonsingular $(2r+s)\times r$ matrices Ω, modulo the equivalence relation $\Omega \sim \Omega A$, is a particular realization of the Grassmannian $Gr(r, 2r+s)$ of r-planes in $\mathbf{C}^{2r+s}$. The first of the two bilinear relations (3.27) exhibits $\check{D}$ as a sub-variety of this Grassmannian. By associating to each nonsingular matrix Ω the *Plücker coordinates*

$$\Omega_{i_1,\ldots,i_r} = \det \begin{vmatrix} \Omega_{i_1,1} & \cdots & \Omega_{i_1,r} \\ \vdots & & \\ \Omega_{i_r,1} & \cdots & \Omega_{i_r,r} \end{vmatrix},$$

one obtains the Plücker embedding of $Gr(r, 2r+s)$, and thereby also of its subvariety $\check{D}$, in the projective space of dimension $\binom{2r+s}{r} - 1$. The set of period matrices forms the total space of a holomorphic principal bundle over $\check{D}$, with structure group $Gl(r, \mathbf{C})$. The character

$$A \longmapsto \det A$$

[14]For arbitrary m, one obtains a collection of $[(m+1)/2]$ period matrices, corresponding to the subspaces F^k, $[m/2]+1 < k < m$.

of $Gl(r, \mathbf{C})$ induces a holomorphic line bundle $\mathbf{L} \to D$, whose space of sections has the Plücker coordinates $\Omega_{i_1,\ldots,i_r}$ as a basis.

Like any principal bundle, the bundle of period matrices has local sections. Hence every holomorphic mapping $f: S \to D$ can be represented locally by a holomorphic, matrix-valued function $\Omega(s)$, $s \in S$, whose values satisfy the two bilinear relations (3.27). The mapping f is horizontal exactly when the column vectors of $d\Omega$ correspond to one-forms with values in $F^1 = F^{2\perp}$; in other words, when

$$ {}^t\Omega(s)\, Q\, d\Omega(s) = 0, \tag{3.28} $$

which looks like an infinitesimal period relation.

We now consider a variation of Hodge structure with base S;[15] or, more concretely, the periods of holomorphic two-forms for a polarized family of algebraic manifolds $\pi: X \to S$. We fix a base point $s_0 \in S$, and we choose a basis $\{e_i\}$ of the fibre of $\mathbf{H_Z}$ over s_0. In the geometric situation, this amounts to choosing a basis for the primitive part of $H_2(V_{s_0}, \mathbf{Z})$. Displacing the e_i horizontally, we obtain a flat frame $\{e_i(s)\}$, $s \in \tilde{S}$, for the pullback of $\mathbf{H_Z}$ to the universal covering $\tilde{S}$ of S. If $\varphi : \pi_1(S) \to G_{\mathbf{Z}} = SO(Q, \mathbf{Z})$ is the representation corresponding to the flat bundle $\mathbf{H} \to S$, one finds that

$$ e_i(\gamma s) = \varphi(\gamma)\, e_i(s), \text{ whenever } \gamma \in \pi_1(S),\ s \in \tilde{S}. \tag{3.29} $$

It may be possible to find a holomorphic frame $\{\sigma_1, \ldots, \sigma_r\}$ of the bundle $\mathbf{F}^2 \to S$, although usually only outside of a subvariety of S. By pairing the σ_i's with the dual frame to $\{e_i(s)\}$, we obtain a holomorphically varying period matrix $\Omega(s)$, $s \in \tilde{S}$, which describes the period mapping.[16] The transformation property (3.29) implies

$$ \Omega(\gamma s) = \varphi(\gamma)\, \Omega(s), \text{ for } \gamma \in \pi_1(S),\ s \in \tilde{S}. \tag{3.30} $$

Equivalently, we may think of $\Omega(s)$ as a *multiple-valued* function on S.

We shall use the language of period matrices to sketch a proof of the theorem on regular singular points. Let $\pi: X \to \Delta^*$ be a family of polarized algebraic manifolds over the punctured disc $\Delta^* =$

[15] The assumptions made at the beginning of this section still apply.

[16] For a general m, this process yields a collection $\{\Omega_i(s)\}$ of $\left[\frac{m+1}{2}\right]$ matrix-valued functions.

$\{z \in \mathbf{C} \mid 0 < |z| < 1\}$, which can be continued to a family $\pi\colon X \to \Delta$ over the entire disc Δ, by inserting a possibly singular fibre over the origin. For any m between 0 and the fibre dimension, we consider the period mapping $f\colon S \to M = \Gamma\backslash D$ which corresponds to the m-th primitive cohomology groups.

(3.31) THEOREM (*Regular singular points*):[17] *The period mapping can be represented by holomorphic, multiple valued period matrices* $\{\Omega_k(t)\}$, $t \in \Delta^*$, *which satisfy the estimate*

$$\|\Omega_k(t)\| \leqslant C|t|^{-\mu}$$

in the slit disc $0 < \arg t < 2\pi$.

REMARK. For a more concrete statement of the theorem, one should mention which choice of a holomorphic frame of the bundles $\mathbf{F}^p \to \Delta^*$ gives period matrices with this property. This will be done in the course of the argument.

SKETCH OF PROOF. We will discuss how one proves the theorem for $m = 2$, i.e. for the periods of the holomorphic two-forms, referring to the reference cited in footnote (17) for the general argument. Our proof will be analytic, but it is worthwhile remarking that a purely algebraic argument has been found by Katz [31].

By assumption, there is a projective embedding $X \subset \mathbf{P}^N$. Standard arguments involving the Lefschetz theorem allows us to replace $V_t = \pi^{-1}(t)$ by a surface lying in V_t[18], and so we shall assume that the V_t are algebraic surfaces. A generic projection $X \to \mathbf{P}^3$ will now realize the V_t as hypersurfaces in $\mathbf{P}^3$ given by a single polynomial equation of fixed degree δ

$$P(x, y, z; t) = 0, \tag{3.32}$$

where the coefficients of P are holomorphic functions of $t \in \Delta$. For this, we may have to shrink the disc Δ. The arguments given by Landman [35] show that we may assume that the surface (3.32) has

[17] The theory of differential equations with regular singular points arising in algebraic geometry is discussed extensively in [15] where several different proofs of the regularity theorem are given.

[18] The point is that, if $\dim V_t = n$, and if $S_t = V_t \cap \mathbf{P}^{n-2}$ is a generic intersection, then there is an injection

$$H^2(V_t, \mathbf{Q}) \hookrightarrow H^2(S_t, \mathbf{Q})$$

at most *ordinary singularities* for $t \neq 0$.[19] The holomorphic 2-forms on V_t are given by

$$\omega = \frac{Q(x, y, z; t)\, dx \wedge dy}{\dfrac{\partial P}{\partial z}(x, y, z; t)}$$

where $Q(x, y, z; t)$ is a polynomial of degree $\delta - 4$ vanishing on the double curve of V_t.[20] Since dim $H^{2,0}(V_t)$ is constant, it follows that we may choose a basis

$$\omega_j(t) = \frac{Q_j(x, y, z; t)\, dx \wedge dy}{\dfrac{\partial P}{\partial z}(x, y, z; t)} \tag{3.33}$$

for $H^{2,0}(V_t)$, where the $Q_j(x, y, z; t)$ are polynomials of degree $\delta - 4$ whose coefficients are holomorphic functions of t in the whole disc Δ, again possibly after shrinking Δ. We may think of the $\omega_j(t)$ as an *algebraic framing* for the vector bundle $\mathbf{F}^2 \to \Delta^*$. In more sheaf-theoretic terms, the $\omega_j(t)$ are rational sections of the *coherent sheaf* over Δ,

$$R^0_{f_*}(\Omega^2_{X/\Delta})$$

which give a basis for each fibre $R^0_{f_*}(\Omega^2_{X/\Delta})_t\, (t \neq 0)$. It is this latter language which forms the natural setting for the generalization of our argument to Hodge structures of arbitrary weight.

Now choose a basis $e_1, \ldots, e_{2r+s}$ for the primitive part of $H_2(V_{t_0}, \mathbf{Q})$. The cycles e_i displace in a multiple-valued fashion to give cycles $e_i(t)$ all fibres $V_t(t \neq 0)$. Moreover, by considering the "collapsing map"

$$V_t \to V_0,$$

the cycles $e_i(t)$ will tend to cycles $e_i(0)$ on V_0, and during this process the *volumes* of the $e_i(t)$ will remain *bounded* (cf. the explicit description of the $e_i(t)$ given by Landman [35]). It is now clear that the

[19] Families of surfaces with ordinary singularities are discussed in Appendix II to [19].

[20] This is classical; c.f. the reference cited in footnote (19).

integrals satisfy the estimate

$$\int_{e_i(t)} \omega_j(t) = 0(|t|^{-\mu})$$

on any sector $0 < \arg t < 2\pi$. The period matrix

$$\Omega(t) = \left(\int_{e_i(t)} \omega_j(t) \right)$$

describes the period mapping, and so we have proved the theorem.

4. Varieties with normal crossings. As was mentioned in the introduction, Deligne has put functorial mixed Hodge structures on the complex cohomology groups of a projective variety [14]. The construction involves a substantial amount of homological algebra. However, the mixed Hodge structures can be described very concretely in one special case, namely that of a variety with normal crossings. According to Hironaka [29], a suitable modification of an arbitrary variety has this form. For many applications, the knowledge of the general existence theorem, together with the concrete description in the case of a variety with normal crossings, suffices already.

(ADDED IN PROOF: Recent notes by M. Anderson at the I.A.S. give a proof of Deligne's general result, extending the methods of this section.)

Thus we consider a compact analytic space V, which can be realized locally as a union of coordinate hyperplanes

$$\{(z_1 \ldots z_{n+1}) \in \mathbf{C}^{n+1} \mid z_1 \cdot z_2 \cdot \ldots \cdot z_k = 0, \ |z_i| < \epsilon\}. \tag{4.1}$$

We assume moreover that globally $V = D_1 \cup \ldots \cup D_N$, where the D are compact Kähler manifolds meeting transversely, as in (4.1).

(4.2) PROPOSITION. *On $H^*(V, \mathbf{C})$, there exists a mixed Hodge structure, which is functorial for holomorphic mappings between compact analytic spaces satisfying the above two conditions.*

As a corollary to the proof, we will find that the weight filtration on $H^m(V, \mathbf{C})$ has the form

$$\{0\} \subset W_0 \subset W_1 \subset \ldots \subset W_{m-1} \subset W_m = H^m(V, \mathbf{C});$$

in particular, $h^{p,q}(H^m(V, \mathbf{C})) = 0$ unless $p, q \geqslant 0$ and $p + q \leqslant m$. The same restrictions apply to the mixed Hodge structures on the cohomology of a general projective variety.

The proof will be given in several steps.

STEP ONE. We recall the spectral sequence of a double complex [17]. Let $A^{**} = \oplus_{p,q \geqslant 0} A^{p,q}$ be a bigraded vector space and assume given

$$\left.\begin{aligned} &d\colon A^{p,q} \to A^{p+1,q},\ d^2 = 0\\ &\delta\colon A^{p,q} \to A^{p,q+1},\ \delta^2 = 0\\ &d\delta + \delta d = 0. \end{aligned}\right\} \tag{4.3}$$

Then one may consider the total differential $D = d + \delta\colon A^k \to A^k$, where $A^k = \oplus_{p+q=k} A^{p,q}$. There is a spectral sequence $\{E_r^{p,q}\}$ with

$$\left.\begin{aligned} &E_1 = H_d^*(A^{**});\\ &E_2 = H_\delta^*(H_d^*(A^{**}));\\ &d_1 = \delta; \text{ and}\\ &E_\infty = H_D^*(A^{**}). \end{aligned}\right\} \tag{4.4}$$

Here the filtration on $H_D^*(A^{**})$, with associated graded E_∞, is induced from the filtration $W_n = \oplus_{p \leqslant n} A^{p,*}$ on A^{**}

STEP TWO. For each index set $I = \{i_1, \ldots, i_q\} \subset \{1, \ldots, N\}$ we set

$$D_I = D_{i_1} \cap \ldots \cap D_{i_q},$$

$$|I| = q,$$

$$D^{[q]} = \bigsqcup_{|I|=q} D_I \qquad \text{(disjoint union)}.$$

Each $D^{[q]}$ is a compact Kähler manifold, and we define

$$A^{p,q} = A^p(D^{[q]}),$$

where $A^*(D^{[q]})$ is the usual de Rham complex. A form $\varphi \in A^{p,q}$ may be written as

$$\varphi = \sum_{|I|=q} \varphi_I;$$

$\varphi_I \in A^p(D_I)$ is the "value" of φ on D_I. Define

$$d: A^{p,q} \to A^{p+1,q} \ (d = \text{exterior derivative}), \text{ and}$$
$$\delta: A^{p,q} \to A^{p,q+1}$$

by the formula

$$(\delta\varphi)_{(j_1 \dots j_{q+1})} = \sum_{l=1}^{q} (-1)^l \varphi_{(j_1 \dots \hat{j}_l \dots j_{q+1})} \Big|_{D(j_1 \dots j_{q+1})} \tag{4.5}$$

The properties (4.3) are immediate, and so $\{A^{**}, d, \delta\}$ is a double complex.

(4.6) Lemma (*de Rham theorem for* V): $H^*_D(A^{**}) \cong H^*(V, C)$.

Proof. There are obvious sheaves $\mathscr{A}^{p,q}$ on V with $\Gamma(V, \mathscr{A}^{p,q}) = A^{p,q}$ and where $H^r(V, \mathscr{A}^{p,q}) = 0$ for $r > 0$ (partition of unity). Setting $\mathscr{A}^n = \oplus_{p+q=n} \mathscr{A}^{p,q}$ we consider the complex of sheaves

$$0 \longrightarrow \mathbf{C}_V \longrightarrow \mathscr{A}^0 \xrightarrow{D} \mathscr{A}^1 \xrightarrow{D} \mathscr{A}^2 \longrightarrow \cdots.$$

The usual sheaf-theoretic proof of the de Rham theorem on manifolds will apply if we can prove the *Poincaré lemma* for D.

This may be done directly, or deduced from the spectral sequence of a double complex as follows: Let $A^{p,q}(U) = \Gamma(U, \mathscr{A}^{p,q})$, where U is the open set (4.1) on V. By the usual d-Poincaré lemma

$$H_d(A^{p,q}) = 0, \quad p > 0.$$
$$H_d(A^{0,q}) \cong H^0 \ (q\text{-fold intersections}).$$

In the spectral sequence for $\{A^{p,q}(U), d, \delta\}$,

$$E_1^{p,q} = 0, \quad p > 0,$$
$$E_1^{0,q} \cong \mathbf{C}^{\binom{n}{q}}.$$

The d_1 map $\delta: E_1^{0,q} \to E_1^{0,q+1}$ is given by the same formula as occurs in the coboundary operator of a simplex, and consequently

$$E_2^{p,q} = 0 \text{ if } p + q > 0,$$
$$E_2^{0,0} \cong \mathbf{C} \cong H^0(U, \mathbf{C}).$$

Thus $E_2 = E_\infty$ and we have the Poincaré lemma for D.

Step three. Returning to the global case, we define a *weight filtration* W_n and *Hodge filtration* F^p on A^{**} by

$$\left.\begin{aligned} W_n &= \oplus_{r \leqslant n} A^{r,*}, \\ F^p &= \oplus_{r,s} F^p(A^{r,s}). \end{aligned}\right| \quad (4.7)$$

These filtrations induce filtrations on $H_D^*(A^{**})$, and by (4.4) the associated graded of the weight filtration on $H_D^*(A^{**})$ is the E_∞ term in the spectral sequence of a double complex. Thus Proposition 4.2 will follow from the

(4.8) LEMMA. *The Hodge filtration* (4.7) *induces a Hodge structure of pure weight m on E_∞^m in the spectral sequence of* (A^{**}, d, δ).

Referring to (4.4) the E_1 term is

$$E_1^m \cong \oplus_q H^m(D^{[q]}),$$

and consequently has a Hodge structure of pure weight m induced by the Hodge filtration (4.7). The d_1 map is

$$\delta : E_1^n \to E_1^n$$

and, by (4.5) is a morphism of Hodge structures of pure weight m. Thus by Corollary 1.15 the E_2^m term has a Hodge structure of pure weight m. Our task will be complete if we can show that

$$E_2 = E_\infty. \quad (4.9)$$

This is accomplished in the last step of our proof.

STEP FOUR. We must show that

$$d_2 = d_3 = \ldots = 0.$$

Suppose that $[\alpha] \in E_2^{m,q}$. Then $[\alpha]$ is represented by a class in $H^m(D^{[q]})$. Decomposing this class into type, we may assume that $[\alpha]$ is represented by a closed C^∞ form α on $D^{[q]}$ which has type (r, s) with $r + s = m$. The condition $d_1[\alpha] = 0$ is that

$$\delta\alpha = d\beta \quad (4.10)$$

for some $\beta \in A^{m-1,q+1}$, and then

$$d_2(\alpha) = \delta\beta$$

viewed as a class in $E_2^{m-1,q+2}$. Applying Lemma 2.13, we may write in two ways

$$\delta\alpha = d\beta'$$

$$\delta\alpha = d\beta''$$

where β' has type $(r, s-1)$ and β'' has type $(r-1, s)$. But $[\delta\beta'] = [\delta\beta'']$ in E_2^{m-1} which has a Hodge structure of pure weight $m-1$. Thus $d_2[\alpha] = 0$ by the principle of two types.

5. The case of non-compact varieties. Let X be a smooth, quasi-projective algebraic variety/**C**. We will prove the following result of Deligne [13]:

(5.1) THEOREM. *The cohomology $H^*(X)$ has a functorial mixed Hodge structure.*

The following will be corollaries of the proof of (5.1):

(5.2) COROLLARY. *The Hodge numbers $h^{p,q}(H^n(X)) = 0$ unless $0 \leqslant p, q \leqslant n$, $n \leqslant p+q \leqslant 2n$.*

(5.3) COROLLARY. *Let $\bar{X}$ be a smooth compactification* (*cf.* §5(a) *below*) *of X. Then the image of $H^n(\bar{X}) \to H^n(X)$ is $W_n\{H^n(X)\}$.*

(a) *The C^∞ log complex.* According to Hironaka [29], we may find a *smooth compactification* $\bar{X}$ of X. Thus X is a smooth, projective variety on which there is a divisor D with *normal crossings*, such that $X = \bar{X} - D$. Locally D is given by

$$\left.\begin{aligned} z = (z_1, \dots, z_n) \in \mathbf{C}^n \\ |z_i| < \epsilon \\ z_1 \dots z_k = 0, \end{aligned}\right\} \qquad (5.4)$$

and it will simplify matters notationally to assume (as we may) that globally

$$D = D_1 \cup \dots \cup D_N,$$

where the D_i are smooth divisors meeting transversely. The general case can be treated with only a slight additional twist.

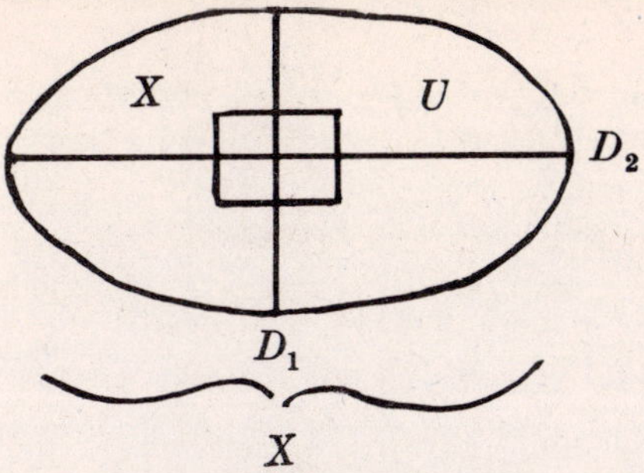

(Fig. 5.5)

By a *neighborhood at infinity* we will mean an open set $U \subset X$ given by (5.4). By $\bar{U}$ we will mean the polycylinder $|z_i| < \epsilon$ so that

$$U = \bar{U} - \bar{U} \cap D.$$

(5.6) DEFINITION. *The 'C^∞ log complex' $A^*(U, \log \langle D \rangle)$ is the complex of C^∞ forms $\varphi \in A^*(U)$ such that*

$$z_1 \dots z_k \varphi$$
$$z_1 \dots z_k \, d\varphi$$

are C^∞ in $\bar{U}$.

Note that $A^*(U, \log \langle D \rangle)$ is closed under d, and it will follow from Lemma 5.7 that the product of two forms in $A^*(U, \log \langle D \rangle)$ is again in the log complex. Thus $A^*(U, \log \langle D \rangle)$ is a sub-complex of the full de Rham complex $A^*(U)$. Note that $A^*(U, \log \langle D \rangle)$ is *not* closed under conjugation.

(5.7) LEMMA. $A^*(U, \log \langle D \rangle) = \dot{A}^*(U) \left\{ \dfrac{dz_1}{z_1}, \dots, \dfrac{dz_k}{z_k} \right\}$.

PROOF. We assume that $k = 1$; the general case is similar. Clearly

$$A^*(\bar{U}) \left\{ \frac{dz_1}{z_1}, \dots, \frac{dz_k}{z_k} \right\} \subset A^*(U, \log \langle D \rangle).$$

Conversely, suppose that $\alpha \in A^*(U, \log \langle D \rangle)$ and write

$$\alpha = \beta \wedge \frac{dz_1}{z_1} + \frac{\gamma}{z_1},$$

where β, γ do not involve dz_1. By the first condition in (5.6), we may assume that β, γ are C^∞ in $\bar{U}$. Now

$$d\alpha = \left(\frac{(d\beta \wedge dz_1}{z_1} + \frac{d\gamma}{z_1}\right) + \frac{\gamma \wedge dz_1}{(z_1)^2},$$

and the second condition in (5.6) gives that $\dfrac{\gamma}{z_1} = \delta \in A^*(\bar{U})$. Then $\alpha = \beta \wedge \dfrac{dz_1}{z_1} + \delta$ lies in $A^*(\bar{U})\left\{\dfrac{dz_1}{z_1}\right\}$.

(5.8) DEFINITION. *The 'C^∞ log complex'* $A^*(X, \log\langle D\rangle)$ *on* X *is the sub-complex of* $A^*(X)$ *consisting of all* φ *which are in* $A^*(U, \log\langle D\rangle)$ *for all neighborhoods* U *at infinity.*

To give the global analogue of (5.7), we consider the line bundles $[D_i] \to \bar{X}$ and choose sections $\sigma_i \in \Gamma(\bar{X}, \mathcal{O}[D_i])$ with $(\sigma_i) = D_i$ and fibre metrics in $[D_i]$. Setting

$$\eta_i = \frac{1}{2\pi\sqrt{-1}}\,\partial \log |\sigma_i|^2,$$

$$\omega_i = \bar{\partial}\eta_i,$$

it follows (cf. §2(b)) that

$$A^*(X, \log\langle D\rangle) = A^*(\bar{X})\{\eta_1, \ldots, \eta_N\}, \tag{5.9}$$

and $\omega_i \in H^2_{DR}(\bar{X})$ is the dual cohomology class of D_i.

It will be proved now that the natural map

$$H^*(A^*(X, \log\langle D\rangle)) \to H^*(A^*(X)) \simeq H^*(X, \mathbf{C})$$

is an isomorphism, and that $A^*(X, \log\langle D\rangle)$ has a weight and Hodge filtration inducing a mixed Hodge structure on $H^*(X)$.

(b) *The weight filtration and the Poincaré residue operator.*

DEFINITION. *On* $A^*(X, \log\langle D\rangle)$ *we define the 'weight filtration'* $W_l = W_l(A^*(X, \log\langle D\rangle)$ *to be those forms* φ *such that locally at infinity*

$$\varphi \in A^*(\bar{U})\left\{\frac{dz_{i_1}}{z_{i_1}}, \ldots, \frac{dz_{i_l}}{dz_{i_l}}\right\}.$$

Informally, φ has weight l if φ involves at most l $\dfrac{dz_i}{z_i}$'s. Clearly

$$\left.\begin{aligned} W_l &\subset W_{l+1} \\ dW_l &\subset W_l \\ W_l \wedge W_{l'} &\subset W_{l+l'}. \end{aligned}\right\} \tag{5.10}$$

Note that the definitions of the log complex and weight filtration are local around a point $z \in X$. Thus we may define *complexes of sheaves on* $\bar{X}$

$$\mathscr{A}^*(\log \langle D \rangle)$$
$$\mathscr{W}_l = W_l\,(\mathscr{A}^*(\log\langle D\rangle))$$

such that $A^*(X, \log\langle D\rangle) = \Gamma(\bar{X}, \mathscr{A}^*(\log\langle D\rangle))$ and similarly for W_l. By the usual partition of unity argument, these sheaves have no higher cohomology.

Given that $D = D_1 \cup \ldots \cup D_N$, we shall use the following notations:

$$I = \{i_1, \ldots, i_k\} \subset \{1, \ldots, N\} \text{ is an } \textit{index set};$$
$$D_I = D_{i_1} \cap \ldots \cap D_{i_k},$$
$$|I| = k;$$

and

$$D^{[k]} = \cup_{|I|=k} D_I.$$

DEFINITION. *The 'Poincaré residue operator'*

$$R^{[k]}: \mathscr{W}_k(\mathscr{A}^*(\log\langle D\rangle)) \to \mathscr{A}^{*-k}(D^{[k]})^{(21)}$$

is defined by

$$R^{[k]}\left(\alpha \wedge \frac{dz_{i_1}}{z_{i_1}} \wedge \ldots \wedge \frac{dz_{i_k}}{z_{i_k}}\right) = \alpha\,|_{D_I}. \tag{5.11}$$

The following lemma is easy to verify

(5.12) LEMMA. (i) *$R^{[k]}$ is well defined and $R^{[k]}(\mathscr{W}_{k-1}) = 0$; and* (ii) *$R^{[k]}$ commutes with d, ∂, and $\bar{\partial}$.*

Not quite so simple is the *Poincaré lemma* in the present context. To explain it, we recall that associated to a complex of sheaves

21 $\mathscr{A}^*(\mathscr{D}^{[k']})$ is the sheaf of C^∞ forms on $D^{[k]}$, sometimes referred to as the *de Rham sheaf.*

$\mathscr{L}^* = \{\ldots \to \mathscr{L}^p \xrightarrow{d} \mathscr{L}^{p+1} \to \ldots\}$ on a space, the *cohomology sheaf*

$$\mathscr{H}^*(\mathscr{L}^*)$$

is that sheaf coming from the presheaf

$$U \longmapsto \text{cohomology of } \{\ldots \Gamma(U, \mathscr{L}^p) \xrightarrow{d} \Gamma(U, \mathscr{L}^{p+1}) \to \ldots\}.$$

Thus, e.g., the statement $\mathscr{H}^q(\mathscr{L}^*) = 0$ for $q > 0$ is nothing other than the Poincaré lemma. In the situations that we shall encounter, $\mathscr{L}^*$ will be a subcomplex of the de Rham sheaf and will be closed under d, ∂, and $\bar{\partial}$. We shall denote the various cohomology sheaves by $\mathscr{H}^*_d(\mathscr{L}^*)$, $\mathscr{H}^*_\partial(\mathscr{L}^*)$, $\mathscr{H}^*_{\bar{\partial}}(\mathscr{L}^*)$.

(5.13) LEMMA. *The induced mappings*

$$R^{[k]}: \mathscr{H}^*_d(\mathscr{W}_k/\mathscr{W}_{k-1}) \to \mathscr{H}^{*-k}_d(\mathscr{A}^*(D^{[k]}))$$
$$R^{[k]}: \mathscr{H}^*_{\bar{\partial}}(\mathscr{W}_k/\mathscr{W}_{k-1}) \to \mathscr{H}^{*-1}_{\partial}(\mathscr{A}^*(D^{[k]}))$$

are isomorphisms.[22]

The proof of (5.13) is essentially the same as the proof of Proposition (2.18), and will be omitted.

(c) *de Rham's theorem for the log complex.*

(5.14) PROPOSITION. *The inclusion* $A^*(X, \log \langle D\rangle) \hookrightarrow A^*(X)$ *induces an isomorphism on cohomology. Thus*

$$H^*(A^*X, \log \langle D\rangle) \xrightarrow{\sim} H^*(X, \mathbf{C}).$$

PROOF. Essentially, this proposition is true because it is true locally at infinity by (5.13). The easiest way to make this precise is using some homological nonsense.

The basic fact we shall utilize is that if $\mathscr{L}^* = \{\ldots \to \mathscr{L}^p \xrightarrow{d} \mathscr{L}^{p+1} \to \ldots\}$ is a complex of sheaves on a space Y and if the cohomology $H^q(Y, \mathscr{L}^p) = 0$ for $q > 0$ then there is a spectral sequence $\{E_r\}$ with

[22] Since $R^{[k]}$ is surjective, the subtlety here is that there are forms $\alpha \in \mathscr{W}_k$ with $R^{[k]}\alpha = 0$ but $\alpha \notin \mathscr{W}_{k-1}$ (e.g. $\bar{z}\,\frac{dz}{z}$ on $\mathbf{C}$.) What must be proved is that these don't matter when we pass to cohomology. The lemma is *false* for ∂-cohomology, essentially because we are using $\frac{dz}{z}$ and not $\frac{d\bar{z}}{\bar{z}}$ for the log complex.

$$E_2 = H^*(Y, \mathscr{H}^*(\mathscr{L}^*))$$

$$E^\infty \Rightarrow \text{cohomology of } \{\ldots \Gamma(Y, \mathscr{L}^p) \xrightarrow{d} \Gamma(Y, \mathscr{L}^{p+1}) \to \ldots\}$$

(cf. [17], pages 176-179). As an application, we see that if $\mathscr{L}^*$, $\mathscr{K}^*$ are two complexes of sheaves with $H^q(Y, \mathscr{L}^*) = 0 = H^q(Y, \mathscr{K}^*)$ for $q > 0$, and if we are given a morphism

$$\mathscr{L}^* \xrightarrow{\psi} \mathscr{K}^*$$

of complexes of sheaves such that cohomology sheaf map

$$\psi_* : \mathscr{H}^*(\mathscr{L}^\bullet) \xrightarrow{\sim} \mathscr{H}^*(\mathscr{K}^\bullet)$$

is an isomorphism, then the global cohomology map

$$\frac{\ker\{\Gamma(Y, \mathscr{L}^p) \to \Gamma(Y, L^{p+1})\}}{d\,\Gamma(Y, L^{p-1})} \to \frac{\ker\{\Gamma(Y, \mathscr{K}^p) \to \Gamma(Y, \mathscr{K}^{p+1})\}}{d\,\Gamma(Y, \mathscr{K}^{p-1})}$$

is also an isomorphism.

We shall apply this principle taking $Y = \bar{X}$, $\mathscr{L}^* = \mathscr{A}^*(\log \langle D \rangle)$, and $\mathscr{K}^* = j_*(\mathscr{A}^*(X))$, where $j : X \hookrightarrow \bar{X}$ is the inclusion mapping. Note that $\Gamma(\bar{X}, \mathscr{A}^*(\log \langle D \rangle)) = A^*(X, \log \langle D \rangle)$ and $\Gamma(\bar{X}, j_*(\mathscr{A}^*(X))) = A^*(X)$. Proposition 5.14 consequently follows from the

LEMMA. *The induced mapping*

$$\mathscr{H}^*(\mathscr{A}^*(\log \langle D \rangle)) \xrightarrow{\sim} \mathscr{H}^*(j_*(\mathscr{A}^*(X)))$$

is an isomorphism.

PROOF. The question is local in a neighborhood U at infinity given by (5.4). Let

$$\mathbf{C}\left\{\frac{dz_1}{z_1}, \ldots, \frac{dz_k}{z_k}\right\} = \mathbf{C}\left\{\left\{\frac{dz_i}{z_i}\right\}\right\}$$

be the free differential graded algebra generated by $\dfrac{dz_1}{z_1}, \ldots, \dfrac{dz_k}{z_k}$ and having differential $d = 0$. There is a commutative triangle

$$\begin{array}{ccc} A^*(U, \log \langle D \rangle) & \longrightarrow & A^*(U) \\ \nwarrow \mu & & \nearrow \nu \\ & \mathbf{C}\left\{\dfrac{dz_1}{z_1}, \ldots, \dfrac{dz_k}{z_k}\right\} & \end{array}$$

of complexes, and ν_* is an isomorphism on cohomology by the usual de Rham theorem. Thus it will suffice to show that μ_* is an isomorphism on cohomology.

There is an obvious weight filtration $W_l\left(\mathbf{C}\left\{\left\{\dfrac{dz_i}{z_i}\right\}\right\}\right)$ such that

$$\mu\left[W_l\left(\mathbf{C}\left\{\left\{\frac{dz_i}{z_i}\right\}\right\}\right)\right] \subset W_l(A^*(U, \log \langle D \rangle)).$$

Moreover, there is a commutative triangle

$$\begin{array}{ccc} \dfrac{W_l(A^*(U, \log \langle D \rangle))}{W_{l-1}(A^*(U, \log \langle D \rangle))} & \overset{R^{[k]}}{\longrightarrow} & A^{*-k}(D^{[k]} \cap U) \\ \nwarrow \alpha & & \nearrow \beta = \text{Poincaré residue} \\ & \dfrac{W_l\left(\mathbf{C}\left\{\left\{\frac{dz_i}{z_i}\right\}\right\}\right)}{W_{l-1}\left(\mathbf{C}\left\{\left\{\frac{dz_i}{z_i}\right\}\right\}\right)} & \end{array}$$

where β_* is obviously an isomorphism on cohomology, and $R_*^{[k]}$ is also an isomorphism by Lemma (5.13). (This is the main step in the proof). Thus α_* is an isomorphism on cohomology. Using this it follows inductively on l that

$$\mu : W_l\left[\mathbf{C}\left\{\left\{\frac{dz_i}{z_i}\right\}\right\}\right] \longrightarrow W_l(A^*(U, \log \langle D \rangle))$$

induces an isomorphism on cohomology. For $l = k$, we obtain our lemma.

REMARK. Lemma 5.13 plus the same method of proof gives the

(5.15) COROLLARY. *The residue mapping* $R^{[k]}: W_k/W_{k-1} \to A^{*-k}(D^{[k]})$ *induces isomorphism on both* d *and* $\bar\partial$ *cohomology.*

(d) *The weight and Hodge filtrations on cohomology.* On the C^∞ log complex $A^*(X, \log\langle D\rangle)$ we have defined the weight filtration, W_l in § 5(b), and we now define the *Hodge filtration* by

$$F^p A^*(X, \log\langle D\rangle) = \oplus_{i \geqslant p} A^{i,*}(X, \log\langle D\rangle). \tag{5.16}$$

The weight and Hodge filtrations induce filtrations on the cohomology $H^*(A^*(\log\langle D\rangle)) \cong H^*(X, \mathbf{C})$, and it is to be proved that (with suitable indexing!) this gives a mixed Hodge structure.

(5.17) PROPOSITION. *The weight filtration* $W_n\{H^*(X, \mathbf{C})\}$ *is defined over* $\mathbf{Q}$.

PROOF. We begin with a preliminary remark. Suppose that $\varphi \in W_l(A^*(X, \log\langle D\rangle))$ is a closed form of filtration level l. Then the Poincaré residue

$$R^{[l]}\varphi \in A^{*-l}(D^{[l]})$$

is a closed form giving a cohomology class in $H^{*-l}(D^{[l]}, \mathbf{C})$. If this class is zero, then by Corollary 5.15 we may find $\theta \in W_l$ such that $\varphi - d\theta \in W_{l-1}$.

To prove the proposition, we take a closed form $\varphi \in A^*(X, \log\langle D\rangle)$. Then

$$R^{[n]}\varphi \in H^*(D^{[n]}, \mathbf{C})$$

is well defined. If $\varphi \in W_{n-1}$, then $R^{[n]}\varphi = 0$, and conversely if $R^{[n]}\varphi$ is zero in cohomology, then by the above remark we may find $\theta \in A^*(X, \log\langle D\rangle)$ such that $\varphi - d\theta \in W_{n-1}$. Repeating this argument we find that

$$W_l\{H^*(X, \mathbf{C})\} = \{\varphi \in H^*(X, \mathbf{C}) : R^{[n]}\varphi = \ldots = R^{[l+1]}\varphi = 0\} \tag{5.18}$$

where, on the right hand side, the $R^{[\nu]}\varphi$ are taken in $H^*(D^{[\nu]}, \mathbf{C})$.

With the description (5.18), it is clear that the weight filtration $W_l\{H^*(X, \mathbf{C})\}$ is defined over $\mathbf{Q}$.

Now consider the filtration

$$W^{-l} = W_l(A^*(X, \log \langle D \rangle)) \tag{5.19}$$

on the log complex. The indices have been reversed in order to give a *decreasing* filtration,

$$\ldots \supset W^{-l} \supset W^{-l+1} \supset \ldots,$$

so that we may consider the spectral sequence (=s.s.) of a filtered complex [17]. Accordingly, there is a s.s. $\{E_r\}$ such that E_∞ is the associated graded to the weight filtration on $H^*(X, \mathbf{C})$ and

$$E_1 = H^*(W^{-l}/W^{-l+1}).$$

(5.20) LEMMA. *The Hodge filtration $\{F^p\}$ induces a Hodge structure of pure weight $k+l$ on $H^k(W^{-l}/W^{-l+1})$.*

PROOF. Using the obvious notations,

$$F^p(W^{-l}/W^{-l+1}) \cong (\oplus_{i \geqslant p} W_l^{i,*})/W_{l-1}.$$

The Poincaré residue operator induces

$$F^p(W^{-l}/W^{-l+1}) \xrightarrow{R^{[l]}} F^{p-l}(A^*(D^{[l]})).$$

Applying Corollary 5.15, it follows that

$$H^k(W^{-l}/W^{-l+1}) \cong H^{k-l}(D^{[l]})$$

is a morphism of Hodge structures of type $(-l, -l)$.

(5.21) LEMMA. *The mapping $d_1 : E_1 \to E_1$ is a morphism of Hodge structures.*

PROOF. Using the isomorphism

$$E_1 \cong \oplus_I H^*(D_I),$$

we shall prove using (2.19) that d_1 is given by Gysin mappings

$$H^*(D_{i_1 \ldots i_l}) \longrightarrow H^{*+2}(D_{i_1 \ldots i_{l-1}}).$$

A class $\varphi \in H^*(W^{-l}/W^{-l+1})$ may be represented by a form $\varphi \in W_l(A^*(X, \log \langle D \rangle))$ such that $R^{[l]}\, d\varphi = d\, R^{[l]}\, \varphi = 0$. Then by definition

$$d_1 \varphi = R^{[l-1]} d\varphi,$$

since d_1 is the coboundary map arising from the exact sequence of complexes

$$0 \longrightarrow W_{l-1}/W_{l-2} \longrightarrow W_l/W_{l-2} \longrightarrow W_l/W_{l-1} \longrightarrow 0.$$

Write $\eta_I = \eta_{i_1} \wedge \cdots \wedge \eta_{i_l}$ and $\varphi = \sum_{|i|=l} \eta_I \wedge \varphi_I$. Then by an obvious computation

$$R^{[l-1]} d\varphi|_{D_{i_1 \ldots i_{l-1}}} = \sum_j \pm \{w_j \wedge \varphi_{i_1 \ldots i_{l-1} j} + \eta_j \wedge d\varphi_{i_1 \ldots i_{l-1} j}\},$$

which is just the Gysin mapping according to (2.19).

(5.22) Corollary. *The weight and Hodge filtrations on* $A^*(X, \log \langle D \rangle)$ *induce a mixed Hodge structure on* E_2.

Proof. This follows from Lemmas 5.20, 5.21, and Corollary 1.15.

The main remaining step in the proof of Deligne's theorem is to show that the spectral sequence in question degenerates at E_2. This will be proved in the next section.

(e) *Completion of the proof of Theorem* 5.1.

(i) *Degeneration of the spectral sequence.*

We continue the discussion of §5(c). In case D is smooth, the weight filtration is just $W_0 \subset W_1$ and consequently $E_2 = E_\infty$. The crucial case is when $D = D_1 \cup D_2$, and we shall check here that $d_2 = 0$—this will suffice to make clear how the general argument goes.

Let α be a class in

$$E_1^{-2} = H^*(W_{-2}/W_1) \cong H^{*-2}(D_1 \cap D_2)$$

(cf. 5.15)). We may assume that α is represented by a closed $C^\infty(p, q)$ form on $D_1 \cap D_2$. Let $\tilde{\alpha}$ be a C^∞ extension of α and η_1, η_2 as in (5.9). Then

$$A' = \eta_1 \wedge \eta_2 \wedge \tilde{\alpha}$$

gives a form in $W_2 \{A^*(X, \log \langle D \rangle)\}$ with

$$R^{[2]} A' = \alpha.$$

If $d_1\alpha = 0$, then, referring to the proof of Lemma 5.21, we may find forms β_i on D_i such that

$$\left.\begin{aligned} R^{[1]}(dA')_{D_1} &= \eta_2 \wedge d\tilde{\alpha} + \omega_2 \wedge \tilde{\alpha}|_{D_2} = d\beta_1, \\ R^{[1]}(dA')_{D_2} &= \eta_1 \wedge d\tilde{\alpha} + \omega_1 \wedge \tilde{\alpha}|_{D_1} = d\beta_2. \end{aligned}\right\} \tag{5.23}$$

Moreover, we may choose the β_i to have type $(p+q+1, 0) + \ldots + (p+1, q)$ (cf. (2.10)). Setting $B' = -(\eta_1 \wedge \tilde{\beta}_1 + \eta_2 \wedge \tilde{\beta}_2)$ where the $\tilde{\beta}_i$ are C^∞ extensions of β_1, β_2, we find the relations

$$\left.\begin{aligned} R^{[2]}(A' + B') &= \alpha \\ R^{[2]}(d(A' + B')) &= R^{[1]}(d(A' + B')) = 0 \\ d(A' + B') &\in F^{p+2} A^*(X, \log \langle D \rangle). \end{aligned}\right\} \tag{5.24}$$

Now we may repeat the same argument using $\bar{\eta}_1$, $\bar{\eta}_2$ and solving the equations (5.23) emphasizing the opposite direction in the Hodge filtration. This leads to A'', B'' satisfying

$$\left.\begin{aligned} R^{[2]}(A'' + B'') &= \alpha \\ R^{[2]}(d(A'' + B'')) &= R^{[1]}(d(A'' + B'')) = 0 \\ d(A'' + B'') &\in F^{q+2} A^*(X, \log \langle D \rangle). \end{aligned}\right\} \tag{5.25}$$

Since $\deg [d(A' + B')] = p + q + 3$, equations (5.24) and (5.25) say exactly that $d_2\alpha \in E_2^{0}$[23] has total degree $p + q + 3$ and is in

$$F^{p+2}(E_2^0)F \cap F^{q+2}(E_2^0) = 0,$$

since E_2 has a mixed Hodge structure by (5.22). Thus $\alpha_2\,\alpha = 0$.

REMARK. As was the case in the proof of Lemma 4.7, this proof is simply an application of the principle of two types as discussed at the end of § 2(a).

(ii) *Tying up loose ends.*

Given X, we have chosen a smooth completion $\bar{X}$ of X, defined the log complex $A^*(X, \log \langle D \rangle)$ with a weight filtration W_n and Hodge filtration F^p, and proved that

[23] Recall that $d_2 \colon E_2^{-2} \to E_2^0$.

$H^*(A^*(X, \log \langle D \rangle)) \cong H^*(X, \mathbf{C})$;

the filtrations W_n, F^p induce a mixed

Hodge structure on $H^*(A^*(X, \log \langle D \rangle))$.

Moreover, Corollaries 5.2 and 5.3 follow immediately from the definitions of W_n and (5.20). It remains, therefore, to prove independence of our construction from the smooth completion $\bar{X}$ and functoriality.

Observe first that, given a diagram

$$\begin{array}{ccc} Y & \hookrightarrow & \bar{Y} \\ f \downarrow & & \downarrow \bar{f} \\ X & \hookrightarrow & \bar{X} \end{array} \tag{5.26}$$

then $\bar{f}^*: A^*(X, \log \langle D \rangle) \to A^*(Y, \log \langle D \rangle)$ commutes with the weight and Hodge filtrations. Given smooth completions $\bar{X}_1$, $\bar{X}_2$ of X, by [29] there exists a smooth completion $\bar{X}_3$ and a diagram

$$\begin{array}{ccccc} & & \bar{X}_3 & & \\ & \swarrow & & \searrow & \\ \bar{X}_1 & & \cup & & \bar{X}_2 \\ & \supset & & \subset & \\ & & X & & \end{array}$$

and independence of the smooth compactification follows. Given a morphism $Y \xrightarrow{f} X$ using [29] again we may find a diagram (5.26) and this implies functoriality.

6. Applications of mixed Hodge structures. (a) *Applications to moduli.* Let X, S be smooth quasi-projective varieties and $f: X \to S$ a smooth, proper mapping. Setting $V_s = f^{-1}(s)$ $(s \in S)$, we may think of X as an algebraic family $\{V_s\}_{s \in S}$ of smooth, projective varieties with algebraic parameter space S. Pick a base point $s_0 \in S$ and set $V = V_{s_0}$. The fundamental group $\pi_1(S, s_0)$ acts on the cohomology $H^n(V, \mathbf{Q})$, and we let

$$I^n = H^n(V, \mathbf{C})^{\pi_1}$$

by the *invariant part* of the cohomology under this action. Note that for each point $s \in S$ there is a well-defined inclusion

$$i_s\colon I^n \hookrightarrow H^n(V_s, \mathbf{C})$$

obtained by "transporting" I^n to V_s along some path from s_0 to s.

(6.1) Proposition. (cf. Corollary 4.1.2 of [13]). *$I^n \subset H^n(V, \mathbf{C})$ has an induced Hodge structure, and the inclusions i_s are all morphisms of Hodge structures.*

Remark. In other words, if φ is an invariant cohomology class and if in each $H^n(V_s, \mathbf{C})$ we decompose φ into (p, q) type

$$\varphi = \sum_{p+q=n} \varphi_{p,q}\ (s),$$

then the $\varphi_{p,q}\ (s)$ are *constant* in s. In particular, if φ is of fixed type (p,q) at one point $s \in S$, then φ is everywhere of the same (p, q) type.

In [19], Proposition 6.1 was proved for an arbitrary variation of Hodge structure $\mathbf{H} \to S$ with compact base space S. Using the results of the second author [41] discussed in §9 below, (6.1) may be proved for an arbitrary variation of Hodge structure $\mathbf{H} \to S$ with an algebraic variety as base space (cf. §10).

Proof. According to Deligne's degeneration theorem discussed in §2(a), the image of

$$H^n(X, \mathbf{C}) \xrightarrow{\rho} H^n(V, \mathbf{C})$$

is exactly I^n. Note that $i_s\ (I^n)$ is then the image of

$$H^n(X, \mathbf{C}) \xrightarrow{\rho_s} H^n(V_s, \mathbf{C})$$

for all $s \in S$. Let $\bar{X}$ be a smooth compactification of X. In the diagram

$$\begin{array}{ccc} H^n(X, \mathbf{C}) & \xrightarrow{\rho} & H^n(V_s, \mathbf{C}) \\ \uparrow & \nearrow_{\bar{\rho}} & \\ H^n(\bar{X}, \mathbf{C}) & & \end{array}$$

we will show that

$$\text{image } \rho = \text{image } \bar{\rho}. \tag{6.2}$$

If this is done, then because $\bar{\rho}$ is a morphism of Hodge structures,

$$I^n \cong H^n(\bar{X}, \mathbf{C})/\ker \bar{\rho}$$

will have an induced Hodge structure. Since $\ker \bar{\rho} = \ker \bar{\rho}_s$ for all s, the inclusion i_s will be morphisms of Hodge structures and our proposition is proved.

According to Theorem 5.1, the cohomology $H^n(X, \mathbf{C})$ has a *functorial* mixed Hodge structure $\{W_n, F^p, H^n(X, \mathbf{C})\}$ with

$$W_n(H^n(X, \mathbf{C})) = \text{image } \{H^n(\bar{X}, \mathbf{C}) \to H^n(X, \mathbf{C})\}. \tag{6.3}$$

Since $H^n(\bar{X}, \mathbf{C})$ and $H^n(V, \mathbf{C})$ have mixed Hodge structures of pure weight n and all maps are morphisms of mixed Hodge structures, our assertion (6.2) follows from (6.3) and the strictness Lemma 1.10.

(b) *A direct proof of a result of Deligne about meromorphic forms on projective varieties.* Let V be a compact Kähler manifold. It is well known that

$$\left.\begin{array}{ll}\text{(i)} & \text{every holomorphic form on } V \text{ is closed;}\\ \text{(ii)} & \text{a non-zero holomorphic form is not exact.}\end{array}\right\} \tag{6.4}$$

Clearly (ii) $\Rightarrow$ (i), and (ii) may be most easily seen as follows: Suppose that $\varphi \not\equiv 0$ is a holomorphic q-form and $\varphi = d\eta$ for a C^∞ form η. Letting ω be the Kähler form, by Stokes' theorem, and because $d\omega = 0$,

$$0 < (\sqrt{-1})^{q^2} \int_V \varphi \wedge \bar{\varphi} \wedge \omega^{n-q} = (\sqrt{-1})^{q^2} \int_V d(\eta \wedge \bar{\varphi} \wedge \omega^{n-q}) = 0,$$

a contradiction. In [13] Deligne proved the following generalization of (6.4):

(6.5) PROPOSITION. *Let $D \subset V$ be a divisor with normal crossings and $\varphi \in \Gamma(V, \Omega^q(\log \langle D \rangle))$ a meromorphic q-form having a logarithmic singularity on D. Then*

(i) *$d\varphi = 0$, and*

(ii) *if $\varphi = 0$ in $H^q(V-D, \mathbf{C})$, then $\varphi \equiv 0$.*

PROOF. We shall give a proof similar to the special case discussed above. Observe that the result is not changed if we replace D by $D + H$, where H is a suitable hypersurface section relative to a projective embedding of V. Thus we are free to make D as ample as we wish.

We first show that (i) $\Rightarrow$ (ii). Setting $u = V - D$, if

$$\varphi \in \Gamma(V, \Omega^q(\log \langle D \rangle))$$

is zero in $H^q(u, \mathbf{C})$, then taking u to be affine, we have by the *algebraic de Rham theorem* [26] that

$$\varphi = d\eta \tag{6.6}$$

where η is meromorphic on V and holomorphic on U. The obstruction to lowering the order of pole of η along D to *one* are in cohomology groups (cf. §10 in [20])

$$H^*(V, \Omega_V^{\bullet} \otimes \mathcal{O}[\mu D]) \qquad (* > 0, \mu > 0),$$

and these may be made zero by making D more ample. Thus we may assume that η has a pole of order one on D, and then

$$\eta \in \Gamma(V, \Omega^{q-1}(\log \langle D \rangle))$$

by (6.6). Applying (i), $0 = d\eta = \omega$ which gives (ii).

We now prove (i). The *Poincaré residue* (cf. §5(b)) $R(\varphi)$ is a holomorphic $q - 1$ form on D, and thus $dR(\varphi) = 0$ by the usual result on Kähler manifolds. Thus

$$R(d\varphi) = d(R\varphi) = 0,$$

and so $d\varphi$ is holomorphic on V. Let T_ϵ be an ϵ-tube around D. Then

$$\begin{aligned} 0 &< (\sqrt{-1})^q \int_V d\varphi \wedge d\bar{\varphi} \wedge \omega^{n-q-1} \text{ (since } d\varphi \text{ is holomorphic on } V) \\ &= (\sqrt{-1})^q \lim_{\epsilon \to 0} \int_{\partial T_\epsilon} \varphi \wedge \overline{d\varphi} \wedge \omega^{n-q-1} \text{ (Stokes' theorem)} \\ &= \frac{(-1)^q}{2\pi\sqrt{-1}} \int_D R(\varphi) \wedge \overline{d\varphi} \wedge \omega^{n-q-1} \text{ (by residue formula)} \\ &= 0, \end{aligned}$$

since $R(\varphi)$ has type $(q-1, 0)$ and $\overline{d\varphi}$ has type $(0, q+1)$.

(c) *Intermediate Jacobians.*[24] Let V be a smooth projective variety of dimension n and

$$T(V) = \otimes_{q=1}^{n} T_q(V)$$

the intermediate Jacobian of V.[25] If $\mathscr{A}(V)$ are the algebraic cycles on V which are algebraically equivalent to zero and taken modulo rational equivalence, then one has the *Abel-Jacobi mapping*

$$\varphi : \mathscr{A}(V) \to T(V).$$

The image $\varphi\{\mathscr{A}(V)\}$ is an abelian subvariety $I^0(V)$, and an outstanding problem is to (i) describe $I^0(V)$ algebro-geometrically and (ii) prove that, up to isogeny, $I_q^0(V)$ is the dual abelian variety $\widehat{I_{n-q+1}^{0}(V)}$ to $I_{n-q+1}^{0}(V)$. Using Deligne's theory, it is possible to do this in one significant new case, which shall now be discussed.

To explain this result, we recall from §1 of [22] the notion of *incidence equivalence*, and let $\mathscr{Q}(V) \subset \mathscr{A}(V)$ be algebraic cycles which are incidence equivalent to zero. The quotient

$$\mathscr{A}(V)/\mathscr{Q}(V) = \mathrm{Pic}^0(V)$$

was termed in [22] the identity component of the (*algebraic*) *Picard variety* of V. *Abel's theorem* (loc. cit, §3) gives a factorization

$$\begin{array}{ccc} \mathscr{A}(V) & \xrightarrow{\varphi} & I^0(V) \\ & \rho\searrow \quad \swarrow\psi & \\ & \mathrm{Pic}^0(V) & \end{array} \tag{6.7}$$

It is conjecturally the case that (i) ψ is an isogeny, and (ii) that the duality relation

$$\mathrm{Pic}_q^0(V) \cong \widehat{\mathrm{Pic}_{n-q+1}^{0}(V)} \tag{6.8}$$

is valid upto isogeny. We shall prove that this is so when

[24] The general reference for this section is [22], whose terminology and notations we shall follow. Intermediate Jacobians are also discussed in [36].

[25] $T_q(V) = F^q H^{2q-1}(V, \mathbf{C}) \backslash H^{2q-1}(V, \mathbf{C})/H^{2q-1}(V, \mathbf{Z})$, so that $T_1(V) \cong H^1(V, \mathcal{O})/H^1(V, \mathbf{Z})$ and $T_n(V) = \mathrm{Alb}(V)$.

$$\left.\begin{array}{c} \dim V = 2m+1 \text{ is odd} \\ H^{2q+1}(V, \mathbf{C}) = 0 \text{ for } q \neq m^{(26)}. \end{array}\right\} \qquad (6.9)$$

Proof. We shall first discuss a special case. Let $\{Z_s\}_{s\in S}$ *be an algebraic family* of m-dimensional subvarieties $Z_s \subset V$ with smooth, projective parameter space S, and which is in *general position* so that the *incidence divisors*

$$D_s = \{s' \in S : Z_{s'} \cap Z_s \neq \varnothing\}$$

are defined. The correspondences

$$\begin{array}{ccc} s & \longrightarrow & Z_s \\ \downarrow & & \\ D_s & & \end{array}$$

induce mappings

$$\begin{array}{ccc} \mathrm{Alb}\,(S) & \xrightarrow{\ \varphi\ } & I^0_m(V) \\ \downarrow{\scriptstyle \eta} & \swarrow{\scriptstyle \psi} & \\ \mathrm{Pic}^0_1\,(S) & & \end{array} \qquad (6.10)$$

where the factorization $\eta = \psi \circ \varphi$ is a consequence of Abel's theorem (§3 of [22]). The homology intersection relation

$$(\varphi_*(\alpha), \varphi_*(\beta))_V = \pm\,(\alpha, \eta_*(\beta))_S \,(\alpha, \beta \in H_1(S, \mathbf{Z})) \qquad (6.11)$$

is easy to verify, and it follows from (6.10) and (6.11) that up to isogeny

$$\left\{\begin{array}{l} \ker\varphi = \ker\eta \\ \text{in } (6.10) \end{array}\right\} \Leftrightarrow \left\{\begin{array}{l} \text{the intersection pairing} \\ \text{is non-singular on} \\ \varphi_*[H_1(S, \mathbf{Q})] \end{array}\right\}. \qquad (6.12)$$

[26] Examples of such V are complete intersections in $\mathbf{P}^N$. The reason for assuming (6.9) is that all of $H^{2m+1}(V, \mathbf{Q})$ is *primitive* (§2), and thus the cup-product $H^{2m+1}(V, \mathbf{Q}) \otimes H^{2m+1}(V, \mathbf{Q}) \rightarrow \mathbf{Q}$ is *non-degenerate* on any sub-Hodge structure of $H^{2m+1}(V, \mathbf{Q})$.

Now $\varphi_*[H_1(S, \mathbf{Q})]$ is a sub-Hodge structure of $H_{2m+1}(V, \mathbf{Q})$, and consequently the intersection pairing is non-singular on $\varphi_*[H_1(S, \mathbf{C})]$, since $H^{2m+1}(V, \mathbf{C})$ is all primitive by the assumption (6.9). Thus $\ker \varphi^0 = \ker \eta^0$ in this case.

In general, the point is that if $\{Z_s\}_{s\in S}$ is any algebraic family of algebraic m-cycles on V, then

$$\varphi_* : H_1(S, \mathbf{C}) \to H_{2m+1}(V, \mathbf{C})$$

is a morphism of mixed Hodge structures of type (m, m) and, as a result, the intersection pairing is non-degenerate on

$$\varphi_*[H_1(\mathscr{A}(V), \ \mathbf{C})] \subset H_{2m+1}(V, \mathbf{C}).$$

Now the same argument as before may be applied to give our desired conclusion.

7. Hyperbolic complex analysis and the period mapping.

(a) *General comments: the Ahlfors lemma.* Hyperbolic complex analysis is the study of holomorphic mappings into *negatively curved complex manifolds*; i.e. complex manifolds M having an Hermitian metric ds_M^2 whose *holomorphic sectional curvatures* $K_M(\zeta)$ ($\zeta \in T(M) =$ holomorphic tangent bundle of M) satisfy

$$K_M(\zeta) \leqslant -A < 0^{(27)}. \tag{7.1}$$

If $N \subset M$ is a complex submanifold with induced metric ds_N^2, then for $\zeta \in T(N)$,

$$K_N(\zeta) = K_M(\zeta) - |S(\zeta)|^2, \tag{7.2}$$

where $S(\zeta)$ is the *second fundamental form* of N in M. In particular, N is negatively curved if M is, and this is one of the two reasons why hyperbolic complex analysis works so well, the other being the Ahlfors Lemma 7.7 below.

Let $\Delta = \{z \in \mathbf{C} : |z| < 1\}$ be the unit disc and define a *pseudo-metric* on Δ to be given by

[27] The general philosophy and uses of hyperbolic complex analysis are discussed in the monograph [34] by Kobayashi, and the paper [48] by Wu, both of which contain the relevant definitions and differential-geometric formulas.

$$h(z)dzd\bar{z}, \tag{7.3}$$

where $h(z)$ is C^∞ on Δ and $h(z) > 0$ on $\Delta - R$, for some discrete subset R of Δ. The *Gaussian curvature* and *Ricci form* are defined by

$$K(h) = -\frac{1}{2h}\frac{\partial^2 \log h}{\partial z \partial \bar{z}}$$

$$\operatorname{Ric}(\omega) = \frac{\sqrt{-1}}{4}\left(\frac{\partial^2 \log h}{\partial z \partial \bar{z}}\right) dz \wedge d\bar{z} = -K(h)\,\omega\,; \tag{7.4}$$

here $\omega = \dfrac{\sqrt{-1}}{2} h\, dz \wedge d\bar{z}$ is the (1.1) form associated to the metric. The rules (cf. [24] for further discussion),

$$\operatorname{Ric}(e^{\mu}\omega) = \frac{\sqrt{-1}}{4}(\partial\bar{\partial}\mu) + \operatorname{Ric}(\omega)$$

$$\operatorname{Ric}(f^*\omega) = f^* \operatorname{Ric}(\omega) \qquad (f\colon \Delta \to \Delta \text{ holomorphic})$$

afford easy manipulation of the Gaussian curvatures.

If $f\colon \Delta \to M$ is a non-constant holomorphic mapping into a negatively curved complex manifold M, then by (7.2) $f^*ds_M^2$ is a pseudo-metric on Δ, whose Gaussian curvature K satisfies

$$K \leqslant -A < 0.^{(28)} \tag{7.5}$$

The Ahlfors lemma compares a general pseudo-metric satisfying (7.5) to the *Poincaré metric*

$$\left.\begin{aligned} ds_P^2 &= \pi(z)\ dzd\bar{z} = \frac{dzd\bar{z}}{(1-|z|^2)^2} \\ \operatorname{Ric}\left(\frac{\sqrt{-1}}{2}\pi(z)\,dz\wedge d\bar{z}\right) &= \frac{\sqrt{-1}}{2}\pi(z)dz\wedge d\bar{z} \end{aligned}\right\} \tag{7.6}$$

Despite its simplicity, it is one of the most powerful and subtle tools available in the study of holomorphic mappings.

[28] The pseudo-metric $\dfrac{1}{A} f^*(ds_M^2)$ has Gaussian curvature $K < -1$, and in this way we may always assume that $A=1$ in (7.1). The points $z_0 \in R$ where $h(z_0) = 0$ should be thought of as having $K(z_0) = -\infty$.

(7.7) LEMMA (Ahlfors). *Given a pseudo-metric $h(z)dz d\bar{z}$ on Δ whose Gaussian curvature satisfies $K(z) \leqslant -1$, then $h(z) \leqslant \pi(z)$.*[29]

PROOF. Let $ds^2_P(\rho) = \pi_\rho(z)\, dz d\bar{z} = \dfrac{\rho^2 dz d\bar{z}}{(\rho^2 - |z|^2)^2}$ be the Poincaré metric on $\Delta_\rho = \{z: |z| < \rho\}$ with Gaussian curvature $K_\rho \equiv -1$. Writing

$$h(z) dz d\bar{z} = \mu_\rho(z)\, ds^2_P(\rho),$$

it will suffice to show that $\mu_\rho \leqslant 1$ for $\rho < 1$ since

$$\lim_{\rho \to 1} \mu_\rho(z) = \mu_1(z) \quad (z \in \Delta \text{ fixed}).$$

The reason for doing this is that $\mu_\rho(z)$ goes to zero as $|z| \to \rho$ for $\rho < 1$, and thus μ_ρ has a maximum at some point $z_0 \in \Delta_\rho$. By the maximum principle and (7.4),

$$0 \geqslant \frac{\partial^2 \log \mu_\rho(z_0)}{\partial z\, \partial \bar{z}} = K_\rho\, \pi_\rho(z_0) - K(z_0)\, h(z_0).$$

Since $K_\rho \equiv -1$ and $K(z_0) \leqslant -1$,

$$h_0(z_0) \leqslant \pi_\rho(z_0),$$

which is the same as $\mu_\rho(z_0) \leqslant 1$.

To conclude this section, there are three little properties of the Poincaré metric we wish to record for future reference.

(i) Let $U = \{z \in \mathbf{C}: z = x + iy,\ y > 0\}$ be the upper half plane and $\rho_U(z, z')$ the distance measured in the Poincaré metric

$$ds^2_P = \frac{dx^2 + dy^2}{y^2}.$$

on U. From the formula for ds^2_P

$$\rho_U(z, z+1) = \frac{1}{\operatorname{Im} z}. \tag{7.8}$$

[29] The invariant form of the *Schwarz lemma* [28], due to Pick, states that if $f: \Delta \to \Delta$ is a holomorphic mapping, then $f^*(ds^2_P) \leqslant ds^2_P$, or equivalently

$$\frac{|f'(z)|^2}{(1 - |f(z)|^2)^2} \leqslant \frac{1}{(1 - |z|^2)^2};$$

this follows from the Ahlfors lemma taking $h(z)\, dz d\bar{z} = f^*(ds^2_P)$. Note that the proof of (7.7) is quite similar to the proof of the Schwarz lemma.

(ii) Let $\Delta^* = \{0 < |z| < 1\}$ be the punctured disc and $\Delta^*_\rho = \{0 < |z| < \rho\}$. Via the universal covering map

$$U \longrightarrow \Delta^*$$
$$z \longrightarrow e^{2\pi\sqrt{-1}\,z}$$

the Poincaré metric induces the metric

$$ds^2_P = \frac{dz d\bar{z}}{|z|^2 (\log |z|)^2} \tag{7.9}$$

on Δ^*. Denoting by $\rho_{\Delta^*}(z, z')$ the distance on Δ^* measured using (7.9), for fixed z and $0 < t \leqslant 1$

$$\rho_{\Delta^*}(z, tz) = 0\,(\log \log \frac{1}{t}). \tag{7.10}$$

PROOF. Using (7.9), we have

$$\rho_{\Delta^*}(z, tz) = 0\left(\int_t^1 \frac{ds}{s \log 1/s}\right) = 0\left(\int_t^1 -\frac{d}{ds}\left(\log \log \frac{1}{s}\right) ds\right)$$
$$= 0\left(\log \log \frac{1}{t}\right).$$

(iii) Finally, it follows from (7.9) that for $\rho < 1$

$$\int_{\Delta^*_\rho} \frac{\sqrt{-1}}{2} \frac{dz \wedge d\bar{z}}{|z|^2 (\log |z|^2)^2} = \int_{\substack{|x| \leqslant 1/2 \\ y > -\log \rho}} \frac{dxdy}{y^2} < \infty \tag{7.11}$$

so that Δ^*_ρ has finite non-Euclidean area for $\rho < 1$.

(b) *Unipotence of the Picard-Lefschetz transformation.* Let $D = G_{\mathbf{R}}/V$ be a classifying space for polarized Hodge structures, $\Gamma = G_{\mathbf{Z}}$ the arithmetic group of integral points in $G_{\mathbf{R}}$, and $\Gamma\backslash D$ the corresponding modular variety. Now the principle of hyperbolic complex analysis does not apply to $\Gamma\backslash D$, but *it does apply relative to those holomorphic mappings which might come from algebraic geometry.* More precisely, from (3.14), (7.2), and (7.7) we have

(7.12) LEMMA. *Let $f: \Delta \to \Gamma\backslash D$ be a locally liftable, holomorphic, horizontal mapping. Then*

$$f^*(ds_D^2) \leqslant ds_\Delta^2,$$

where ds_D^2 is the metric on $\Gamma\backslash D$ induced from the $G_{\mathbf{R}}$-invariant metric on D and ds_Δ^2 is the Poincaré metric.

A beautiful and simple application of (7.12) to the Picard-Lefschetz transformation (§3(b)) has been given by Borel. Let

$$f\colon \Delta^* \to \Gamma\backslash D \tag{7.13}$$

be a locally liftable, holomorphic, horizontal mapping of the punctured disc into $\Gamma\backslash D$. Letting $U = \{z = x + iy,\ y > 0\}$ be the universal covering of Δ^*, we obtain from (7.13) a diagram

$$\begin{array}{ccc} U & \xrightarrow{\ F\ } & D \\ \downarrow & & \downarrow \\ \Delta^* & \xrightarrow{\ f\ } & \Gamma\backslash D \end{array} \tag{7.14}$$

where $F\colon U \to D$ is a holomorphic horizontal mapping which covers f. In particular

$$F(z+1) = T\cdot F(z), \tag{7.15}$$

where $T \in \Gamma$ is the Picard-Lefschetz transformation associated to f.

(7.16) Proposition (Borel). *The eigenvalues of T are roots of unity.*[30]

Proof. According to a theorem of Kronecker, an algebraic integer, all of whose conjugates have absolute value one, must be a root of unity. Since $T \in G_{\mathbf{Z}}$ is an integral matrix, it will therefore suffice to show that all eigenvalues of T have modulus one. Now $V \subset G_{\mathbf{R}}$ is a *compact* matrix group, and thus the eigenvalues of all $h \in V$ are of absolute value one. Thus it will be enough to find a

[30] When $f\colon \Delta^* \to \Gamma\backslash D$ "comes from algebraic geometry", i.e. when there is a family

$$X \xrightarrow{\ \pi\ } \Delta^*$$

of polarized projective varieties with $f(t) =$ "Hodge structure on $P^n(\pi^{-1}(t))$", then (7.16) is part of the so-called *monodromy theorem* (cf. Landman [35] for the original proof plus further references). Matrices T all of whose eigenvalues have finite order are said to be *quasi-unipotent*. For a suitable positive integer N, $T^N - I$ is then nilpotent. The exact position of T in $G_{\mathbf{C}}$ has been determined in [41] (cf. §10).

sequence $\{g_n\} \subset G_{\mathbf{R}}$ such that

$$g_n^{-1} T g_n \longrightarrow V.$$

Let $\{z_n\} \subset U$ be a sequence of points such that $\operatorname{Im} z_n \to \infty$. Denoting by ρ_D and ρ_U the distances associated to ds_D^2 and ds_U^2, we have by (7.15), (7.12), and (7.8)

$$\begin{aligned} \rho_D(F(z_n), T\cdot F(z_n)) &= \rho_D(F(z_n), F(z_n+1)) \\ &\leqslant \rho_U(z_n, z_n+1) \\ &\leqslant \frac{C}{\operatorname{Im} z_n} \end{aligned}$$

which tends to zero as $n \to \infty$. On the other hand, since $G_{\mathbf{R}}$ acts transitively on D, we may write $F(z_n) = g_n \cdot p_0$ for some $g_n \in G_{\mathbf{R}}$(p_0= reference point in D), and

$$\rho_D(p_0, g_n^{-1} T g_n p_0) = \rho_D(g_n p_0, T g_n p_0) \longrightarrow 0$$

since ds_D^2 is $G_{\mathbf{R}}$ invariant. Thus $g_n^{-1} T g_n \to V$, and we are done.

Further applications of the Ahlfors' lemma to variation of Hodge structure will be discussed in §§8, 9 below, and are also given in §9 and Appendix D of [19].

8. Applications of Nevanlinna theory to the period mapping.

(a) *A preliminary result from Nevanlinna theory.* The general philosophy of hyperbolic complex analysis perhaps finds its deepest manifestation in the *Nevanlinna theory* (cf. pages 247-260 in [38] and [24]). We want to give two applications of this theory to variation of Hodge structure, and as a preliminary we shall prove a proposition which is a sort of *big Picard theorem.* Our method is similar to that in §9 (b) of [24].

Let V be a projective algebraic variety and $U \subset V$ an open subset consisting of smooth points.[31] We wish to find conditions under which a holomorphic mapping

$$f: \Delta^* \longrightarrow U$$

[31] Both cases where U is a Zariski open set in V and where $U = D$ is a classifying space for Hodge structures and $V = \check{D}$ is the compact dual will be used.

extends across the origin $z = 0$, to a holomorphic mapping

$$f : \Delta \longrightarrow V.^{(32)}$$

On U we assume given a negatively curved ds_U^2 with associated (1,1) form ω_U. On V we assume given an algebraic line bundle $\mathbf{L} \to V$ and holomorphic sections $\sigma_0, \ldots, \sigma_N$ such that the rational map

$$[\sigma_0, \ldots, \sigma_N] : V \to \mathbf{P}^N$$

is a holomorphic embedding on U. The ratios $\varphi_\alpha = \sigma_\alpha / \sigma_0$ are rational functions on V, the pullbacks $f^* \varphi_\alpha$ are meromorphic functions on Δ^*, and clearly

$$\begin{aligned} &f(z) \text{ extends across } z = 0 \Leftrightarrow \text{the meromorphic} \\ &\text{functions } f^*\varphi_\alpha \text{ do not have an essential} \\ &\text{singularity at } z = 0. \end{aligned} \tag{8.1}$$

Relating the metric on U to the algebraic geometry on V we assume that there exists a fibre metric in the restriction $L|_U \to U$ which satisfies the two conditions

$$\left.\begin{aligned} &\text{(i)} \quad dd^c \log \frac{1}{|\sigma|^2} \leqslant C\omega_U (\sigma \in \Gamma(U, \mathcal{O}(L)))^{(33)} \\ &\text{(ii)} \quad |\sigma_i(f(z))| = 0\left(\frac{1}{|z|^N}\right) (z \in \Delta^*) \end{aligned}\right\} \tag{8.2}$$

(8.3) PROPOSITION. *Under the above conditions, any holomorphic mapping* $f : \Delta^* \to U$ *extends to* $f : \Delta \to V$.[34]

PROOF. A meromorphic function $\varphi(z)$ on Δ^* has an inessential singularity at $z = 0 \Leftrightarrow$ for some A the equation

$$\varphi(z) = a \quad (a \in \mathbf{P}^1, z \in \Delta^*)$$

[32] We are interested in the possible singularity of $f(z)$ at $z = 0$, and not on the boundary circle $|z| = 1$. Thus we shall assume that f extends to the slightly larger punctured disc $0 < |z| < 1 + \epsilon$.

[33] For any holomorphic section σ, the C^∞ (1,1) form $dd^c \log \frac{1}{|\sigma|^2}$ represents the Chern class of $\mathbf{L}|_U$ computed from the curvature of the given metric (cf. §0 (a) of [24]).

[34] This proposition implies the usual big Piccard theorem by taking $V = \mathbf{P}^1$, $U = \mathbf{P}^1 - \{0, 1, \infty\}$, $\mathbf{L} \to \mathbf{P}^1$ to be the standard bundle with usual sections and metric, and ds_U^2 the metric constructed in §2 of [24].

has $\leqslant A$ solutions for *all* points $a \in \mathbf{P}^1$; this follows from the *Casorati-Weierstrass theorem*. Given a linear combination

$$\sigma = \sum_{i=0}^{N} a_i \sigma_i \quad (a_i \in \mathbf{C})$$

of our sections σ_i, we consider the section

$$f^* \sigma \in \Gamma(\Delta^*, \mathcal{O}(f^* \mathbf{L})).$$

Denote by $n(\sigma, r)$ the number of zeroes of $f^*\sigma$ in $A_r = \left\{\frac{1}{r} \leqslant |z| \leqslant 1\right\}$.[35]

It will suffice to show that we have a uniform estimate

$$n(\sigma, r) \leqslant A \tag{8.4}$$

for all r and σ.

Now the problem of estimating the *counting function* $n(\sigma, r)$, both from *above* and from *below*, is the basic problem of Nevanlinna theory (cf. the Introduction to [24]). In the present situation, we shall simply apply the First Main Theorem (FMT) of Nevanlinna theory to the data at hand to prove the upper bound (8.4) (The Second Main Theorem is the tool for proving more subtle lower bounds.)

To give the FMT, we let $\mu(z) \geqslant 0$ be a C^∞ function on $0 < |z| \leqslant 1$ such that, around any point z_0,

$$\mu(z) = |z - z_0|^{2\mu_{z_0}} \mu_0(z). \tag{8.5}$$

where μ_0 is C^∞ and $\mu_0(z_0) > 0$. We call μ_{z_0} the *multiplicity* of z_0 and let

$$D = \sum \mu_{z_0} \cdot z_0$$

be the *divisor* of μ. Denote by

$$n(\rho) = \text{degree}(D \cap A_\rho)$$

$$N(r) = \int_1^\rho n(\rho) \frac{d\rho}{\rho}$$

[35] Of course, we assume that $f^*\sigma \not\equiv 0$. Recall also that f is assumed to extend across the circle $|z| = 1$.

the counting functions associated to D. Clearly

$$n(\rho) \leqslant A \tag{8.6}$$

is equivalent to an estimate

$$N(r) \leqslant A \log r + B. \tag{8.7}$$

Taking $\mu(z) = |f^*(\sigma)(z)|^2$, $D = D_\sigma$ is the divisor of $f^*\sigma$ and (8.4) follows from (8.7) with a uniform A independent of σ. The following *Jensen-type formula* is the FMT in the present context.

(8.8) LEMMA. *For $\mu(z)$ as above, we have*

$$N(r) = \frac{1}{2\pi}\int_0^{2\pi} \log \mu\left(\frac{1}{r}e^{i\theta}\right) d\theta - \int_1^r \left(\int_{A\rho} dd^c \log \mu\right) \frac{d\rho}{\rho} + 0(1).$$

PROOF. Make the change of variables $w = \frac{1}{z}$ so that A_ρ is given by $1 \leqslant |w| \leqslant \rho$. In case

$$\mu|_{A_r} = \prod_\alpha |w - w_\alpha|^{2\mu_\alpha}$$

(8.8) is the usual Jensen formula. In case $\mu|_{A_r}$ is everywhere > 0, $\log \mu$ is C^∞ and Stokes' theorem applies to give

$$\begin{aligned}\int_1^r \left(\int_{A\rho} dd^c \log \mu\right)\frac{d\rho}{\rho} &= \int_1^r \left(\int_{|w|=\rho} d^c \log \mu\right)\frac{d\rho}{\rho} \\ &= \int_1^r \rho\frac{\partial}{\partial\rho}\left(\frac{1}{2\pi}\int_0^{2\pi} \log \mu(\rho e^{i\theta})\, d\theta\right)\frac{d\rho}{\rho} \\ &= \frac{1}{2\pi}\int_0^{2\pi} \log \mu(re^{i\theta})\, d\theta - \frac{1}{2\pi}\int_0^{2\pi} \log \mu(e^{i\theta}) d\theta,\end{aligned}$$

which is the desired formula. The general case follows by writing

$$\mu|_{A_r} = \left(\prod |w - w_\alpha|^{2\mu_\alpha}\right). \mu_0,$$

where $\mu_0 > 0$ on A_r.

The estimate (8.7) follows immediately from the Ahlfors' lemma (7.7), (7.11), our assumptions (8.2), and the integral formula (8.8):

$$\frac{1}{2\pi}\int_0^{2\pi} \log \mu\left(\frac{1}{r}e^{i\theta}\right) d\theta \leqslant A_1 \log r + B_1 \qquad \text{by (ii) in (8.2);}$$

$$\int_{A\rho} dd^c \log \frac{1}{\mu} \leqslant C \int_{A\rho} f^*\omega_U \qquad \text{by (8.2)}$$

$$\leqslant A_2 < \infty \qquad \text{by (7.7) and (7.11),}$$

and so $$N(r) \leqslant A \log r + B\ (A = A_1 + A_2) \qquad \text{by (8.8).}$$

REMARK. The estimate

$$\int_0^r \left(\int_{A_\rho} f^*\omega_U\right)\frac{d\rho}{\rho} = 0\,(\log r) \tag{8.9}$$

may be proved by (8.8) without using the Ahlfors lemma as follows: Write

$$f^*\omega_U = \frac{\sqrt{-1}}{2\pi}\,\mu\; dz \wedge d\bar{z}$$

and apply (8.8) to μ, using the negative curvature assumption in the form

$$dd^c \log \mu \geqslant \frac{\sqrt{-1}}{2\pi}\,\mu\; dz \wedge d\bar{z}, \tag{8.10}$$

to obtain:

$$T(r) = \int_0^r \left(\int_{A_\rho} f^*\omega_U\right)\frac{d\rho}{\rho} \qquad \text{, by definition}$$

$$= \int_1^r \left(\int_0^\rho \left(\frac{1}{2\pi}\int_0^{2\pi} \mu\,(te^{i\theta})\, d\theta\right) tdt\right)\frac{d\rho}{\rho} \qquad \text{(polar coordinates)}$$

$$\leqslant \int_0^r \left(\int_{A_\rho} dd^c \log \mu\right)\frac{d\rho}{\rho} \qquad \text{, by (8.10)}$$

$$\leqslant \frac{1}{2\pi}\int_0^{2\pi} \log \mu\,(re^{i\theta})\,d\theta + 0(1) \qquad \text{by (8.8)}$$

$$\leqslant \log\left(\frac{1}{2\pi}\int_0^{2\pi} \mu\,(re^{i\theta})d\theta\right) + 0(1) \qquad \text{concavity of log}$$

$$= \log\left[\frac{d^2\,T(r)}{dr^2}\right] + 0\,(\log r); \quad \text{i.e.}$$

$$T(r) \leqslant \log\left[\frac{d^2\,T(r)}{dr^2}\right] + 0\,(\log r). \tag{8.11}$$

It is a calculus lemma that "(8.11) ⇒ $T(r)$ is 0 (log r)" (cf. [38], page 253). This proves (8.9).

(b) *Borel's extension theorem.* Let D be a bounded, symmetric domain, $\Gamma \subset \text{Aut}\,(D)$ an arithmetically defined discrete group of automorphisms, and $U = \Gamma\backslash D$. Borel and Baily [2] have constructed a *compactification* V of U, where V is a projective variety in which U appears as a Zariski open set. From our point of view, V may be best described as follows:

> Let $K \to D$ be the *canonical line bundle* (*canonical factor of automorphy*). The Γ-invariant sections of $\mathbf{K}^\mu \to D$ are called *automorphic forms of weight* μ and induce sections $\sigma \in \Gamma(U, \mathcal{O}(\mathbf{K}^\mu))$. Now the graded ring $\oplus_{\mu\geqslant 0}\Gamma(U, \mathcal{O}(\mathbf{K}^\mu))$ is of finite type,[36] and for sufficiently large μ the sections in $\Gamma(U, \mathcal{O}(\mathbf{K}^\mu))$, induce an embedding $U \subset \mathbf{P}^N$ in which U appears as a Zariski open set in its Zariski closure V.

(8.12) PROPOSITION (Borel). *A holomorphic mapping f: $\Delta^* \to U$ extends to f: $\Delta \to V$.*

PROOF. It is well known that $\mathbf{K} \to D$ has an invariant metric whose Chern class ω is a negatively-curved ds^2_D on D. This then gives the ds^2_U in §8(a), and we may take $\mathbf{L} = \mathbf{K}^\mu$ and (i) in (8.2) is satisfied. For our sections $\sigma_i \in \Gamma(U, \mathcal{O}(\mathbf{K}^\mu))$, we take the so-called *cusp forms*:

[36] cf. Andreotti-Grauert [1] for a function-theoretic proof. The proofs of these statements require general information about the fundamental domains for Γ.

these are automorphic forms which, so to speak, vanish at infinity [2]. For such σ_i, the length

$$\sup_{p\in U} |\sigma_i(p)| < +\infty$$

is bounded, and for large μ there are sufficiently many cusp forms to induce a projective embedding of U. Proposition (8.3) now applies to yield a proof of Borel's result.

REMARKS. (i) Properly speaking, what we have proved is that, given f: $\Delta^* \to \Gamma\backslash D$, the ratio σ/σ' of two cusp forms of the same weight pulls back to give a meromorphic function $f^*(\sigma/\sigma')$ having an inessential singularity at $z = 0$. With a little work, the same could be proved for general automorphic forms using the full strength of condition (ii) in (8.2).

(ii) The result of Borel [5] is stronger, in that he shows that if

$$f\colon (\Delta^*)^k \times \Delta^l \to U$$

is a holomorphic mapping of a *punctured polycylinder* into $\Gamma\backslash D$, then f extends holomorphically to

$$f\colon \Delta^k \times \Delta^l \to V.$$

In both his proof and the later proof by Kobayashi-Ochiai, extensive use is made of the detailed description of Siegel sets. The strongest result along these lines is due to Kierman-Kobayashi [32], who show that

$$F\colon \Gamma\backslash D \to \Gamma'\backslash D'$$

extends continuously to the compactifications on each side.

(c) *A Riemann extension theorem for variation of Hodge structures.* As a second application of Nevanlinna theory, we shall prove the following (cf. [19], [41])

(8.13) PROPOSITION. *Let D be a classifying space for variation of Hodge structure and f: $\Delta^* \to D$ a holomorphic, horizontal mapping. Then f extends to $f : \Delta \to D$.*

PROOF. Let $\check{D}$ be the compact dual to D. We will first prove that f extends to $f : \Delta \to \check{D}$, where possibly $f(0) \in \partial D$. For this we want to apply Proposition 8.3 when $U = D$, $V = \check{D}$ and $\mathbf{L} \to D$ is a standard ample *homogeneous line bundle* (cf. §3(b)). Writing

$$D = G_{\mathbf{R}}/V$$
$$\check{D} = M/V,$$

the line bundle $\mathbf{L} \to \check{D}$ has an M-invariant metric $|\quad|_M$ and $\mathbf{L} \to D$ has a $G_{\mathbf{R}}$-invariant metric $|\quad|_G$. The ratio

$$\chi = \frac{|\sigma|_G^2}{|\sigma|_M^2}$$

is a positive C^∞ function on D. Let $o \in D$ be the reference point and denote by $\rho_D(p, q)$ the $G_{\mathbf{R}}$-invariant distance on D. From Lemmas 3.19 and 3.23 in §3(b) we have

$$\chi(p) = 0(\exp \rho_D(o, p)). \tag{8.14}$$

We are now ready to verify the hypotheses of Proposition 8.3. The sections $\sigma_i \in \Gamma(\check{D}, \mathcal{O}(L))$ will be chosen as a basis for this vector space; obviously

$$\sup_{q \in D} |\sigma_i(q)|_M \leqslant C < \infty. \tag{8.15}$$

For the metric in $\mathbf{L}|_U$, we take $|\quad|_G$, then clearly

$$dd^c \log \frac{1}{|\sigma|_G^2} = 0(\omega_{\check{D}}),$$

since both forms are $G_{\mathbf{R}}$-invariant. Finally

$$\begin{aligned} |\sigma_i(f(z))|_G &= |\sigma_i(f(z))|_M \, \chi(f(z)) \\ &\leqslant C\,\chi(f(z)) && \text{, by (8.15)} \\ &= 0(\exp \rho_D(f(z_0), f(z))) && \text{, by (8.14)} \\ &= 0\left(\log \frac{1}{|z|}\right) && \text{, by (7.10).} \end{aligned}$$

Thus we may apply (8.3) to have $f\colon \Delta \to \check{D}$ extending our original mapping.

In proving that $f(0) \in D$, we shall limit ourselves to the case of Hodge structures of weight two. The general case can be treated similarly, and this will also provide an alternative point of view for the proof of (8.13). We shall use the notation of §3(c). In particuler, we make the identifications $H \cong \mathbf{C}^{2r+s}$, $H_{\mathbf{Z}} \cong \mathbf{Z}^{2r+s}$, and the bilinear from Q corresponds to a $(2r + s) \times (2r + s)$ symmetric, nonsingular, rational matrix Q. The Grassmann variety $Gr(r, 2r + s)$ of r-planes in $H \simeq \mathbf{C}^{2r+s}$ will be realized as the set of nonsingular $(2r + s) \times r$ matrices Ω, modulo the equivalence relation

$$\Omega \sim \Omega A, \text{ if } A \in Gl(r, \mathbf{C}). \tag{8.16}$$

The subvariety $\check{D} \subset Gr(r, 2r + s)$ is described by the equation

$$^t\Omega\, Q\, \Omega = 0, \tag{8.17}$$

and the points of D correspond to those Ω which satisfy, in addition to (8.17),

$$-{}^t\Omega\, Q\, \bar{\Omega} > 0. \tag{8.18}$$

The line bundle $\mathbf{L} \to \check{D}$ which we shall use is the one induced by the character $A \longmapsto \det A$ of $Gl(r, \mathbf{C})$. Given an index set $I = \{1 \leqslant i_1 < i_2 < \ldots < i_r \leqslant 2r + s]$, we let Ω_I be the corresponding minor of Ω. As was discussed in §3 (c), the space of sections of $\mathbf{L} \to \check{D}$ is spanned by the Plücker coordinates

$$\sigma_I = \det \Omega_I.$$

The M-invariant and G_R-invariant metrics on $\mathbf{L}$ are given by, respectively,

$$|\sigma_I|_M^2 = \frac{|\det \Omega_I|^2}{\|\Omega\|^2} \text{ and } |\sigma_I|_G^2 = \frac{|\det \Omega_I|^2}{\det({}^t\Omega Q \bar{\Omega})},$$

where $\|\Omega\|^2 = \Sigma_I |\det \Omega_I|^2$. Thus the comparison function is

$$\chi(\Omega) = \frac{\|\Omega\|^2}{\det({}^t\Omega Q \bar{\Omega})}. \tag{8.19}$$

We are now ready to prove that given a horizontal holomorphic mapping

$$f: \Delta \longrightarrow \check{D} \text{ such that}$$
$$f: (\Delta^*) \subset D$$

then $f(0) \in D$. Represent $f(z)$ by a holomorphic matrix $\Omega(z)$ having rank k for all $z \in \Delta$. Then

$$-{}^t\Omega(z) Q\, \overline{\Omega(z)} = H(z) > 0$$

for $z \in \Delta^*$, and we want to show that $H(0) > 0$. In any case $H(0) \geqslant 0$, and if the inequality is not strict, then

$$\det H(z) \leqslant C\,|z| \tag{8.20}$$

for small $|z|$. Since $\|\Omega(z)\|^2 = 0(1)$, it follows from (8.19) and (8.20) that for $|z| < \epsilon$

$$\chi(\Omega(z)) \geqslant \frac{A}{|z|}. \tag{8.21}$$

On the other hand, by (8.14) and (7.10),

$$\chi(\Omega(z)) \leqslant B\left(\log \frac{1}{|z|}\right)^{\mu}. \tag{8.22}$$

The inequalities (8.21) and (8.22) cannot both hold, and thus $f(0) \in D$.

9. Asymptotic analysis of the period mapping. Recently the second author has been able to give a detailed analysis of an arbitrary variation of Hodge structure over the punctured disc [41]. In this section, we shall discuss the two main theorems from [41], giving in the "geometric case"[37] an alternate proof of the first result, the *nilpotent orbit theorem*, and then presenting a heuristic discussion of the motivation and proof of the second result, the SL_2-*orbit theorem*.

(a) *The nilpotent orbit theorem.* Let D be a classifying space for variation of Hodge structure and

$$f\colon \Delta^* \longrightarrow \Gamma\backslash D \tag{9.1}$$

a locally liftable, holomorphic, horizontal mapping (§3(b)). Denote by $U = \{w = u + iv\colon v > 0\}$ the upper half plane and let

$$U \longrightarrow \Delta^*$$

$$w \longmapsto z = e^{2\pi\sqrt{-1}\,w}$$

be the universal covering mapping. Then (9.7) induces

[37] The "geometric case" means that we assume given a complex manifold X and a proper holomorphic mapping $\pi\colon X \to \Delta$ such that π is smooth outside $\pi^{-1}(0)$ and such that there is a projective embedding $X \to \mathbf{P}^N$. The fibers $V_z = \pi^{-1}(z)$ are smooth, projective varieties for $z \neq 0$, but V_0 may have singularities. As explained in §3, this situation generates a holomorphic *period mapping* $f\colon \Delta^* \to \Gamma\backslash D$. The results in [41] are proved for an arbitrary locally liftable, holomorphic, horizontal mapping f.

$$\begin{array}{ccc} U & \xrightarrow{\;F\;} & D \\ \downarrow & & \downarrow \\ \Delta^* & \xrightarrow{\;f\;} & \Gamma\backslash D \end{array} \tag{9.2}$$

$$F(w+1) = T \cdot F(w)$$

where T is the *Picard-Lefschetz transformation* (cf. §7(b)). Using (7.16) and passing to a finite cyclic covering of Δ^* if necessary, we may assume that T is unipotent with *index of unipotency* l.[38] Define

$$\left.\begin{aligned} N = \log T = (T-I) - \frac{(T-I)^2}{2} + \ldots + (-1)^l \frac{(T-I)^l}{l} \\ G(w) = \exp(-wN)\, F(w) \in \check{D}. \end{aligned}\right\} \tag{9.3}$$

From (9.2) we see that $G(w+1) = G(w)$, so that G induces a mapping

$$g: \Delta^* \longrightarrow \check{D}.$$

(9.4) PROPOSITION. *In the geometric case, this mapping extends to a holomorphic mapping* $g: \Delta \to \check{D}$.

PROOF. We consider the case of Hodge structures of weight two. Then $f(z)$ is given by a period matrix (cf. §3(c))

$$\Omega(z) = \left(\int_{\gamma_\mu} \omega_i(z) \right),$$

where $\omega_i(z)$ are a basis for $H^{2,0}(V_z)$ and the γ_μ are a basis for the primitive part of $H_2(V_2, \mathbf{Q})$. According to the theorem on regular singular points (§3(c)), the $\omega_i(z)$ may be chosen so that

$$\left| \int_{\gamma_\mu} \omega_i(z) \right| = 0(|z|^{-K}) \qquad (0 \leqslant \arg z < 2\pi).$$

Now $g(z)$ is the point in the Grassmannian given by the matrix

[38] Thus l is the smallest integer such that $(T-I)^{l+1} = 0$. It is a consequence of the results in [41] that $l < m$ where m is the weight of the Hodge structures classified by D (cf: [35] for the geometric case).

$$\left(e^{-\frac{\log z}{2\pi\sqrt{-1}}N}\right)\Omega(z) = \Psi(z).$$[39]

Since $N^{l+1} = 0$, it follows that

$$\|\Psi(z)\| = 0(|z|^{-K}). \tag{9.5}$$

Consider the composed mapping

$$\begin{array}{ccc} \Delta^* & \xrightarrow{\ g\ } & Gr(r, 2r+s) \\ & \searrow^{h} & \downarrow p \\ & & \mathbf{P}^{\binom{2r+s}{r}-1} \end{array}$$

where p is the Plücker embedding. Using (9.5), it is clear that $h(z)$ is given by a homogeneous vector

$$h(z) = [h_1(z), \ldots, h_{\binom{n}{k}}(z)]$$

where the $h_i(z)$, being the $k \times k$ minors of $\Psi(z)$, are meromorphic at $z = 0$. Taking a common factor z^σ out of all $h_i(z)$, we may arrange that the $h_i(z)$ are holomorphic at $z = 0$ and some $h_j(0) \neq 0$. Then g extends across the origin as desired.

REMARK. It seems likely that this proposition could be proved in general, using Nevanlinna theory and arguments similar to those in §8(b).

We set $g(0) = p_0 \in \check{D}$ and consider the *nilpotent orbit*

$$\left\{ \begin{array}{ll} \mathcal{O}(w) = \exp(wN)\, p_0 & (w \in U) \\ \mathcal{O}(w+1) = T\mathcal{O}(w). & \end{array} \right\} \tag{9.6}$$

(9.7) THEOREM (NILPOTENT OROBIT THEOREM): (i) *For* $\operatorname{Im} w \geqslant C$, *the orbit* $\mathcal{O}(w) \in D$ *and* $w \longmapsto \mathcal{O}(w)$ *is a horizontal mapping*; *and* (ii) *given* $\epsilon > 0$,

$$\rho_D(F(w), \mathcal{O}(w)) \leqslant A(\epsilon) \exp(-2\pi(1-\epsilon)\operatorname{Im} w)$$

[39] That is to say, $g(z)$ is the k-plane in $\mathbf{C}^{2r+s}$ spanned by the columns of $\Psi(z)$.

for $\operatorname{Im} w \geqslant C.$[40]

PROOF. We continue discussing the case of Hodge structures of weight two. Accordingly we may represent $f(z)$ by a period matrix $\Omega(z)$ satisfying (8.16) and the relations

$$
\begin{aligned}
{}^t\Omega(z)Q\Omega(z) &= 0\\
-{}^t\Omega(z)Q\overline{\Omega(z)} &> 0\\
\Omega(z)Q\Omega'(z) &= 0\\
\Omega(e^{2\pi i} z) &= T\,\Omega(z),
\end{aligned}
$$

where $\Omega(e^{2\pi i} z)$ is the result of analytically continuing $\Omega(z)$ around $z=0$ and $T=e^N$ is the Picard-Lefschetz transformation. Now define

$$l(z) = \frac{\log z}{2\pi i}$$

$$\Psi(z) = e^{-l(z)N}\Omega(z).$$

Then $\Psi(e^{2\pi i}z) = \Psi(z)$, and indeed $\Psi(z)$ is the period matrix representing $g(z)$ in (9.4), so that we assume $\Psi(z)$ extends across $z=0$ as a mapping into $\check{D}$. Set.

$$\Psi(0) = \Psi_0 = p_0 \in \check{D}.$$

The orbit mapping $\mathcal{O}(z)$ is represented by the matrix

$$\Theta(z) = e^{l(z)N}\Psi_0.$$

It follows from the nilpotency of N that

$$\Omega(z) = \Theta(z) + z\,\Xi(z, \log z) \tag{9.9}$$

where $\Xi(z, \log z)$ is a polynomial in $\log z$ whose coefficients are holomorphic functions of z. Our theorem will follow by looking closely at (9.9) and using the Ahlfors Lemma 7.7, together with the metric comparison Lemmas 3.19 and 3.23.

We first prove that $\Theta(z)$ is horizontal. Differentiating (9.9) gives

(40) By the second condition in (9.6) and (i) in the theorem, $\mathcal{O}$ induces a locally, liftable holomorphic, horizontal mapping

$$\mathcal{O} : \Delta_\rho^* \to \Gamma\backslash D;$$

(ii) says that $\mathcal{O}(z)$ and $f(z)$ are asymptotic in the sense of the estimate.

$$\Omega'(z) = \frac{1}{2\pi i}\,\frac{N}{z}\,\Theta(z) + \Xi_1(z, \log z) \tag{9.10}$$

where Ξ_1 $(z, \log z)$ is as before. Plugging (9.10) into the third relation in (9.8) and looking at the coefficient of $\frac{1}{z}$ gives

$${}^t\Psi_0\, Q\, N\, \Psi_0 = 0,$$

which implies that $\Theta(z)$ is horizontal.

Next we want to prove that for $0 < |z| < \rho$

$$-{}^t\Theta(z)\, Q\, \overline{\Theta(z)} > 0, \tag{9.11}$$

this being the condition that $\Theta(z) \in D$. By (9.9)

$$0 < -{}^t\Omega(z)\, Q\, \overline{\Omega(z)} = -{}^t\Theta(z)\, Q\, \overline{\Theta(z)} + |z|^{1-\epsilon}\,\Gamma(z) \tag{9.12}$$

where $\Gamma(z)$ is bounded near $z = 0$. The idea now is to show, using the Ahlfors lemma, that

$$-{}^t\Omega(z)\; Q\; \overline{\Omega(z)} \geqslant c(\log \frac{1}{|z|})^K . I \tag{9.13}$$

for some K; (9.11) then follows from (9.12) and (9.13). We begin by looking at

$$h(z) = \det\,\{-{}^t\Omega(z)\, Q\, \overline{\Omega(z)}\} = \sum_{\alpha=0}^{m} h_\alpha(z)\left(\log \frac{1}{|z|}\right)^{\alpha}$$

$$j(z) = \det\,\{-{}^t\Theta(z)\, Q\, \overline{\Theta(z)}\} = \sum_{\beta=0}^{m} j_\beta(z)\left(\log \frac{1}{|z|}\right)^{\beta} \tag{9.14}$$

where $h_\alpha(z)$, $j_\beta(z)$ are C^∞ functions on $|z| < \rho$. By (9.12),

$$0 < h(z) = j(z) + 0\,(|z|^{1-\epsilon}). \tag{9.15}$$

Suppose for a moment we can show that

$$h(z) \geqslant C\left(\log \frac{1}{|z|}\right)^{\sigma} \tag{9.16}$$

for some (possibly negative) σ. From (9.16) it follows first that not all $h_\alpha(0) = 0$. If $h_m(0) = \ldots = h_{\tau+1}(0) = 0$, $h_\tau(0) > 0$, then by (9.15), $j_m(0) = \ldots = j_{\tau+1}(0) = 0$ and $j_\tau(0) > 0$, so that $j(z) > 0$ for $|z| < \rho$.

To prove (9.16), we use (8.19) and (8.22) to obtain

$$\frac{1}{h(z)} = \frac{\chi(\Omega(z))}{\|\Omega(z)\|^2} \leqslant \frac{C\left(\log \frac{1}{|z|}\right)^{\mu}}{\|\Omega(z)\|^2}$$

which gives (9.16), since $\|\Omega(z)\|^2 \geqslant C\left(\log \frac{1}{|z|}\right)^{\theta}$ for some θ.

Now we were able to prove that $j(z) > 0$ for $0 < |z| < \rho$ using (9.12) and Lemmas 3.19 and 3.23 relating the M- and $G_{\mathbf{R}}$-invariant metrics in the homogeneous line bundle $\mathbf{L} \to D$. Applying the same argument to the homogeneous vector bundle $\mathbf{E} \to D$, whose fibre E_{Ω} is the vector space spanned by the columns of Ω, gives (9.13) and subsequently

$$-{}^t\Theta(z)\, Q\, \overline{\Theta(z)} > 0 \text{ for } 0 < |z| < \rho.$$

The estimate

$$\rho_D(f(z), \mathcal{O}(z)) = 0(|z|^{1-\epsilon})$$

is proved by a similar argument, and will therefore be omitted.

(b) *The SL_2-orbit Theorem.* Theorem (9.7) asserts that the period mapping of a degenerating family of Hodge structures is asymptotic to a mapping of a very special nature. By itself, this result is not particularly useful. It merely allows one to reduce questions about the original mapping to questions about the approximating nilpotent orbit. To get further, one needs a description of the nilpotent orbits which can come up. Such a description is the content of the second main theorem of [41], the SL_2-orbit theorem.

In the remarks preceeding (9.7), the Picard-Lefschetz transformation T is an element of the arithmetic group $G_{\mathbf{Z}}$, which has been made unipotent, if necessary, by going to a finite covering of Δ^*. This makes $N = \log T$ a rational, nilpotent element of the Lie algebra $\mathfrak{g}_0$ of $G_{\mathbf{R}}$. Thus we have the following data : a point $p_0 \in D$, a rational and nilpotent element $N \in \mathfrak{g}_0$, and a constant C, such that

$$\left.\begin{array}{l} \exp(wN)\cdot p_0 \in D, \text{ for } w \in \mathbf{C},\ \mathrm{Im}\, w \geqslant C; \\ \text{and } w \longmapsto \exp(wN)\cdot p_0 \text{ is a horizontal map.} \end{array}\right\} \quad (9.17)$$

Let us first look at the simplest possible case, when D is the ordinary upper half plane U, $\check{D}$ the Riemann sphere $\mathbf{P}^1 = \mathbf{C} \cup \{\infty\}$, and $G_{\mathbf{R}}$ the group $SL(2, \mathbf{R})$. The Lie algebra $\mathfrak{sl}(2, \mathbf{R})$ contains exactly two conjugacy classes of nonzero nilpotent elements, namely those of

$$\begin{pmatrix} 0 & 1 \\ 0 & 0 \end{pmatrix} \text{ and } \begin{pmatrix} 0 & 0 \\ 1 & 0 \end{pmatrix}.$$

If N is one of these two, $\exp(wN) \cdot p_0$ equals $p_0 + w$ in the first instance, and $\left(\dfrac{1}{p_0} + w\right)^{-1}$ in the second. The condition (9.17) eliminates the latter possibility. Hence, up to an automorphism of U, the approximating nilpotent orbit takes the form $w \longmapsto p_0 + w$. The period mapping of a family of elliptic curves which acquire an ordinary double point has this kind of singularity.

In very general terms, the SL_2-orbit theorem, combined with the nilpotent orbit theorem, says the following: given any one-parameter family of Hodge structures which becomes singular, one can equivariantly embed a copy of the upper half plane in the classifying space, such that the period mapping asymptotically approaches a mapping into this upper half plane, of the type described just above. The copy of the upper half plane arises as an orbit of the group $SL(2, \mathbf{R})$, and this may serve to explain the name of the theorem.

For a precise statement, some more notation will be necessary. We consider the basis

$$z = \begin{pmatrix} 0 & -\sqrt{-1} \\ \sqrt{-1} & 0 \end{pmatrix}, X_+ = \tfrac{1}{2}\begin{pmatrix} -\sqrt{-1} & 1 \\ 1 & \sqrt{-1} \end{pmatrix}, X_- = \tfrac{1}{2}\begin{pmatrix} \sqrt{-1} & 1 \\ 1 & -\sqrt{-1} \end{pmatrix} \tag{9.18}$$

of the Lie algebra $\mathfrak{sl}(2, \mathbf{C})$, and we set

$$A = \begin{pmatrix} -1 & 0 \\ 0 & 1 \end{pmatrix}, F = \begin{pmatrix} 0 & 1 \\ 0 & 0 \end{pmatrix}.$$

As usual, U will stand for the upper half plane; we also identify $\mathbf{P}^1$ with the Riemann sphere $\mathbf{C} \cup \{\infty\}$.

(9.19) THEOREM (SL_2-ORBIT THEOREM): *Under the hypotheses* (9.17), *there exist*

(i) *a homomorphism of algebraic groups* $\psi: SL(2, \mathbf{C}) \to G_{\mathbf{C}}$, *defined over* $\mathbf{R}$,

(ii) *a holomorphic, horizontal embedding* $\tilde{\psi}: U \to D$, *which is* $SL(2, \mathbf{R})$*-equivariant with respect to* ψ,

(iii) *and a holomorphic mapping* $w \longmapsto g(w)$ *of a neighborhood* $\mathscr{V}$ *of* ∞ *in* $\mathbf{P}^1$ *into* $G_{\mathbf{C}}$,

with all of the following properties:

(a) $\exp(wN) \circ p_0 = g(-\sqrt{-1}w) \circ \tilde{\psi}(w)$ *for* $w \in U \cap \mathscr{V}$:

(b) $g(\infty) = e$, *and* $g(v) \in G_{\mathbf{R}}$ *for* $\sqrt{-1}\, v \in \mathscr{V} \cap \sqrt{-1}\, \mathbf{R}^+$;

(c) $N = \psi_*(F)$;

(d) *with respect to the Hodge structure corresponding to the point* $\psi(\sqrt{-1}) \in D$, *the linear transformations* $\psi_*(X_+)$, $\psi_*(Z)$, $\psi_*(X_-)$ *are of Hodge type* $(-1, 1)$, $(0, 0)$ *and* $(1, -1)$ *respectively*;

(e) *if* $g(w) = 1 + g_1 w^{-1} + g_2 w^{-2} + \ldots$ *is the power series expansion of the matrix-valued function* $g(w)$ *around* $w = \infty$, *then* $(\operatorname{ad} N)^{k+1}\, g_k = 0$;

(f) $\operatorname{ad} \psi_*(A)$ *operates semisimply, with integral eigenvalues*; *let* g_k^l *be the component of* g_k *in the* l*-eigenspace*; *then* $g_k^l = 0$ *unless* $l \leqslant k - 1$, *for* $k \geqslant 1$.

Moreover, when the condition $g(\infty) = e$ *is weakened to* $g(\infty) \in \exp(\text{image}\{\operatorname{ad} N: \mathfrak{g}_0 \to \mathfrak{g}_0\} \cap \text{kernel}\{\operatorname{ad} N: \mathfrak{g}_0 \to \mathfrak{g}_0\})$, *one can arrange that the homomorphism* ψ *is defined over* $\mathbf{Q}$.

According to (a) and (b), the two mappings $w \to \exp(wN) \circ p_0$ and $w \longmapsto \tilde{\psi}(w)$ take the same value at $w = \infty$; the conditions (e), (f), when looked at more closely, actually say that the two mappings are asymptotic as $\operatorname{Im} w \to \infty$. In various applications (cf. §10), (c), (e), and (f) allow one to reduce problems about the nilpotent orbit to questions about $\tilde{\psi}$. Because of (d), the one-parameter family of Hodge structures parameterized by the SL_2-orbit $\tilde{\psi}$ degenerates in a very simple fashion. To be more precise, the upper half plane U classifies polarized Hodge structures of weight 1 on $\mathbf{C}^2$; from this universal family, by the operations of symmetric products, tensor products with constant Hodge structures, and direct sums, the family parametrized by $\tilde{\psi}$ can be built up.

Although the proof of the theorem is technical, its basic idea can be described in simple terms. We shall do so below, in the hope that this may motivate and clarify the statement of the theorem.

Let $\{H_0^{p,q}\}$ be the reference Hodge structure on H, corresponding to the base point of D. It induces a Hodge structure of weight zero on $\mathrm{Hom}(H, H)$ which in turn determines a Hodge structure $\{\mathfrak{g}^{p,-p}\}$ the Lie algebra $\mathfrak{g}$ of $G_{\mathbf{C}}$. We identify the real subspace $\mathfrak{g}_0$ of $\mathfrak{g}$ with the Lie algebra of $G_{\mathbf{R}}$. Then

$\mathfrak{v}_0 = \mathfrak{g}^{0,0} \cap \mathfrak{g}_0$ is the Lie algebra of the subgroup $V \subset G_{\mathbf{R}}$; $\mathfrak{b} = \oplus_{p \geqslant 0} \mathfrak{g}^{p,-p}$ is the Lie algebra of $B_{\mathbf{C}} \subset G_{\mathbf{C}}$; under the natural isomorphism between $\oplus_{p<0} \mathfrak{g}^{p,-p} \cong \mathfrak{g}/\mathfrak{b}$ and the holomorphic tangent space of $\check{D} \cong G_{\mathbf{C}}/B_{\mathbf{C}}$ at the origin, $\mathfrak{g}^{-1,1}$ corresponds to the subspace of horizontal vectors. (9.20)

According to the hypotheses (9.17),

$$v \longmapsto \exp(\sqrt{-1}\, v N) \quad p_0, \text{ with } v \in \mathbf{R},\ v > C, \tag{9.21}$$

represents a smooth, real curve in $D \cong G_{\mathbf{R}}/V$. The Ad V invariant splitting

$$\mathfrak{g}_0 = \mathfrak{v}_0 \oplus (\oplus_{p \neq 0} \mathfrak{g}^{p,-p} \cap \mathfrak{g}_0)$$

defines a $G_{\mathbf{R}}$-invariant connection on the principal bundle

$$V \longrightarrow G_{\mathbf{R}} \longrightarrow D \cong G_{\mathbf{R}}/V.$$

Hence there exists an essentially unique lifting $v \longmapsto h(v) \in G_{\mathbf{R}}$ of the curve (9.21) to $G_{\mathbf{R}}$, which is tangential to this connection.

We now introduce the three $\mathfrak{g}_0$-valued functions

$$A(v) = -2h^{-1}(v)h'(v),\ F(v) = \mathrm{Ad}\, h(v)^{-1}N,\ E(v) = -C_0F(v) \tag{9.22}$$

(the apostrophe stands for differentiation, and C_0 denotes the Weil operator of the Hodge structure $\{\mathfrak{g}^{p,-p}\}$ on $\mathfrak{g}$), which are defined for $v \in \mathbf{R}$, $v > C$. By construction of $h(v)$, $A(v)$ takes values in $\oplus_{p \neq 0} \mathfrak{g}^{p,-p}$. Because of the horizontal nature of the mapping

$$v \longmapsto h(v) \circ o = \exp(\sqrt{-1}\, v\, N) \cdot p_0 \in D \cong G_{\mathbf{R}}/V,$$

combined with the last statement in (9.20), $A(v)$ actually lies in $(\mathfrak{g}^{-1,1} \oplus \mathfrak{g}^{1,-1}) \cap \mathfrak{g}_0$. When the holomorphic tangent space of $\check{D} \cong G_{\mathbf{C}}/B_{\mathbf{C}}$ at the origin is identified with $\mathfrak{g}/\mathfrak{b}$, the image of $F(v)$ in $\mathfrak{g}/\mathfrak{b}$ represents the tangent vector in the $\dfrac{\partial}{\partial u}$ direction of the orbit $w \longmapsto \exp(wN) \circ p_0$ at $w = \sqrt{-1}\, v$, translated back to the origin by $h(v)^{-1}$. Similarly, $-\frac{1}{2} A(v)$ represents the tangent vector in the $\dfrac{\partial}{\partial v}$ direction. Since the orbit is a holomorphic mapping, we find that $A(v) + 2\sqrt{-1}\, F(v) \in \mathfrak{b}$. Thus

$$\left.\begin{aligned} &E(v),\ F(v) \in (\mathfrak{g}^{-1,1} \oplus \mathfrak{g}^{0,0} \oplus \mathfrak{g}^{1,-1}) \cap \mathfrak{g}_0, \\ &A(v) \in (\mathfrak{g}^{-1,1} \oplus \mathfrak{g}^{1,-1}) \cap \mathfrak{g}_0, \\ &A(v) + 2\sqrt{-1}\, F(v) \in \mathfrak{g}^{1,-1} \oplus \mathfrak{g}^{0,0}. \end{aligned}\right| \tag{9.23}$$

The functions $A(v)$, $E(v)$, $F(v)$ satisfy the system of differential equations

$$\left.\begin{aligned} 2E'(v) &= -[A(v), E(v)] \\ 2F'(v) &= [A(v), F(v)] \\ A'(v) &= -[E(v), F(v)] \end{aligned}\right| \tag{9.24}$$

Indeed, the second equation is obtained by differentiating the equation which defines $F(v)$, the first equation follows from the second by applying the Weil operator on both sides, and the third is a formal consequence of the preceding two, if one uses the information in (9.23).

For the moment, let us assume that

$h(v)$ has a Laurent series expansion in powers of $v^{-1/\alpha}$ near $v = \infty$, for some $\alpha \in \mathbf{N}$, which converges and represents $h(v)$ for all sufficiently large $y \in \mathbf{R}$. (9.25)

The functions $A(v)$, $E(v)$, $F(v)$ will then share this property. According to the discussion of the Ahlfors lemma and its consequences in §7, the mapping

$$w \longmapsto \exp((w + (\sqrt{-1}\, C)\, N) \circ p_0,\ w \in U,$$

is distance-decreasing, relative to the Poincaré metric on U. Consequently $\| A(v) \| = 0(v^{-1})$ as $y \to \infty$. With some additional work, one obtains the same estimate for $E(v)$ and $F(v)$. Hence, still under the assumption that (9.25) holds, the functions (9.22) have series expansions

$$\left.\begin{aligned} A(v) &= A_0\, v^{-1} + A_1\, v^{-1-1/\alpha} + \ldots \\ E(v) &= E_0\, v^{-1} + E_1\, v^{-1-1/\alpha} + \ldots \\ F(v) &= F_0\, v^{-1} + F_1\, v^{-1-1/\alpha} + \ldots \end{aligned}\right] \qquad (9.26)$$

The equations (9.24) now give recursive relations on the coefficients of the series. In particular,

$$[A_0, E_0] = 2E_0, \quad [A_0, F_0] = -2F_0, \quad [E_0, F_0] = A_0.$$

Also, if $A_0 = E_0 = F_0$, all three series must vanish identically, which can happen only if $N = 0$. We may disregard this special case. Hence A_0, E_0, F_0, span a subalgebra $\mathfrak{sl}(2, R) \in \mathfrak{g}_0$; this observation is the key to the entire proof. The recursion relations which follow from (9.24), when analyzed in terms of the representation theory of $\mathfrak{sl}(2, R)$, limit the possibilities for the coefficients A_i, E_i, F_i very much. For example, if one defines a $G_{\mathbf{R}}$-valued function $g(v)$ by

$$h(v) = g(v) \exp(-\tfrac{1}{2} \log v\, A_0),$$

$g(v)^{-1}\, g'(v)$ turns out to have a convergent power series expansion in integral powers of v^{-1} near $v = \infty$, starting with a term of order v^{-2}. Thus $g(v)$ must be regular near $v = \infty$, and $g(v)$ becomes defined for complex values of its variable v. Similarly, one obtains the various other ingredients and conclusions of the theorem.

It remains to justify the assumption (9.25). To begin with, by elementary arguments in linear algebra, one constructs a lifting $v \longmapsto h_1(v)$ of the curve (9.21) to $G_{\mathbf{R}}$, which need not be tangential to the connection used to define $h(v)$, but which does have the property (9.25). The two liftings $h(v)$, $h_1(v)$ are then related by a

V-valued function. This function satisfies a linear differential equation with (at worst) a regular singular point at $v = \infty$. As a result, $h(v)$ has a series expansion in fractional powers of v^{-1} and integral powers of log v. An algebraic argument, which depends on the structure theory of a semisimple Lie algebra, then excludes the presence of logarithmic terms, proving (9.25).

10. Some applications. (a) *Monodromy and the weight filtration.* We consider a family of polarized algebraic manifolds $\pi: X \to \Delta^*$, parametrized by the punctured disc Δ^*, and we let

$$T: P^m(V_t, \mathbf{C}) \to P^m(V_t, \mathbf{C})$$

denote the Picard-Lefschetz transformation; thus T is the action of the canonical generator of $\pi_1(\Delta^*)$ on the mth primitive cohomology group of a typical fibre $V_t = \pi^{-1}(t)$, $t \in \Delta^*$.

According to Landman's monodromy theorem [35],[41] some power T^k of T is unipotent, and T^k has index of unipotency at most m (i.e. $(T^k - 1)^{m+1} = 0$). In (7.16), we gave Borel's simple proof of the first part of the statement. Conjecture (8.4) of [21] suggested a somewhat sharper bond on the index of unipotency, which has been proven by Katz [31]. As a direct consequence of the SL_2-orbit theorem, one obtains an additional slight improvement of the bound:

(10.1) PROPOSITION. *Let l be the largest number of successive nonzero Hodge subspaces of $P^m(V_t, \mathbf{C})$; then $(T^k - 1)^l = 0$.*

PROOF. We set $N = \log T^k$; since $T^k = \exp N$, it suffices to show that $N^l = 0$. In the notation of (9.19), $N = \psi_*(F)$. Any two nonzero nilpotent elements of $\mathfrak{sl}(2, \mathbf{C})$ are conjugate, and thus $\psi_*(F)$, $\psi_*(X_-)$ are conjugate under some $g \in G_{\mathbf{C}}$. By part (d) of (9.19), in the Hodge structure corresponding to the point $\psi(\sqrt{-1}) \in D$, $\psi_*(X_-)$ shifts the indices of the Hodge subspaces exactly by one, so that $\psi_*(X_-)^l = 0$. Hence also $N^l = \psi_*(F)^l = 0$.

41 The theorem, and various proofs of it, are discussed in [21]. It should also be mentioned that all the "geometric" proofs depend on the existence of a continuation of the family to the entire disc, if a suitable, possibly singular, fibre is inserted over the origin. Moreover, some arguments require that the family should come from a global, algebro-geometric family.

With slightly more care, Conjecture 8.4′ in the appendix of [21] can also be deduced from the SL_2-orbit theorem. The index of unipotency is of course not the only significant information which one can give about a nilpotent linear transformation. More subtle properties of the Picard-Lefschetz transformation are implicit in Deligne's conjecture (9.17) in [21], concerning the limit of a degenerating family of Hodge structures, which we shall now discuss.

We again look at a family $\pi: X \to \Delta^*$, as described above, and we set $N = \log T^k$. Then $N^{m+1} = 0$, according to the monodromy theorem. From pp. 255-6 of [21], we recall the existence of the *monodromy weight filtration*

$$0 \subset W_0 \subset W_1 \subset \ldots \subset W_{2m-1} \subset W_{2m} = P^m(V_t, \mathbf{C}), \qquad (10.2)$$

which is characterised uniquely by the following properties:

$$\left.\begin{array}{l} N: W_l \subset W_{l-2}, \text{ for all } l, \text{ and} \\ N^l: W_{m+l}/W_{m+l-1} \to W_{m-l}/W_{m-l-1} \text{ is an isomorphism.} \end{array}\right\} \qquad (10.3)$$

Since N is defined over $\mathbf{Q}$, so is the filtration. The vector spaces $P^m(V_t, \mathbf{C})$ are the fibres of a flat bundle $\mathbf{P}^m \to \Delta^*$ (cf. 3(a)), and with respect to the flat structure, N becomes independent of t. Hence (10.2) defines a filtration of $\mathbf{P}^m$ by flat subbundles

$$0 \subset \mathbf{W}_0 \subset \ldots \subset \mathbf{W}_{2m-1} \subset \mathbf{W}_{2m} = \mathbf{P}^m. \qquad (10.4)$$

The pullback of $\mathbf{P}^m$ by the universal covering $U \to \Delta^*$, $z \longmapsto e^{2\pi\sqrt{-1}z}$, becomes canonically trivial, so that we may talk of *the* fibre H of this pullback. Now (10.4) corresponds to a rationally defined filtration

$$0 \subset W_0 \subset W_1 \subset \ldots \subset W_{2m-1} \subset W_{2m} = H; \qquad (10.5)$$

(10.3) remains valid in this context. To each point $z \in U$, there corresponds a Hodge filtration $\{F_z^p\}$ on H, such that $TF_z^p = F_{z+1}^p$ (cf. §3 (a)). According to Deligne's Conjecture (9.17) in [21], for every $z \in U$ with sufficiently large imaginary part, the two filtrations $\{F_z^p\}$ and $\{W_l\}$ were to give a mixed Hodge structure on H. Deligne has since pointed out that this is more than should be expected.

Instead, the conjecture holds "in the limit", as can be deduced from the two theorems of §9. Full details can be found in §6 of [41]; here we shall only give a precise statement of the result and a brief indication of the proof. For every $z \in U$, we consider the filtration $\{\exp(-\frac{z}{k}N) \circ F_z^p\}$, which is invariant under $z \longmapsto z + k$, because $\exp N = T^k$. As a consequence of the nilpotent orbit theorem, the limit

$$F_\infty^p = \lim_{\operatorname{Im} z \to \infty} \exp\left(-\frac{z}{k}N\right) \circ F_z^p \tag{10.6}$$

exists. Indeed, the filtration $\{F_\infty^p\}$ corresponds to the point $p_0 \in \check{D}$.

(10.7) THEOREM. *The two filtrations* $\{F_\infty^p\}$, $\{W_l\}$ *constitute a mixed Hodge structure on* H. *With respect to it,* $N: H \to H$ *is a morphism of type* $(-1, -1)$.

Conjecture (9.17) in [21] also contains some statements about the interaction of the polarization form and N; these again hold in the limit: the quotients W_l/W_{l-1} carry suitably defined bilinear forms, which polarize the Hodge structures of pure weight on the quotients.

As for the proof of (10.7), the nilpotent orbit theorem allows us to assume that the period mapping is one of the special orbits to which theorem (9.19) applies. Because of (9.19), and part (e) in particular, the filtration $\{F_\infty^p\}$ and the filtration corresponding to the point $\tilde{\psi}(\sqrt{-1}) \in D$ induce the same filtrations on the quotients W_l/W_{l-1}. Hence the nilpotent orbit can be replaced by the SL_2-orbit $\tilde{\psi}(U)$. As was mentioned above (9.19), the Hodge structures corresponding to the points $\psi(z)$ degenerate in a very simple manner as $\operatorname{Im} z \to \infty$, and in this situation, (10.7) can be verified by an explicit computation.

When a one-parameter family of algebraic manifolds degenerates to a singular variety, the limiting mixed Hodge structure has geometric significance. We shall take this up in (b) below.

The SL_2-orbit theorem also leads to a description of the monodromy weight filtration, in terms of order of growth of cohomology classes. We assume that the total space X of the family $\pi: X \to \Delta^*$ lies as an immersed submanifold in some projective space. The standard metric of this projective space restricts to a Kähler metric on each of the fibres $V_t = \pi^{-1}(t)$. With respect to the Kähler metrics, one can measure the length of a cohomology class $c \in P^m(V_t, \mathbf{C})$, $t \in \Delta^*$. In other words, the flat bundle $\mathbf{P}^m \to \Delta^*$ inherits a Hermitian metric. Now let $c \in H$ be given.[42] By flat translation, one obtains a multiple-valued, flat section of $\mathbf{P}^m \to \Delta^*$. The length of its values in the various fibres of $\mathbf{P}^m$ is a multiple-valued, real function, which we denote by $t \longmapsto \|c\|_t$, $t \in \Delta^*$. Over any radial ray or proper angular sector in Δ^*, one can choose a single-valued determination of this function.

(10.8) Theorem. *An element $c \in H$ belongs to W_l if and only if*

$$\|c\|_t = 0((-\log|t|)^{(l-m)/2}), \qquad \text{as } t \to 0,$$

over some, or equivalently any, radial sector in Δ^*.

A detailed proof can be found in §6 of [41]. As a first step, the nilpotent orbit theorem allows one to assume, in effect, that the pullback of the period mapping to the universal covering $U \to \Delta^*$ is one of the special nilpotent orbits. By (9.19), in particular part (f), one may further replace the orbit by the embedding $\tilde{\psi}: U \to D$. This situation can be treated by an easy, explicit computation.

If a cohomology class $c \in P^m(V_t, \mathbf{C})$ is invariant under the action of the fundamental group, i.e. if $Tc = c$, it must lie in the kernel of N. In view of the second statement in (10.3), W_m contains the kernel of N. Hence $\|c\|_t = 0(1)$, as $t \to 0$.

(10.9) Corollary. *An invariant cohomology class has bounded length, near the puncture of* Δ^*.

Remark. Although we have stated the preceding results only for families of algebraic manifolds, they carry over immediately to the case of an arbitrary variation of Hodge structure, parametrized by Δ^*.

[42] As before, H denotes the fibre of the pullback of $\mathbf{P}^m$ to the universal covering $U \to \Delta^*$.

In [19], the first author studied global properties of the period mapping of a variation of Hodge structure with compact parameter space. The main technical result was (a slightly more general version of) the "theorem of the fixed part" (cf. §6 (a)), from which the properties of the period mapping were essentially deduced as corollaries. Roughly speaking, the argument went as follows: Let σ be a flat section of the total bundle **H** of the variation of Hodge structure. If σ takes values in the subbundle $\mathbf{F}^p \subset \mathbf{H}$, for some p, then the length of the $(p, m-p)$-Hodge component of σ is a plurisubharmonic function on the parameter space S, as follows from the curvature properties of the Hodge bundles. Moreover, the $(p, m-p)$-component of σ is flat exactly when the length function is constant. Since a compact analytic space does not admit nonconstant plurisubharmonic functions, the $(p, m-p)$-component must be flat, and one can now apply induction on p, peeling off one Hodge component at a time.

If S, instead of being compact, is only Zariski open in some compact analytic space, it may carry nonconstant plurisubharmonic functions, but all of these are unbounded. Hence the arguments of [19] carry over to this more general situation, as soon as one knows the boundedness of the length functions which come up in the proof. Corollary (10.9) gives just the needed information. Section 7 of [41] describes in more detail how the results of [19] can be extended. We should also add that the "theorem of the fixed part" was proved by Deligne for algebraic families, as described in §6(a).

(b) *Degeneration of algebraic manifolds.* In this section, we shall summarize some results of H. Clemens and the second author about the topology of a family of algebraic manifolds which acquire singularities [10]. Let $\bar{X}$ be an immersed submanifold of some projective space, $\pi\colon \bar{X} \to \Delta$ a proper holomorphic map onto the unit disc Δ, with connected fibres, which has maximal rank at each point of $X = \pi^{-1}(\Delta^*)$ ($\Delta^* = \Delta - \{0\}$). Thus $\pi\colon X \to \Delta^*$ is a family of polarized algebraic manifolds, as defined in §3, and the central fibre $V_0 = \pi^{-1}(0)$ has the structure of projective variety. We may think of V_0 as a specialization of the typical fibre $V_t = \pi^{-1}(t)$, $t \in \Delta^*$. It is natural to ask to what extent the topology of the general fibre determines that of the singular fibre (or vice versa).

In some sense, the question is not really well posed. One can perform operations on the family, such as blowing up or down along a subvariety of V_0, going to a finite covering of the base, again blowing up or down, etc., which affect the singular fibre, but not the regular fibre. Thus, in order to get as concrete results as possible, one should bring the singular fibre into some kind of "normal form". According to Hironaka, one can arrange that V_0 is a divisor in $\bar{X}$, with no singularities other than normal crossings. Recently Mumford showed that in addition by repeated blowing up and change of the base parameter, all components of V_0 can be made to have multiplicity one. In this case, V_0 can be covered by coordinate systems with local holomorphic coordinates $z_1, z_2, \ldots, z_{n+1}$, such that

$$\pi(z_1, \ldots, z_{n+1}) = z_1 \cdot z_2 \cdot \ldots \cdot z_l$$

for some l, depending on the coordinate system. Unless we say otherwise, we shall assume that this simplification has been made. We also assume, as we may, that the components of V_0 have no self-intersection. Some of the final conclusions, like the solution of the "local invariant cycle problem", do not depend on these assumptions, whereas others do.

After shrinking the base Δ, if necessary, the family $\pi\colon \bar{X} \to \Delta$ will continue to a neighborhood of the closure of Δ. The boundary $\partial X = \pi^{-1}(\partial\Delta)$ is then a C^∞ fibre bundle over $\partial\Delta$, with fibre V_t ($t \in \partial\Delta$). Let

$$T\colon H^*(V_t) \longrightarrow H^*(V_t)$$

be the Picard-Lefschetz transformation, i.e. the action of the canonical generator of $\pi_1(\Delta^*) = \pi_1(\partial\Delta)$ on the cohomology[43] of the typical fibre V_t. One then has the exact sequence of a fibre bundle with base S^1,

[43] Here, as in the following, the homology and cohomology groups have complex coefficients.

$$\longrightarrow H^m(\partial X) \longrightarrow H^m(V_t) \xrightarrow{\;T-1\;} H^m(V_t) \longrightarrow H^{m+1}(\partial X) \longrightarrow . \tag{10.10}$$

Under the hypotheses we made, T itself is already unipotent; in the notation of part (a), $k = 1$. This follows from the "geometric" proofs of the monodromy theorem, e.g. Landman's [35]. We set $N = \log T$. Since N and $(T-1)$ have the same kernels and cokernels, we may replace $(T-1)$ by N in (10.10), which gives the exact sequence

$$\longrightarrow H^m(\partial X) \longrightarrow H^m(V_t) \xrightarrow{\;N\;} H^m(V_t) \longrightarrow H^{m+1}(\partial X) \longrightarrow . \tag{10.11}$$

The total space $\bar{X}$ has the central fibre V_0 as a strong retract (cf. [8]), so that $H^*(\bar{X}) \simeq H^*(V_0)$. Taking into account Poincaré duality, the exact cohomology sequence of the pair $(\bar{X}, \partial X)$ therefore becomes

$$\longrightarrow H^{m-1}(\partial X) \longrightarrow H_{2n-m+2}(V_0) \longrightarrow H^m(V_0) \longrightarrow H^m(\partial X) \longrightarrow \tag{10.12}$$

$(n = \dim_{\mathbf{C}} V_t)$. The exact sequences (10.11) and (10.12) can be combined into the diagram

$$\begin{array}{ccccccccc}
\longrightarrow H^m(V_0) & & & & H^m(V_t) & \longrightarrow & H^m(V_t) & & \nearrow \\
 & \searrow & & \nearrow & & & & \searrow & \\
 & & H^m(\partial X) & & & & & & H^{m+1}(\partial X) \\
 & \nearrow & & \searrow & & & & \nearrow & & \searrow \\
\longrightarrow H^{m-1}(V_t) & & & & H_{2n-m+1}(V_0) & \longrightarrow & H^{m-1}(V_0) & &
\end{array} \tag{10.13}$$

Except for a shift in the indices by one, the two rows are identical. Let us look at one of them, with the missing arrows filled in:

$$\longrightarrow H_{2n-m+2}(V_0) \xrightarrow{\;\mu\;} H^m(V_0) \xrightarrow{\;\nu\;} H^m(V_t) \xrightarrow{\;N\;} H^m(V_t) \xrightarrow{\;\psi\;} H_{2n-m}(V_0) \longrightarrow . \tag{10.14}$$

Under the identifications $H_{2n-m+2}(V_0) \simeq H^m(\bar{X}, \partial X)$, $H^m(V_0) \simeq H^m(\bar{X})$, μ corresponds to the natural mapping $H^m(\bar{X}, \partial X) \to H^m(\bar{X})$; ν is the mapping on cohomology induced by the "collapsing map"

$V_t \to V_0$; and finally ψ is dual to ν, with a shift in indices, and relative to Poincaré duality on $H^*(V_t)$.

According to Deligne [14], $H^*(V_0)$ carries a canonical mixed Hodge structure, which was described explicitly in §4. By duality, $H_*(V_0)$ inherits a mixed Hodge structure, too. Theorem 10.7 gives the existence of a limiting mixed Hodge structure on the primitive part of the cohomology of the nonsingular fibres. Since the Lefschetz decomposition and the Picard-Lefschetz transformation commute, one can deduce the analogous statement about the full cohomology. To be more precise, let $\mathbf{H}^m \to \Delta^*$ be the flat bundle of the mth cohomology groups of the fibres, and let H^m be *the* fibre of the pullback of $\mathbf{H}^m$ to the universal covering $U \to \Delta^*$, which is canonically trivial. On H^m, there is a unique rational filtration $\{W_l\}$ with the properties (10.3). A limiting Hodge filtration $\{F^p_\infty\}$ on H^m can be defined as in (10.6). Then $\{F^p_\infty\}$ and $\{W_l\}$ give a mixed Hodge structure on H^m.

For each $t \in \Delta^*$, one can choose various natural identifications $j: H^m \xrightarrow{\sim} H^m(V_t)$, which are indexed by the fibre of $U \to \Delta^*$ over t. Any two of them are related by a power of $T = \exp N$. Hence $j^{-1} \cdot \nu$, $\psi \cdot j$, and of course $j^{-1} \cdot N \cdot j$ become independent of the particular choice of j. We may therefore replace $H^m(V_t)$ in (10.14) by H^m, for the sake of simplicity, we shall not change the symbols for the maps which occur:

$$\to H_{2n-m+2}(V_0) \xrightarrow{\mu} H^m(V_0) \xrightarrow{\nu} H^m \xrightarrow{N} H^m \xrightarrow{\psi} H_{2n-m}(V_0) \to \cdot \tag{10.15}$$

All of the vector spaces in (10.15) now carry canonical mixed Hodge structures. The main results of [10] can be summarized by the formal statement

(10.16) THEOREM. *The sequence* (10.15) *is exact. The mappings* μ, ν, N, ψ *are morphisms of mixed Hodge structures, of types* $(n+1, n+1)$, $(0, 0)$, $(-1, -1)$, *and* $(-n, -n)$, *respectivly.*

The local invariant cycle problem, listed as Conjecture (8.1) in [21], asks whether the image of $H^m(V_0)$ in $H^m(V_t)$, $t \in \Delta^*$, consists of all $\pi_1(\Delta^*)$-invariant cohomology classes. For surfaces, an affirmative answer was given by Katz, and in general, by Deligne (unpublished); both arguments depend on having the family come from a global algebraic family. Theorem (10.16) now also asnwers this problem.[44]

Because of the mixed Hodge structures which are present, and with the help of Lemma (1.13), quite a bit of information beyond the local invariant cycle problem can be squeezed out of the theorem. We do not want to go into details here, and instead refer the reader to [10].

APPENDIX

The current status of the problems and conjectures listed in [21].

(a) *Torelli-type questions* (§7)[45]. Conjecture 7.2 has been proved by Piatetskii-Shapiro and Shafarevich [40]. Very briefly, they deduce the global Torelli theorem from the local result by verifying the global statement for "sufficiently many" polarized $K3$ surfaces. It seems to be still unknown whether every point in $\Gamma\backslash D$ comes from a polarized $K3$ surface having at most rational singularities [37].

Problem 7.3 has been settled in the affirmative by Tjurin [43], and independently by Clemens and the first author [9]. A recent paper by Tjurin [42] discusses the general problem of the global Torelli theorem for the *Fano threefolds*.

The local Torelli problem 7.1 for simply connected surfaces of general type seems to remain open. In their recent work on canonical embeddings of surfaces of general type, Bombieri and Kodaira [3] have found surfaces whose canonical series exhibit various kinds of extreme behavior (generalizing that of hyperelliptic curves), and Cornalba has verified the local Torelli theorem for these.

[44] One can show that going to a finite covering, blowing up, etc. does not affect the statement in question.

[45] The numbering of sections, problems, etc. is that of [21].

(b) *Local monodromy and the behavior of the periods at infinity* (§§8, 9). As was mentioned already in §10(b) above, the local invariant cycle problem 8.1 has been solved by Katz for families of surfaces, and by Deligne in general; see also [10]. Problem 8.2 about the local monodromy around an isolated singularity was settled by counterexamples of A'Campo and Deligne. Conjecture 8.4 was proven by Katz [31], and the refined version 8.4′ in the appendix is a consequence of the results of [41] (cf. §9, 10 (a) above).

For the classifying spaces of the periods of holomorphic two-forms, Cattani [6] has constructed a partial compactification, as suggested by conjecture 9.2. The general case is still open, as is conjecture 9.5, even in the special case of Hodge structures of weight two. Conjecture 9.17 is discussed in detail in §10 (a) above.

(c) *Uniformization of periods*; *automorphic cohomology* (§§10, 11). As far as we know, there has been no substantial progress on conjecture 10.1, problem 10.6, and problem 11.1. Sommese (Harvard thesis, to appear) seems close to a solution of conjecture 10.5, although some details remain unsettled at the time of this writing.

(d) *General uniformization* (§12). Problem 12.1 has been answered in the affirmative by the first author [23]. However, the result lacks the symmetry exhibited by the classical uniformization theorem; basically, this is due to the breakdown of the theory of conformal mappings in the case of several variables. (Consequently, the theorem in [23] should be thought of only as a partial result, and we refer to the above paper for a further discussion and specific open problems, having to do with the general uniformization question for algebraic varieties).

We want to conclude by stating a new problem. Let $\pi: X \to S$ be an *algebraic* family of polarized algebraic manifolds, and

$$\Gamma \subset \operatorname{Hom}(H^m(V_{s_0}), H^m(V_{s_0})), \quad s_0 \in S,$$

the global monodromy group, i.e. the action of $\pi_1(S)$ on the mth cohomology group of a typical fibre $V_{s_0} = \pi^{-1}(S_0)$.

Problem. *Is the monodromy group an arithmetic group?*

To be more precise, we recall that Γ lies in the automorphism group $G_{\mathbf{R}}$ of the appropriate classifying space of Hodge structures, and that $\Gamma \subset G_{\mathbf{Z}}$ (cf. §3). The Zariski closure over $\mathbf{Q}$ of Γ in $G_{\mathbf{R}}$, which we denote by $G_{\mathbf{R}}(\Gamma)$, is the group of real points of an algebraic $\mathbf{Q}$-group. We are asking whether Γ is arithmetic in this group; equivalently, whether

$$[\Gamma : G_{\mathbf{Z}} \cap G_{\mathbf{R}}(\Gamma)] < \infty.$$

The period mapping can be taken to have values in the quotient $\Gamma\backslash D$ (rather than $G_{\mathbf{Z}}\backslash D$), and the volume of its image in $\Gamma\backslash D$ can be shown to be finite [19]. This at least lends some plausibility to the conjecture.

REFERENCES

1. A. Andreotti and H. Grauert: Algebraische Körper von automorphen Funktionen, *Göttinger Nachrichten*, (1961), 39-48.
2. W. Baily and A. Borel: Compactification of arithmetic quotients of bounded symmetric domains, *Ann. of Math.*, 84 (1966), 442-528.
3. E. Bombieri: The pluricanonical map of a complex surface, Several Complex Variables I, *Lecture Notes in Mathematics*, 155, Springer, (1970).
4. A. Borel: *Linear Algebraic Groups*, Benjamin, (1969).
5. A. Borel: Some metric properties of arithmetic quotients of symmetric spaces and an extension theorem, *J. Diff: Geom.*, 6 (1972), 543-560.
6. E. Cattani: *On the partial compactification of the "modular variety" associated to a Kähler surface*: Thesis, St. Louis, (1972).
7. S. S. Chern: A generalization of Kähler geometry, *Symposium in honor of S. Lefschetz*: Princeton Univ. Press, (1957). 103-121,
8. H. Clemens: Picard-Lefschetz theorem for families of nonsingular algebraic varieties acquiring ordinary singularities, *Trans. Amer. Math. Soc.*, 136 (1969), 93-108.

9. H. Clemens and P. Griffiths: The intermediate Jacobian of the cubic threefold, *Ann. of Math.*, 96 (1972), 281-356.

10. H. Clemens and W. Schmid: Degeneration of algebraic manifolds, to appear.

11. P. Deligne: Le travaux de Griffiths, *Séminaire Bourbaki* Exp. 376.

12. P. Deligne: Théorie de Hodge I, *Actes du Congrès International des Mathematiciens*, Nice, (1970).

13. P. Deligne: Théorie de Hodge II, *Publ. Math. I.H.E.S.*, 40(1972), 5-57.

14. P. Deligne: Théorie de Hodge III, to appear in *Publ. Math. I.H.E.S.*

15. P. Deligne: Equations différentielles à points singuliers réguliers, *Lecture Notes in Mathematics*, 163, Springer, (1970).

16. V. Gugenheim and D. C. Spencer: Chain homotopy and the Poincaré lemma, *Proc. Amer. Math. Soc.*, 7 (1956), 144-152.

17. R. Godement: Topologie algebriques et theorie des faisceaux, *Publ. Inst. Math. Univ. Strasbourg*, XIII, Paris, Hermann, (1958).

18. P. Griffiths: Periods of integrals on algebraic manifolds I, II, *Amer. J. Math.*, 90 (1968), 568-626, 805-865.

19. P. Griffiths: Periods of integrals on algebraic manifolds III, *Publ. Math. I.H.E.S.*, 38 (1970), 125-180.

20. P. Griffiths: On the periods of certain rational integrals, I, II, *Ann. of Math.*, 90 (1969), 460-541.

21. P. Griffiths: Periods of integrals on algebraic manifolds: summary of main results and discussion of open problems, *Bull. Amer. Math. Soc.*, 76 (1970), 228-296.

22. P. Griffiths: Some transcendental methods in the study of algebraic cycles, Proceedings of the Maryland Conference on Complex Analysis, *Lecture Notes in Mathematics*, 185, Springer, (1971).

23. P. Griffiths: Complex analytic properties of certain Zariski open sets on algebraic varieties, *Ann. of Math.*, 94 (1971), 21-51.

24. P. Griffiths and J. King: Nevanlinna theory and holomorphic mappings between algebraic varieties, to appear in *Acta Math.*, 130 (1973).

25. P. Griffiths and W. Schmid: Locally homogeneous complex manifolds, *Acta Math.*, 123 (1969), 253-302.

26. A. Grothendieck: On the de Rham cohomology of algebraic varieties, *Publ. Math. I.H.E.S.*, 29 (1966), 95-103.

27. S. Helgason: *Differential Geometry and Symmetric Spaces*, Academic Press, New York, (1962).

28. E. Hille: *Analytic Function Theory, vol. II*, Ginn and Company, New York, (1962).

29. H. Hironaka: Resolution of singularities of an algebraic variety over a field of characteristic zero I, II, *Ann. of Math.*, 79 (1964), 109-326.

30. W. V. D. Hodge: *The Theory and Application of Harmonic Integrals*, 2nd ed., Cambridge Univ. Press, (1952).

31. N. Katz: Nilpotent connections and the monodromy theorem; applications of a result of Turritin, *Publ. Math. I.H.E.S.*, 39 (1971), 175-232.

32. P. Kiernan and S. Kobayashi: Satake compactification and extension of holomorphic mappings, to appear in *Inventiones Math.*

33. K. Kodaira and D. C. Spencer: On deformations of complex analytic structures III, *Ann. of Math.*, 71 (1960).

34. S. Kobayashi: *Hyperbolic Manifolds and Holomorphic Mappings*: Marcel Dekker, New York, (1970).

35. A. Landman: On the Picard-Lefschetz transformations, to appear in *Trans. Amer. Math. Soc.*

36. D. Lieberman: Intermediate Jacobians, *Proc. Oslo Conf.*, (1970).

37. A. Mayer: Families of $K3$ surfaces, *Nagoga J. Math.*, 48 (1972).

38. R. Nevanlinna: *Analytic Functions*, Springer, (1970).

39. H. Nickerson: On the complex form of the Poincaré lemma, *Proc. Amer. Math. Soc.*, 9 (1958), 183-188.

40. I. I. Piatetskii - Shapiro and I. R. Shafarevich: Theorem of Torrelli on algebraic surfaces of type $K3$, *Izvestia Akad. Nauk*, 35 (1971).

41. W. Schmid: Variation of Hodge structure—the singularities of the period mapping, to appear in *Inventiones Math.*

42. A. N. TJURIN: The geometry of the Fano surface of a nonsingular cubic $F \subset P^4$ and Torelli's theorem for Fano surfaces and cubics. *Izvestia Akad. Naùk*, 35 (1971), 498-529.

43. A. N. TJURIN: Five lectures on three-dimensional varieties, *Uspeki Mat. Nauk*, 27 (1972), 4-50.

44. F. WARNER: *Foundations of differentiable manifolds and Lie groups*: Scott, Foresman and Co., (1971).

45. A. WEIL: Variétés Kählériennes, *Publ. Inst. Math. Univ. Nancago*, VI, Hermann, Paris (1958).

46. A. WEIL: Un théorème fondamental de Chern en géométric riemannienne, *Séminaire Bourbaki*, Exp. 239.

47. R. WELLS: *Differential analysis on complex manifolds*, to appear.

48. H. WU: Normal families of holomorphic mappings, *Acta. Math.*, 119 (1967), 193-233.

ON THE COHOMOLOGY OF DISCRETE ARITHMETICALLY DEFINED GROUPS*

By G. HARDER

INTRODUCTION. In this paper I want to come back to the questions which I discussed in [7]. These questions arise from the study of the cohomology of discrete arithmetically defined groups Γ. To investigate the cohomology of Γ one considers the action of Γ on the corresponding symmetric space X and makes use of the fact that $H^{\cdot}(X/\Gamma,\mathbf{R})=H^{\cdot}(\Gamma,\mathbf{R})$. If this quotient X/Γ is compact then the Hodge theory (comp. [12] §31) is a powerful tool for the investigation of these cohomology groups. But in general the quotient X/Γ is not compact and my concern in this paper are those phenomena which are due to this noncompactness. The Hodge theory fails in this case and I want to find a substitute for it or in other words I want to get control of the deviation from Hodge theory. The basic idea is to make use of Langlands' theory of Eisenstein series (comp. [8]) to describe the cohomology at "infinity".

In this paper I mainly consider the case that Γ is of rank one, i.e. the semi-simple group G/k which defines Γ is of rank one over the algebraic number field k. The main result is Theorem 4.6. This theorem is to some extent a generalization of Theorem 2.1 in [7] which is stated there without proof. On the other hand the results in [7] are still much more precise, because in the case of $G = SL_2/k$ the intertwining operators $c(s)$ (comp. §3) are accessible for explicit computations. I hope to come back to these problems later.

§1. Basic notions and results on the cohomology of Γ. Let G_∞ be a real Lie group acting transitively from the right on a contractible C^∞-manifold X. Let $x_0 \in X$ and let us assume that the stabilizer of x_0 is a compact subgroup K. Therefore we have $K\backslash G_\infty = X$.

*Supported by the "Sonderforschungsbereich für theoretische Mathematik an der Universität Bonn" and NSF grant GP-36418X.

Let $\Gamma \subset G_\infty$ be a discrete subgroup without torsion, let us assume that X/Γ is a manifold; then Γ is the fundamental group of X/Γ and the higher homotopy groups of X/Γ are trivial, i.e., X/Γ is a $K(\Gamma, 1)$-space.

Let E be a finite dimensional vector space over the complex numbers and let $\rho_\infty \colon G_\infty \to GL(E)$ be a representation of G_∞. We define an action of G_∞ on the functions $f : X \to E$ by

$$T_g(f)(x) = \rho_\infty(g^{-1})(f(xg)) \text{ for } x \in X \text{ and } g \in G_\infty.$$

We restrict the action of G_∞ on X and E to the subgroup Γ, then we can construct the induced bundle $\tilde{E}_{\rho_\infty} \to X/\Gamma$ and it follows from the definition of the induced bundles that the C^∞-sections of $\tilde{E}_{\rho_\infty} \to X/\Gamma$ are precisely the Γ-invariant C^∞-functions $f : X \to E$. The pull back of the bundle $\tilde{E}_{\rho_\infty}$ to X is the trivial bundle $E \times X$ over X and the trivial connection on $E \times X$ induces a flat connection $\tilde{\nabla}$ on $\tilde{E}_{\rho_\infty}$, which will be called the canonical connection on $\tilde{E}_{\rho_\infty}$. By means of this connection we define the sheaf of locally constant sections in $\tilde{E}_{\rho_\infty}$, and the cohomology with coefficients in this sheaf will be denoted by

$$H^{\cdot}(X/\Gamma, \tilde{E}_{\rho_\infty}) = H^{\cdot}(\Gamma, \rho_\infty, E)$$

These cohomology groups can also be defined via the de Rham-cohomology: Let $\Omega^p(X/\Gamma, E)$ be the vector space of C^∞-p-differential forms on X/Γ with values in $\tilde{E}_{\rho_\infty}$. Making use of the canonical connection we define the exterior derivative $d \colon \Omega^p(X/\Gamma, E) \to \Omega^{p+1}(X/\Gamma, E)$ which is given by the following formula : If $P_1, P_2, \ldots, P_{p+1}$ are C^∞-vector fields on X/Γ and if $\omega \in \Omega^p(X/\Gamma, E)$ then

$$d\omega(P_1, \cdot, P_{p+1}) = \sum_{i=1}^{i=p+1} (-1)^{i+1} \tilde{\nabla}_{P_i} \omega(P_1, \cdot, \hat{P}_i, \cdot, P_{p+1}) +$$

$$+ \sum_{1 \leq i < k \leq p+1} (-1)^{i+j} \omega([P_i, P_j], P_1, \ldots, \hat{P}_i, \cdot, \hat{P}_j, \cdot, P_{p+1}) \quad (1.1)$$

Since $\tilde{\nabla}$ is flat we have $d^2 = 0$ and it is well known that the cohomology groups of the de Rham complex $(\Omega^{\cdot}(X/\Gamma, E), d)$ are canonically isomorphic to the cohomology groups $H^{\cdot}(X/\ \ , \tilde{E}'_{\rho_\infty})$ (comp. [10], §1 Prop. 1.1).

We may look at our bundle $\tilde{E}_{\rho'_\infty}$ from a different point of view. We consider the fibration

$$q: G_\infty/\Gamma \to X/\Gamma$$

which is principal with structure group K. Therefore the restriction of ρ_∞ to K defines another induced bundle E_{ρ_∞} whose global C^∞-sections are given by the C^∞-functions

$$f: G_\infty/\Gamma \to E$$

which satisfy

$$f(kg) = \rho_\infty(k)(f(g)) \text{ for all } g \in G_\infty, k \in K.$$

LEMMA 1.2. *For any C^∞-section f of E_{ρ_∞} we define a section $\tilde{f}$ of $\tilde{E}_{\rho_\infty}$ by*

$$\tilde{f}(x) = \tilde{f}(x_0 g) = \rho_\infty(g^{-1}) f(x_0) \quad x = x_0 g$$

Then the mapping $f \to \tilde{f}$ induces an isomorphism from E_{ρ_∞} to $\tilde{E}_{\rho_\infty}$

The proof is either trivial or can be found in [10], Prop. 3.1.

By means of this identification we can define a connection ∇ on E_{ρ_∞} and we shall derive a formula for this connection in terms of the bundle E_{ρ_∞} itself.

Let $\mathfrak{G}$ be the Lie algebra of right invariant vector fields on G_∞. Let $\mathfrak{K}$ be the Lie algebra of K and we put $\mathfrak{p} = \mathfrak{G}/\mathfrak{K}$. Let T^G (resp. T^X) be the tangent bundle of G_∞/Γ (resp. X/Γ). The projection q defines at any point $g \in G$ a surjection

$$D_{g,q}: T^G_g = \mathfrak{G} \longrightarrow T^X_x, \; x = x_0 g$$

and this yields isomorphisms

$$\bar{D}_{g,q}: \mathfrak{p} \xrightarrow{\sim} T^X_x.$$

This shows that $q^*(T^X) = G_\infty/\Gamma \times \mathfrak{p}$ and that the C^∞-vector fields on X/Γ can be identified with the C^∞-functions

$$P: G_\infty/\Gamma \to \mathfrak{p}$$

which satisfy

$$P(kg) = ad_{\mathfrak{p}}(k)P(g) \qquad g \in G_\infty, k \in K, \tag{1.3}$$

i.e., the tangent bundle of X/Γ is the bundle induced by adjoint representation $ad_{\mathfrak{p}}$ of K on $\mathfrak{p}$.

Now we can give our formula for the connection ∇.

LEMMA 1.4. *Let* $f : G_\infty/\Gamma \to E$ *be a* C^∞*-section of* E_{ρ_∞} *and let* $P : Kg \to \mathfrak{p}$ *be a tangent vector at* $x = x_0\, g = Kg$. *Then*

$$\nabla_{P(g)}(f)\,|_g = \tilde{P}(g) f\,|_g - \rho_\infty(\tilde{P}(g))(f(g))$$

where $\tilde{P}(g)$ *is a representative for* $P(g)$ *in* $\mathfrak{G}$, *and where* $\tilde{P}(g) f\,|_g$ *is the ordinary derivative of the function* f *with respect to the tangent vector* $\tilde{P}(g)$ *at* g.

This lemma follows from direct calculations and it is also essentially stated in [10], Prop. 4.1.

REMARK. The bundles E_{ρ_∞} and $\tilde{E}_{\rho_\infty}$ are only two different realizations of the same thing. We need $\tilde{E}_{\rho_\infty} \to X/\Gamma$ for the definition of the connections $\tilde{\nabla}$ and ∇, but from now on we shall work with the realization $E_{\rho_\infty} \to X/\Gamma$.

Let us now assume that G_∞ is the group of real points of a reductive algebraic group over **R**. Then the Lie algebra splits

$$\mathfrak{G}' = \mathfrak{Z} \oplus \mathfrak{G}$$

where $\mathfrak{Z}$ is the centre of $\mathfrak{G}'$ and where $\mathfrak{G}$ is the semisimple part of $\mathfrak{G}$. Let us assume moreover that $K \subset G_\infty$ is a maximal compact subgroup, then $X = K\backslash G_\infty$ is homeomorphic to an euclidean space. Let

$$B : \mathfrak{G} \times \mathfrak{G} \longrightarrow \mathbf{R}$$

be a symmetric nondegenerate bilinear form on $\mathfrak{G}$ which has the following properties

(1) B is invariant under the adjoint action of G_∞ on $\mathfrak{G}$.

(2) The restriction of B to $\mathfrak{K}$ is negative definite and if $\mathfrak{p}$ is the orthogonal complement of $\mathfrak{K}$ with respect to B, then $B\backslash\mathfrak{p}$ is positive definite.

The condition (1) implies that $\mathfrak{G} = \mathfrak{Z} \oplus \mathfrak{G}'$ is orthogonal with respect to B.

By means of the identification $\bar{D}_{e,q} : \mathfrak{p} \to T^X_{x_0}$ we define as usual a Riemannian metric on X.

We assume now that ρ_∞ is the complexification of a real representation $\rho_0: G \to GL(E_0)$. Following Matsushima and Murakami in [10] we call an euclidean bilinear form

$$\langle , \rangle : E_0 \times E_0 \to \mathbf{R}$$

admissible if

(1) $\langle , \rangle$ is invariant under $\rho(K)$,

(2) $\langle \rho_\infty(P)e, f \rangle = \langle e, \rho_\infty(P) f \rangle$ for all $P \in \mathfrak{p}$, $e. f \in E_0$.

We extend $\langle , \rangle$ to a hermitian form on $E = E_0 \otimes \mathbf{C}$. The existence of such an admissible metric on E_0 is proved in [10], Lemma 3.1, if K is connected. In the general case we have to average over the different connected components of K.

Let us fix an admissible metric on E. It induces a $\mathbf{C}$-linear isomorphism

$$\# : E \xrightarrow{\sim} E^* = \mathrm{Hom}_{\mathbf{C}}(E, \mathbf{C})$$

which is defined by $(\# a)(b) = \langle b, \bar{a} \rangle$ for all $a,b \in E$. Because of condition (1) this extends to an isomorphism $\#: E_{\rho_\infty} \xrightarrow{\sim} E^*_{\rho_\infty}$ This yields a map

$$\# : \Omega^p(X/\Gamma, E) \xrightarrow{\sim} \Omega^p(X/\Gamma, E^*), \tag{1.5}$$

If dim $X = N$ and if Γ preserves an orientation of X then we define the operator

$$* : \Omega^p(X/\Gamma, X) \longrightarrow \Omega^{N-p}(X/\Gamma, E) \tag{1.6}$$

(comp. [10], §2) as usual and it is easy to see that $*$ and $\#$ commute.

If $\omega \in \Omega^p(X/\Gamma, E)$ and $\omega' \in \Omega^{N-p}(X/\Gamma, E^*)$ then $\omega \wedge \omega'$ takes its values in $E \otimes E^*$ and the evaluation map $tr : E \otimes E^* \to \mathbf{C}$ brings us back to the complex numbers. We define

$$(\omega, \omega') = tr(\omega \wedge \omega') \in \Omega^N(X/\Gamma, \mathbf{C}) \tag{1.7}$$

We now define as usual a hermitian inner product on the C^∞-p-forms with compact support

$$\langle \omega, \omega' \rangle = \int_{X/\Gamma} (\omega, \overline{* \circ \# \omega'})$$

The operator $\delta : \Omega^{p+1} (X/\Gamma, E) \to \Omega^p (X/\Gamma, E)$ is defined as the adjoint of d with respect to this metric. We know from [10], §2 that this operator can also be defined by

$$\delta = (-1)^{p+1} *^{-1} \circ \#^{-1} d \# \circ * .$$

We define the Laplacian

$$\Delta = d\delta + \delta d.$$

This Laplacian operator does not depend on the choice of an orientation on X and therefore we can define Δ also if Γ does not preserve the orientation.

If the quotient X/Γ is compact then one knows that a harmonic p-form ω, i.e., a form that satisfies $\Delta\omega = 0$, is closed and coclosed, i.e., it satisfies $d\omega = 0$ and $\delta\omega = 0$. Therefore we get a mapping from the space of harmonic forms $\mathbf{H}^p (X/\Gamma, E)$ to the cohomology

$$\mathbf{H}^p (X/\Gamma, E) = \{ \omega \in \Omega^p (X/\Gamma, E) \mid \Delta\omega = 0 \} \longrightarrow H^p (X/\Gamma, E)$$

and a straightforward generalization of the Hodge theory (comp. [12], §31, Cor. 4) tells us that this map is an isomorphism.

To conclude this section we shall explain the relation of Δ to the Casimir operator C. The bundle E_{ρ_∞} is the induced bundle of the representation $\rho_\infty | K$, and we have seen that the tangent bundle T^X is the bundle induced by $ad_{\mathfrak{p}}$. Therefore we have a natural identification between the space $\Omega^p (X \Gamma, E)$ and the functions

$$\varphi : G_\infty/\Gamma \to \mathrm{Hom}(\Lambda^p_{\mathfrak{p}}, E)$$

which satisfy

$$\varphi(kg) = \Lambda^p ad_{\mathfrak{p}}{}^*(k) \otimes \rho_\infty (k)(\varphi(g)) \quad g \in G_\infty, \ k \in K \qquad (1.8)$$

where $ad_{\mathfrak{p}}{}^*$ is the dual representation to $ad_{\mathfrak{p}}$.

The enveloping algebra $U(\mathfrak{G})$ is operating as an algebra of differential operators on the C^∞-functions on G_∞/Γ with values in $\mathrm{Hom}\ (\Lambda^p_{\mathfrak{p}}, E)$ (Comp. [8], Chap. I, §2). Our metric on $\mathfrak{G}$ defines the Casimir operator $C \in \mathfrak{Z}\ (\mathfrak{G})$ = centre of $U(\mathfrak{G})$, and this is an operator of order 2 which sends the functions φ satisfying (1.8) into itself. We can extend the representation ρ_∞ of $\mathfrak{G}$ in End (E, E) to a representation ρ_∞ of $U(\mathfrak{G})$ in End (E, E). Then we have for all C^∞-p-forms φ: $G_\infty/\Gamma \to \mathrm{Hom}(\Lambda_{\mathfrak{p}}^P, E)$ the formula

$$\Delta\varphi = -C\varphi + \rho_\infty(C)\varphi. \tag{1.9}$$

This is the lemma of Kuga and it is proved in [10] §6 for semisimple groups. The generalization to the reductive case is easy.

§2. The cohomology of 'parabolic' discrete groups. Let k be an algebraic number field and let G/k be a semisimple algebraic group. We define $k_\infty = k \otimes_{\mathbf{Q}} \mathbf{R} = \mathbf{R}^{r_1} \oplus \mathbf{C}^{r_2}$ and we put $G_\infty = G_{k_\infty}$. Let $\Gamma \subset G_k$ be an arithmetically defined subgroup without torsion. Let $\rho_k : G \to GL(V)$ be a rational representation of G/k. From this we get a representation ρ_∞ of G_∞ on the vector space $E_0 = V \otimes_{\mathbf{Q}} \mathbf{R}$. Let us fix an admissible metric on E_0 with respect to a fixed maximal compact subgroup $K \subset G_\infty$. Again we put $E = E_0 \otimes \mathbf{C}$. This is the situation to which we shall apply the results of §1.

Let $P \subset G/k$ be a parabolic subgroup. In this section we shall investigate the cohomology groups

$$H^\cdot(\Gamma \cap P_k, \rho'_\infty, E)$$

where $\Gamma \cap P_k = \Gamma \cap P_\infty \subset P_\infty$ is the "parabolic" discrete group in the heading of this section.

Let us assume that G/k is simply connected. Then the character module $X(P) = \mathrm{Hom}(P, G_m)$ is generated by fundamental dominant weights $\chi_1, \ldots, \chi_r$. These characters χ_i induce homomorphisms

$$|\chi_i| : P_\infty \longrightarrow (\mathbf{R}^+)^*$$
$$|\chi_i| : p \longrightarrow N_{k_\infty/\mathbf{R}}(\chi_i(p))|$$

The intersection of the kernels of these homomorphisms $|\chi_i|$ is called $P_\infty(1)$, and it is well known that $P_\infty(1) \cap \Gamma = P_\infty \cap \Gamma$. We put $X = K\backslash G_\infty$ where K is our fixed maximal compact subgroup and $x_0 = Ke \in X$. We define

$$X(1) = X_P(1) = x_0 P_\infty(1).$$

This is a s-codimensional subspace of X which depends on P. We drop the index P if it is clear with respect to which P this subspace is defined. Since $X(1)$ is contractible we have

$$H^{\cdot}(X(1)/\Gamma \cap P_\infty, E) = H^{\cdot}(X/\Gamma \cap P_\infty, E) = H^{\cdot}(\Gamma \cap P_\infty, \rho'_\infty, E)$$

where ρ'_∞ is of course the restriction of ρ_∞ to P_∞.

Let us denote the unipotent radical of P by U. Then $M = P/U$ is a reductive algebraic group over k, and $M_\infty(1)$ is the group of real points of an algebraic subgroup of $M \times_{\mathbf{Q}} \mathbf{R}$. Let K_M be the image of $K \cap P_\infty$ in $M_\infty(1)$ then the quotient

$$X_M = K_M \backslash M_\infty(1)$$

is again a symmetric space. As metric on X_M we take the one which is induced by the restriction of the Killing form to the Lie algebra of $M_\infty(1)$, this restriction satisfies obviously the conditions (1) and (2) in §1. Let Γ_M be the image of $\Gamma \cap P_\infty$ in $M_\infty(1)$—this is again an arithmetically defined group. The map

$$\pi : X(1)/\Gamma \cap P_\infty \to X_M/\Gamma_M$$

is easily seen to be a fibration with fiber $U_\infty/U_\infty \cap \Gamma$. Therefore we have a spectral sequence

$$H^p(X_M/\Gamma_M, H^q(U_\infty/U_\infty \cap \Gamma, E)) \Rightarrow H^n(X(1)/\Gamma \cap P_\infty, E) \qquad (2.1)$$
$$p + q = n.$$

We shall see that this spectral sequence collapses and that the cohomology $H^{\cdot}(X(1)/\Gamma \cap P_\infty, E)$ is actually the direct sum of the E_2-terms on the left hand side.

First we apply the considerations of §1 to the cohomology of the fiber. Since $K \cap U_\infty = e$ the tangent bundle and the bundle E_{ρ_∞} restricted to $U_\infty/U_\infty \cap \Gamma$ are trivial and $\Omega^q(U_\infty, E)$ is the space of C_∞-functions $\omega : U_\infty \to \mathrm{Hom}(\Lambda^q u, E)$ where u is the Lie algebra of right invariant vector fields on U_∞. The group U_∞ is acting on the q-forms by right translations, and $\Omega^q(U_\infty/U_\infty \cap \Gamma, E)$ consists exactly of those forms which are invariant under $U_\infty \cap \Gamma$. Therefore we have an embedding

$$\mathrm{Hom}(\Lambda^{\cdot} u, E) = \Omega^{\cdot}(U_\infty, E)^{U_\infty} \xhookrightarrow{j} \Omega^{\cdot}(U_\infty/U_\infty \cap \Gamma, E).$$

We have also a map in the opposite direction

$$h : \omega \to \int_{U_\infty/U_\infty \cap \Gamma} \omega(u)\, du.$$

Here, du is a Haar measure on U_∞ such that $\mathrm{vol}_{du}(U_\infty/U_\infty \cap \Gamma) = 1$.

THEOREM 2.2. (van Est, [4]). *The maps j and h induce isomorphisms on the cohomology which are inverse to each other. We have*

$$H^{\cdot}(U_\infty/U_\infty \cap \Gamma, E) = H^{\cdot}(\underline{u}, E)$$

where the expression on the right hand side denotes the cohomology of the Lie algebra $\underline{u}$ *with coefficients in the* $\underline{u}$*-module E.*

The first assertion is essentially a special case of Theorem 8 in van Est's paper, and the second one follows from his Theorem 9; but at this place we have to make a few remarks about signs. Let us assume for the moment that U_∞ is any Lie group and that $\underline{u}$ is the Lie algebra of right invariant vector fields. If ρ_∞ is still our representation of U_∞ in $GL(E)$ then we define as usual for $T \in \underline{u}$

$$\rho_\infty(T)\, f = \lim_{t \to 0} \frac{\exp(tT)\, f - f}{t}$$

and with this definition E becomes a right $\underline{u}$-module, i.e., we have

$$(\rho_\infty(T_1)\, \rho_\infty(T_2) - \rho_\infty(T_2)\, \rho_\infty(T_1))f = \rho_\infty([T_2, T_1])f \text{ for } T_1, T_2 \in \underline{u},\ f \in E$$

where of course $[T_1 T_2] = T_1 T_2 - T_2 T_1$. Especially for the adjoint representation this yields $\mathrm{ad}(T_1)\,(T_2) = [T_2, T_1]$. For a right $\underline{u}$-module E the cohomology groups are computed from the following complex (comp. [3], Chap. XIII, §8).

$$\longrightarrow \mathrm{Hom}(\Lambda^{q-1}\underline{u}, E) \xrightarrow{d} \mathrm{Hom}(\Lambda^{q}\underline{u}, E) \xrightarrow{d} \mathrm{Hom}(\Lambda^{q+1}\underline{u}, E) \longrightarrow$$

where for $f \in \mathrm{Hom}(\Lambda^q \underline{u}, E)$ we have

$$df(T_1, \cdot, T_{q+1}) = \sum_{i=1}^{i=q+1} (-1)^i \rho_\infty(T_i) f(T_1, \cdot, \hat{T}_i, \cdot, T_{q+1}) +$$

$$+ \sum_{1 \leqslant i < j \leqslant q+1} (-1)^{i+j} f([T_i, T_j], T_1, \cdot, \hat{T}_i, \cdot, \hat{T}_j \cdot T_{q+1}). \quad (2.3)$$

Actually the second statement of Theorem 2.2 is easy consequence of the definitions, Lemma 1.4, and formula (2.3).

In the spectral sequence (2.1) we have to take the cohomology of X_M/Γ_M with coefficients in the local system of the cohomology of the fiber $H^{\cdot}(U_{\infty}/U_{\infty} \cap \Gamma, E) = H^{\cdot}(\underline{u}, E)$. This local system is determined by the operation of the fundamental group Γ_M on $H^{\cdot}(\underline{u}, E)$. On the other hand, the group $P_{\infty}(1)$ acts in a natural way on the cochain complex which defines $H^{\cdot}(\underline{u}, E)$ and this gives us a representation of $M_{\infty}(1)$ on $H^{\cdot}(\underline{u}, E)$. We claim that the restriction of the latter action to Γ_M is equal to the first one. To see this we have to consider the pull back of the fibration π over π itself, we get a diagram

$$\begin{array}{ccc} X(1)/\Gamma \cap P_{\infty} & \xrightarrow{\pi} & X_M/\Gamma_M \\ \uparrow & & \uparrow \pi \\ Y & \xrightarrow{\pi'} & X(1)/\Gamma \cap P_{\infty} \end{array}$$

and the fiber bundle $Y \xrightarrow{\pi'} X(1)/\Gamma \cap P_{\infty}$ is induced by the operation of $\Gamma \cap P_{\infty}$ on $U_{\infty}/U_{\infty} \cap \Gamma$. But then the corresponding assertion for the fibration π' is clear, and this proves our original claim.

This tells us that we can apply the results of §1 to the E_2-terms of the spectral sequence. The next thing we shall do is to construct for a fixed q an imbedding

$$\Omega^p(X_M/\Gamma_M, H^q(\underline{u}, E)) \hookrightarrow \Omega^{p+q}(X(1)/\Gamma \cap P_{\infty}, E)$$

and we shall see that this imbedding commutes with the exterior derivatives.

The Cartan involution corresponding to K is denoted by θ. It induces the well-known Cartan decomposition $\mathfrak{G} = \mathfrak{k} \oplus \mathfrak{p}$. Moreover it is well known that the intersection

$$\tilde{M}_{\infty} = P_{\infty} \cap P_{\infty}^{\theta}$$

is a real Levi subgroup of P_{∞} and that the restriction of θ to $\tilde{M}_{\infty}$ is again a Cartan involution. We introduce a positive definite metric on $\mathfrak{G}$ by

$$B_{\theta}(N, N') = -B(N, \theta(N'))$$

where B is the Killing form. In $\tilde{M}_{\infty}$ we fix the maximal compact subgroup $K \cap \tilde{M}_{\infty} = K_M$.

Lemma 2.4. *The metric B_θ restricted to $\underline{u}$ is admissible with respect to the adjoint representation of $\tilde{M}_\infty$ on $\underline{u}$.*

Proof. Obvious.

This metric on $\underline{u}$ extends to an admissible metric on $\mathrm{Hom}(\Lambda^q\underline{u}, E)$ since we have already chosen an admissible metric on E. Following Kostant [9] we construct the adjoint operator $\delta : \mathrm{Hom}(\Lambda^p\underline{u}, E) \rightarrow \mathrm{Hom}(\Lambda^{p-1}\underline{u}, E)$ to the coboundary operator d. Since our metric is invariant under the action of K_M the operators d and δ commute with the action of K_M. We define the Laplacian by $L = d\delta + \delta d$, and Kostant has proved (comp. [9], Prop. 2.1 and 3.5.4)

$$H^q(\underline{u}, E) \xrightarrow{\sim} \{\zeta \in \mathrm{Hom}\,(\Lambda^q\underline{u}, E) \mid L\zeta = 0\} = \mathbf{H}^q(\underline{u}, E).$$

This is an M_∞ invariant subspace of $\mathrm{Hom}(\Lambda^q\underline{u}, E)$ since our metric is admissible.

Let us put $\underline{P} = \mathrm{Lie}(P_\infty)$, $\underline{P}_1 = \mathrm{Lie}(P_\infty(1))$, $\underline{m} = \mathrm{Lie}(\tilde{M}_\infty)$ and $\underline{m}_1 = \mathrm{Lie}(\tilde{M}_\infty(1))$. The Cartan decomposition of $\underline{m}$ (resp. $\underline{m}_1$) with respect to the restriction of θ to $\tilde{M}_\infty$ (resp. $\tilde{M}_\infty(1)$) is denoted by

$$\underline{m} = \mathfrak{K}_M \oplus \bar{\mathfrak{p}}$$

(resp.)

$$\underline{m}_1 = \mathfrak{K}_M \oplus \bar{\mathfrak{p}}_1$$

With respect to B_θ we have the following orthogonal decomposition (comp.[6], Prop. 1.1.1)

$$\underline{P} = \underline{m} \oplus \underline{u}$$

and this induces

$$\underline{P}_1 = \underline{m}_1 \oplus \underline{u} = \mathfrak{K}_M \oplus \bar{\mathfrak{p}}_1 \oplus \underline{u}.$$

Our general considerations in §1 yield that the tangent bundle of $X(1)$ (resp. X_M) is canonically isomorphic to the bundle which is induced by the adjoint operation of K_M on $\bar{\mathfrak{p}}_1 \oplus \underline{u}$ (resp. $\bar{\mathfrak{p}}_1$). Let us introduce the notation $\underline{r}_1 = \bar{\mathfrak{p}}_1 \oplus \underline{u}$. Then we have the decomposition of K_M-modules

$$\Lambda^n\underline{r}_1 = \Lambda^n(\bar{\mathfrak{p}}_1 \oplus \underline{u}) = \bigoplus_{p+q=n} \Lambda^p\bar{\mathfrak{p}}_1 \oplus \Lambda^q\underline{u}$$

We have interpreted an element in $\Omega^p(X_M/\Gamma_M, H^q(\underline{u}, E))$ as a C^∞-function

$$\varphi : M_\infty(1)/\Gamma_M \longrightarrow \mathrm{Hom}(\Lambda^p \bar{\mathfrak{p}}_1, \mathbf{H}^q(\underline{u}, E))$$

which satisfies

$$\varphi(kg) = \Lambda^p \,\mathrm{ad}^*_{\bar{\mathfrak{p}}_1}(k) \otimes \Lambda^q \,\mathrm{ad}^*_{\underline{u}}(k) \otimes \rho_\infty(k)\,(\varphi(g))$$

(comp. (1.8)). But since $\mathbf{H}^q(\underline{u}, E) \subset \mathrm{Hom}(\Lambda^q\underline{u}, E)$ we have $\mathrm{Hom}(\Lambda^p\bar{\mathfrak{p}}_1, \mathbf{H}^q(\underline{u}, E)) \subset \mathrm{Hom}(\Lambda^p\bar{\mathfrak{p}}_1, \mathrm{Hom}(\Lambda^q\underline{u}, E)) \subset \mathrm{Hom}(\Lambda^{p+q}\underline{\mathfrak{r}}_1, E)$ we may as well consider φ as an element of $\Omega^{p+q}(X(1)/\Gamma \cap P_\infty, E)$. Therefore we have constructed an embedding where we consider q as fixed and p as variable. Let us denote by d_M the exterior derivative of the complex $\Omega^\cdot(X_M/\Gamma_M, \mathbf{H}^q(\underline{u}, E))$ and respectively by d the exterior derivative of $\Omega^\cdot(X(1)/\Gamma \cap P_\infty, E)$.

LEMMA 2.6. *We have*

$$i_q \circ d_M = d \circ i_q.$$

PROOF. Let us take a form $\varphi \in \Omega^p(X_M/\Gamma_M, \mathbf{H}^q(\underline{u}, E))$, we put $\tilde{\varphi} = i_q(\varphi)$. We pull these two forms back to $P_\infty(1)/\Gamma \cap P_\infty$, and carry out the computations upstairs, the forms upstairs will be denoted by the same letters. By $M_1, M_2, \cdot, M_i$, (resp. $U_1, U_2, \ldots$) we denote the right invariant vector fields on $P_\infty(1)$ which are obtained from elements $M_i \in \underline{m}_1$ (resp. $U_i \in \underline{u}$). (There is no confusion in the notation since the indices are always less than infinity in M_∞). We get from our construction

$$\varphi(M_1, M_2, \cdots, M_p)(U_1, \cdots, U_q) = \tilde{\varphi}(M_1, M_2, \cdot, M_p, U_1, \cdot, U_q)$$

and moreover it is clear that φ is bihomogeneous of degree (p, q), i.e., we have $\varphi(.M_i\cdot, \cdot U_j\cdot) = 0$ unless the number of M's (resp. U's) is equal to p (resp. q). This implies that

$$d\tilde{\varphi}(M_1, \cdots, M_p, U_1, \cdots, U_{q+1}) = 0$$

because the terms involving ∇_{M_i}, $[M_i, M_j]$, and $[M_i, U_j]$ are zero and the terms involving ∇_{U_i} and $[U_i, U_j]$ add up to zero since φ takes its values in $\mathbf{H}^q(\underline{u}, E)$.

Therefore we have only to compute $d\varphi(M_1, \cdot, M_{p+1}, U_1, \cdot, U_q)$.

This breaks up into several terms; we use Lemma 1.4 repeatedly to get

$$d\tilde{\varphi}(M_1, \cdots, M_{p+1}, U_1, \cdot, U_q) =$$

$$\sum_{i=1}^{i=p+1} (-1)^{i+1} M_i \tilde{\varphi}(M_1, \cdot, \hat{M}_i, \cdot, M_{q+1}, U_1, \cdot, U_q) \qquad \text{(I)}$$

$$- \sum_{i=1}^{i=p+1} (-1)^{i+1} \rho_\infty(M_i) \tilde{\varphi}(M_1, \cdot, \hat{M}_i, \cdot, M_{p+1}, U_1, \cdot, U_q) \qquad \text{(II)}$$

$$+ \sum_{1 \leqslant i < j \leqslant p+1} (-1)^{i+j} \tilde{\varphi}([M_i, M_j], M_1, \cdot, \hat{M}_i, \cdot, \hat{M}_j, \cdot, M_{p+1}, U_1, \cdot, U_q) \qquad \text{(III)}$$

$$+ \sum_{i=1}^{i=p+1} \sum_{j=1}^{j=q} (-1)^{i+p+1+j} \tilde{\varphi}([M_i, U_j], M_1, \cdot, \hat{M}_i, \cdot, M_{p+1}, + $$

$$+ U_1, \cdot, \hat{U}_j, \cdot, U_q) \qquad \text{(IV)}$$

$+2$ other terms which vanish because $\tilde{\varphi}$ is bihomogeneous of degree (p, q).

On the other hand we have

$$d_{\mathcal{M}} \varphi(M_1, \cdots, M_{p+1}) =$$

$$\sum_{i=1}^{i=p+1} (-1)^{i+1} M_i \varphi(M_1, \cdot, \hat{M}_1, \quad, M_{p+1}) \qquad \text{(I}')$$

$$- \sum_{i=1}^{i=p+1} (-1)^{i+1} (\rho_\infty(M_i) + \Lambda^q \mathrm{ad}^*_{\underline{u}}(M_1)) \varphi(M_1, \cdot, \hat{M}_i, \cdot, M_{p+1}) \qquad \text{(II}')$$

$$+ \sum_{1 \leqslant i < j \leqslant p+1} (-1)^{i+j} \varphi([M_i, M_j], M_1, \cdot, \hat{M}_i, \cdot, \hat{M}_j, \cdot, M_{p+1}). \qquad \text{(III}')$$

We have to evaluate this on $(U_1, \cdot, U_q)$ and we see that (I) = (I′) and (III) = (III′). We claim that II′ = II + IV. We see that the typical term in IV is equal to

$$(-1)^i \tilde{\varphi}(M_1, \cdot, \hat{M}_i, \cdot, M_{p+1}, U_1, \cdot, [M_i, U_j], \cdot, U_q).$$

We see that (II′) naturally breaks up into two terms, one of them is obviously equal to (II) and the other one is

$$\sum_{i=1}^{i=p+1} (-1)^i (\Lambda^q \mathrm{ad}_{\underline{u}}(M_i) \varphi(M_1, \cdot, \hat{M}_i, , M_{p+1}))(U_1, \cdot, U_q).$$

If we take into account that $\mathrm{ad}_{\underline{u}}(M_i)(U_j) = [U_j, M_i]$ (comp. remark following Theorem 2.2.) then we see that this second term is equal to (IV).

We introduced an admissible metric on $\mathrm{Hom}(\Lambda^q \underline{u}, E)$ which induces an admissible metric on $\mathbf{H}^q(\underline{u}, E)$. From this metric we get a Laplacian operator

$$\Delta_M : \Omega^p(X_M/\Gamma_M, \mathbf{H}^q(\underline{u}, E)) \longrightarrow \Omega^p(X_M/\Gamma_M, \mathbf{H}^q(\underline{u}, E)).$$

On the other hand we have a Laplacian operator Δ_1 on the complex $\Omega^\cdot(X(1)/\Gamma \cap P_\infty, E)$ which is defined by means of the admissible metric on E.

LEMMA 2.7. *We put* $s = \dim(\underline{u})$. *Then we have*

$$i_q \circ \Delta_M = \Delta_1 \circ i_q.$$

PROOF. Let us denote by $M_\infty^0(1)$ (resp. K_M^0) the connected components of $M_\infty(1)$ (resp. K_M). Since X_M is connected we know that $X_M = K_M \backslash M_\infty(1) = K_M^0 \backslash M_\infty^0(1)$ and we may assume that $\Gamma_M \subset M_\infty^0(1)$, since the Laplacian operator does not depend on an orientation. If we put $s = \dim \underline{u}$, then K_M^0 acts trivially on $\Lambda^s \underline{u}$, we choose an isomorphism $\Lambda^s \underline{u} \xrightarrow{\sim} \mathbf{R}$. The nondegenerated pairing

$$\mathrm{Hom}(\Lambda^q \underline{u}, E) \times \mathrm{Hom}(\Lambda^{s-q} \underline{u}, E^*) \longrightarrow \mathbf{C} = \mathrm{Hom}(\Lambda^s \underline{u}, \mathbf{C})$$

induces a nondegenerated pairing

$$\mathbf{H}^q(\underline{u}, E) \times \mathbf{H}^{s-q}(\underline{u}, E^*) \longrightarrow \mathbf{C}$$

which is compatible with the action of K_M^0.

To prove the assertion in our lemma, we compare the operators δ_1 and δ_M. The operator δ_M is constructed by means of the mapping (comp. §1)

$$\#_M : \mathbf{H}^q(\underline{u}, E) \longrightarrow \mathbf{H}^q(\underline{u}, E)^* = \mathbf{H}^{s-q}(\underline{u}, E^*)$$

and δ_1 is constructed by means of the operator

$$\#_1 : \Omega^{p+q}(X(1)/\Gamma \cap P_\infty, E) \xrightarrow{\sim} \Omega^{p+q}(X(1)/\Gamma \cap P\ \ , E^*)$$

which is constructed by means of the admissible metric on E. The $*$-operator on the forms on $X(1)/\Gamma \cap P_\infty$ is defined by means of the restriction of B_θ to $\underline{r}_1 = \bar{\mathfrak{p}}_1 \oplus \underline{u}$ where this decomposition is orthogonal. If we restrict B_θ to $\underline{u}$ we get the admissible metric on $\underline{u}$ which induces the admissible metric on $\mathbf{H}^q(\underline{u},E)$. Therefore we get the following commutative diagram

$$\begin{array}{ccc} \Omega^p(X_M/\Gamma_M, \mathbf{H}^q(\underline{u},E)) & \xrightarrow{i_q} & \Omega^{p+q}(X(1)/\Gamma \cap P_\infty, E) \\ \downarrow {\scriptstyle (-1)^{(n-p)q} *_M \circ \#_M} & & \downarrow {\scriptstyle *_1 \circ \#_1} \\ \Omega^{n-p}(X_M/\Gamma_M, \mathbf{H}^{s-q}(\underline{u}, E^*)) & \xrightarrow{i_{s-q}} & \Omega^{s+n-p-q}(X(1)/\Gamma \cap P_\infty, E^*) \end{array}$$

where $n = \dim X_M$. A simple calculation yields

$$i_q \circ \delta_M = \delta_1 \circ i_q$$

and now the lemma is obvious.

THEOREM 2.8. *If X_M/Γ_M is compact then the mapping i induces an isomorphism for the cohomology*

$$i^*: \bigoplus_{p+q=m} H^p(X_M/\Gamma_M, \mathbf{H}^q(\underline{u},E)) \xrightarrow{\sim} H^m(X(1)/\Gamma \cap P_\infty, E).$$

PROOF. The Lemma 2.6 tells us that i^* induces a homomorphism on the cohomology. The Hodge theory tells us that the cohomology groups on both sides are isomorphic to the corresponding spaces of harmonic forms with respect to Δ_M and Δ_1. This implies that i^* is injective and the surjectivity follows then from the spectral sequence.

The group $\tilde{M}_\infty$ is the group of real points of an algebraic group over the real numbers. Let us denote by T the connected component of the identity of the centre of that group. This is a torus over $\mathbf{R}$ and let us denote its group of real points by T_∞. The adjoint action induces an operation of T_∞ on $\mathbf{H}^q(\underline{u}, E)$ which is semisimple, and the eigenvalues of this action are induced by algebraic characters of T. The metric B_θ induces a decompositian

$$\left.\begin{aligned} T_\infty &= T_\infty(1)\cdot A \\ \Phi: A &\xrightarrow{\sim} ((\mathbf{R}^+)^*)^r \\ \Phi: a &\xrightarrow{\sim} (|\chi_1|(a), \cdot, |\chi_r|(a)) \end{aligned}\right\} \qquad (2.9)$$

where we shall use Φ to identify $A = ((\mathbf{R}^+)^*)$ and we shall write $a = (t_1, t_2, \cdot, t_r)$. If $r = 1$ we write $a = t$. We now restrict our above action to A, Then A acts by algebraic characters

$$\lambda\colon a \longrightarrow a^\lambda = t_1^{\lambda_1} \qquad t_r^{\lambda_r} \qquad \lambda_i \in \mathbf{Z}$$

on $\mathbf{H}^q(\underline{u}, E)$ and we denote the corresponding decomposition by

$$\mathbf{H}^q(\underline{u}, E) = \bigoplus_\lambda \mathbf{H}^q_\lambda(\underline{u}, E) \tag{2.10}$$

and we shall call $\mathbf{H}^q_\lambda(\underline{u}, E)$ the space of classes of weight λ.

The group A acts on Hom $(\Lambda^s\underline{u}, \mathbf{C})$ by the character $a \to a^{-2\rho}$ where 2ρ is the sum of the positive roots. This implies

$$\#_M\colon \mathbf{H}^q_\lambda(\underline{u}, E) \longrightarrow \mathbf{H}^{s-q}_{-2\rho-\lambda}(\underline{u}, E^*).$$

Since the action of A on $\mathbf{H}^q(\underline{u}, E)$ commutes with the action of $\tilde{M}_\infty$ we get a decomposition

$$\begin{aligned} H^\cdot(X(1)/\Gamma \cap P_\infty, E) &= \bigoplus_\lambda H^\cdot_\lambda(X(1)/\Gamma \cap P_\infty, E) \\ &= \bigoplus_\lambda H^\cdot(X_M/\Gamma_M, \mathbf{H}^\cdot_\lambda(\underline{u}, E)) \end{aligned}$$

§3. Extension of differential forms and Eisenstein series. We assume that G/k is of k-rank one. If $P \subset G/k$ is a minimal parabolic subgroup then we have only one fundamental dominant weight $\chi\colon P \longrightarrow G_m$. We define the function (comp. 2.9)

$$h : X \longrightarrow (\mathbf{R}^+)^*$$
$$h : x = x_0 p \longrightarrow |\chi|(p) = t.$$

This function is equivariant with respect to the action of P_∞. Let us denote $h^*\left(\frac{dx}{x}\right) = \frac{dt}{t}$, this is a P_∞-invariant 1-form on the symmetric space X.

We define $X(t) = \{x \in X \mid h(x) = t\}$. Let us define T as the vector field on X which is orthogonal to the vector fields along the slices $X(t)$ and for which $\frac{dt}{t}(T) = 1$.

The metric B_θ induces a decomposition

$$P_\infty = A \cdot P_\infty(1)$$

where A centralizes $K \cap P_\infty$ and where $|\chi| : A \xrightarrow{\sim} (R^+)^*$ is the identification we introduced at the end of §2. Now the mapping

$$m_t : X(1) \xrightarrow{\sim} X(t)$$

$$m_t : x = x_0 p \overset{\sim}{\longmapsto} x_0 t p \qquad p \in P_\infty(1)$$

is a diffeomorphism which is compatible with the action of $P_\infty(1)$. Therefore it induces a diffeomorphism

$$\bar{m}_t : X(1)/\Gamma \cap P_\infty \longrightarrow X(t)/\Gamma \cap P_\infty$$

and we find a diffeomorphism $X/\Gamma \cap P_\infty \xrightarrow{\sim} (X(1)/\Gamma \cap P_\infty) \times \mathbf{R}$. The tangent bundle T^X of $X/\Gamma \cap P_\infty$ has the orthogonal decomposition

$$T^X = T_0^X \oplus \langle T \rangle$$

where T_0^X is the bundle of tangent vectors along the slices $X(t)$ and where $\langle T \rangle$ is the subbundle spanned by the vector field T. In the language of induced bundles this corresponds to the decomposition

$$\mathfrak{r} = \mathfrak{r}_1 \oplus \mathbf{C}\, T$$

where T now is considered as an element of Lie (A) and where $\mathfrak{r} = \mathfrak{p} \oplus \mathfrak{u}$ (comp. §2.). The group K_M acts trivially on T.

We now want to give some preparations which are necessary for the definition of Eisenstein series. Let us start with a cohomology class

$$[\varphi] \in H^m(X(1)/\Gamma \cap P_\infty, E).$$

This cohomology class has a unique harmonic representative $\Omega^m(X(1)/\Gamma \cap P_\infty, E)$ which may be viewed as a function

$$\varphi: P_\infty(1)/\Gamma \cap P_\infty(1) \longrightarrow \mathrm{Hom}\,(\Lambda^m \mathfrak{r}_1, E)$$

which satisfies for $k \in K_M$, $p \in P_\infty(1)$, $u \in U_\infty$

$$\varphi(kpu) = \Lambda^m \mathrm{ad}^*_{\mathfrak{r}_1}(k) \otimes \rho_\infty(k)(\varphi(p))$$

(comp. 1.8). The right invariance with respect to the action of U_∞ follows from Theorem 2.8. We have a K_M-invariant embedding $\mathrm{Hom}(\Lambda^m \mathfrak{r}_1, E) \hookrightarrow \mathrm{Hom}\,(\Lambda^m \mathfrak{r}, E)$. On the bigger space the group K

is acting, but the smaller space is not necessarily invariant under this action of K. We now extend our function φ to a function

$$\varphi_s : G_\infty/\Gamma \cap P_\infty \longrightarrow \mathrm{Hom}\,(\Lambda^m \underline{r}, E)$$

by

$$\varphi_s(ktp) = \Lambda^m \mathrm{ad}^*_{\underline{r}}(k) \otimes \rho_\infty(k)\,(\varphi(p)\,)\,t^{-\rho-s}$$

where $p \in P_\infty(1)$, $k \in K$, $t \in A$, and $s \in \mathbf{C}$. This function can be considered as a differential p-form on $X/\Gamma \cap P_\infty$. In this case ρ is of course an integer.

We now denote the exterior derivative (resp. exterior coderivation, resp. Laplacian operator) on the complex $\Omega^\cdot(X/\Gamma \cap P_\infty, E)$ by d (resp. δ, resp. Δ). Then

LEMMA 3.1. *If* $[\varphi]$ *is of weight* λ, *then*

$$d\varphi_s = (-\rho - s - \lambda)\,\frac{dt}{t} \wedge \varphi_s$$

$$\delta\varphi_s = 0$$

$$\Delta\varphi_s = (s^2 - (\rho+\lambda)^2)\varphi_s.$$

PROOF. We pull φ_s back to P_∞. We have to evaluate $d\varphi_s$ on $m+1$-tuples of right invariant vector fields. Since φ itself is closed the only nonzero terms are

$$d\varphi_s\,(T, M_1,\cdot, M_a, U_1,\cdot, U_b) \qquad \begin{array}{l} M_i \in \bar{\mathfrak{p}}_1,\ U_j \in \underline{u} \\ a+b = m \end{array}$$

Since $\varphi_s\,(T, \cdot\,, M, \cdot\,, U\cdot) = 0$ we get for this last expression

$$\nabla_T\varphi_s(M_1,\cdot,M_a,U_1,\cdot,U_b) + \sum_{1\leqslant j\leqslant b}(-1)^{1+p+1+j}\varphi_s([T,U_j],M_i\cdot,\cdot U_j)$$

$$= (-\rho - s)\varphi_s(M_1,\cdot,M_a,U_1,\cdot,U_b) - \rho_\infty(T)\varphi_s(M_1,\cdot,M_a,U_1,\cdot,U_b) - $$

$$- \sum_{1\leqslant j\leqslant b}\varphi_s(M_1,\cdot,M_a,U_1,\cdot,[T, U_j],\cdot, U_b).$$

Since $\mathrm{ad}(T)(U_j) = [U_j, T]$ (comp. remark following Theorem 2.2) we see that the last two terms add up to $-\lambda\varphi_s(M_1,\cdot, M_a,U_1,\cdot,U_b)$. To prove the second formula we compare the operators $*$ and $\#$ on $\Omega^\cdot(X/\Gamma \cap P_\infty, E)$ to the corresponding operators $*_1$ and $\#_1$ on

$\Omega^{\cdot}(X(1)/\Gamma \cap P_\infty, E)$. It is clear that $\#_1$ is the restriction of $\#$ to $\Omega^*(X(1)/\Gamma \cap P_\infty, E)$ and since $\frac{dt}{t}$ is orthogonal to the vector fields along the slices $X(t)$ we get

$$\left.\begin{aligned} *\varphi_s &= (-1)^p \frac{dt}{t} \wedge (*_1\varphi)_s \\ *\left(\frac{dt}{t} \wedge \varphi_s\right) &= ({*}_1\varphi)_s \end{aligned}\right\} \tag{3.2}$$

Since φ is harmonic we find

$$\delta\varphi_s = (-1)^{p+1} *^{-1} \circ \#^{-1} d \# \circ *\varphi_s = - *^{-1} \circ \#^{-1} d \left(\frac{dt}{t} \wedge (\#_1 \circ *_1\varphi)_s\right) = 0$$

because of our previous formula.

To prove the formula for the Laplacian operator, we have to observe that it follows from our considerations at the end of §2 that $*_1 \circ \#_1 \varphi = *_M \circ \#_M \varphi$ is of weight $-2\rho - \lambda$. Then the formula becomes obvious.

We want to define the Eisenstein series $E(\varphi, s) \in \Omega^p(X/\Gamma; E)$ which is associated to $[\varphi]$. To do this we have to recall the general context in which the Eisenstein series are defined.

We start with a representation

$$\sigma : K \longrightarrow GL(V)$$

where $\dim V < +\infty$. Let $\eta \in \mathrm{Hom}_{K_M}(V, V)$, we introduce the vector space

$$\mathscr{A}(M_\infty(1)/\Gamma_M, \sigma, \eta) = \left\{\psi : M_\infty(1)/\Gamma_M \to V \;\middle|\; \begin{aligned} &\psi(km) = \sigma(k)\,\psi(m) \text{ for } k \in K_M \\ &(C_M \psi)(m) = \eta(\psi(m)) \end{aligned}\right\}$$

where C_M is the Casimir operator. This vector space is of finite dimension since $M_\infty(1)/\Gamma_M$ is compact and since the Casimir operator induces an elliptic operator on the bundle $V_\sigma \to X_M/\Gamma_M$ which is induced by σ. If $\psi \in \mathscr{A}(M_\infty(1)/\Gamma_M, \sigma, \eta)$ we put, following Harish-Chandra, $\psi_s(g) = \psi_s(ktp) = \sigma(k)\,\psi(p) \cdot t^{-\rho-s}$ (comp. [7], Chap. II, §2) and we define for $s \in \mathbf{C}$, $\mathrm{Re}(s) > \rho$

$$E(g, \psi, s) = E(\psi, s) = \sum_{a \in \Gamma/\Gamma \cap P_\infty} \psi_s(ga).$$

It is known that this series converges for $\mathrm{Re}(s) > \rho$ and that it has meromorphic continuation into the entire s-plane (comp. [8], Chap. IV).

If P' is another minimal parabolic subgroup then the constant Fourier coefficient of $E(g, \psi, s)$ along P' is defined by

$$E^{P'}(g, \psi, s) = \int_{U_\infty / U'_\infty \cap \Gamma} E(gu', \psi, s)\, du'$$

where U' is the unipotent radical of P' and where

$$\mathrm{vol}_{du'}(U'_\infty / U'_\infty \cap \Gamma) = 1.$$

We shall state briefly some of the results concerning these constant Fourier coefficients which are proved in [8], Chap. II. We put

$$\epsilon = \begin{cases} 0 \text{ if } P \text{ and } P' \text{ are not } \Gamma\text{-conjugate} \\ 1 \text{ if } P \text{ and } P' \text{ are } \Gamma\text{-conjugate.} \end{cases}$$

If $\epsilon = 1$ we assume that $P = P'$. Then

$$E^{P'}(\psi, s) = \epsilon\, (\psi)_s + (c(s)\psi)_{-s}$$

where $c(s)$ is a linear mapping

$$c(s)\colon \mathscr{A}(M_\infty(1)/\Gamma_M, \sigma, \eta) \longrightarrow \mathscr{A}(M'_\infty(1)/\Gamma_M, \sigma, \eta')$$

which is meromorphic in the variable s. We claim that this follows from [8], Chap. II, Theorem 5 and the results in Chap. IV on analytic continuation. To see this we choose an element $y \in K$ which conjugates $\tilde{M}_\infty$ to $\tilde{M}'_\infty$ and which conjugates P into the opposite $\overline{P}'$ of P' with respect to $\tilde{M}'_\infty$, i.e., $\overline{P}'_\infty \cap P'_\infty = \tilde{M}'_\infty$. This map sends the Casimir operator on $\tilde{M}_\infty(1)$ to the Casimir operator on $M'_\infty(1)$ and it sends $\eta \in \mathrm{Hom}_{K_M}(V, V)$ into an element

$$\eta' = \mathrm{ad}(y)(\eta) \in \mathrm{Hom}_{K_{M'}}(V, V).$$

Then our statement above is a slight modification of Lemma 36 in [8], Chap. II.

Now we come back to the cohomology. We have seen that the cohomology group $H^m(X(1)/\Gamma \cap P_\infty, E)$ can be identified with the space of harmonic forms

$$\varphi : M_\infty(1)/\Gamma_M \longrightarrow \bigoplus_{p+q=m} \mathrm{Hom}(\Lambda^p \underline{P}_1, \mathbf{H}^q(\underline{u}, E)) \text{ for which } \Delta_M\varphi = 0$$

(comp. Theorem 2.8.). We have the inclusions

$$\bigoplus_{p+q=m} \mathrm{Hom}(\Lambda^p \bar{\mathfrak{p}}_1, \mathbf{H}^q(\underline{u}, E)) \subset \bigoplus_{p+q=m} \mathrm{Hom}(\Lambda^p \bar{\mathfrak{p}}_1, \mathrm{Hom}(\Lambda^q \underline{u}, E))$$
$$= \mathrm{Hom}(\Lambda^m \underline{r}_1, E) \subset \mathrm{Hom}(\Lambda^m \underline{r}, E).$$

The biggest space is invariant under K; the other spaces are in general only K_M-invariant. The Laplacian operator Δ_M acts on the space of functions

$$\varphi : M_\infty(1)/\Gamma_M \longrightarrow \bigoplus_{p+q=m} \mathrm{Hom}(\Lambda^p \bar{\mathfrak{p}}_1, \mathrm{Hom}(\Lambda^q \underline{u}, E))$$

which satisfy (1.8), and it follows from the lemma of Kuga that (comp. (1.9))

$$\Delta_M\varphi = -C_M\varphi + \lambda(C_M)\varphi$$

where $\lambda(C_M) = \bigoplus_{p+q=m} \lambda_{p,q}(C_M)$ and $\lambda_{p,q}(C_M)$ is the linear transformation induced by the Casimir on

$$\mathrm{Hom}(\Lambda^p \mathfrak{p}_1, \mathrm{Hom}(\Lambda^q \underline{u}, E)).$$

Because of our identifications

$$\lambda(C_M) \in \mathrm{End}(\mathrm{Hom}(\Lambda^m \underline{r}_1, E))$$

and since

$$\Lambda^m \underline{r} = \Lambda^m \underline{r}_1 \oplus \Lambda^{m-1} \underline{r}_1 \otimes \mathbf{C}T$$

we may extend $\lambda(C_M)$ trivially to $\Lambda^{m-1}\underline{r}_1 \otimes \mathbf{C}T$ and therefore we may consider $\lambda(C_M)$ also as an element of

$$\mathrm{End}_{K_M}(\mathrm{Hom}(\Lambda^m \underline{r}, E)).$$

Our considerations show that

$$H^m(X(1)/\Gamma \cap P_\infty; E) \hookrightarrow \mathscr{A}(M_\infty(1)\Gamma_M, \Lambda^m \mathrm{ad}^* \otimes \rho_\infty, \lambda(C_M))$$

and therefore we can associate to any cohomology class an Eisenstein series. Let us denote

$$\mathscr{A}(M_\infty(1)/\Gamma_M, \Lambda^m \mathrm{ad}^*_{\underline{r}} \otimes \rho_\infty, \lambda(C_M)) = \mathscr{H}_M^{(m)}.$$

It is clear that in this case we have

$$c(s) : \mathscr{H}_M^{(m)} \longrightarrow \mathscr{H}_{M'}^{(m)}$$

since the representation $\mathrm{ad}_{\underline{r}_1}|K_M$ extends to the normaliser of K_M in K.

Remark. The space $\mathscr{H}_M$ plays only an auxiliary role; we need it, because we want to have a home for our $c(s)\varphi$ and we want to have some space of functions which is invariant under the transformations $c(s)$. It is not clear that the $c(s)\varphi$ is again a cohomology class if φ was one.

The Eisenstein series will now be used in the following context: We know that X/Γ is up to homotopy a compact manifold V with boundary ∂V. The theory of Eisenstein series will help us to associate to any cohomology class $[\varphi] \in H^m(\partial V; E)$ an Eisenstein series $E(\varphi, s) \in \Omega^m(X/\Gamma: E)$.

Before we can do this we must recall some facts from reduction theory. Let $P_1, P_2, \dots, P_d$ be a set of representatives for the Γ-conjugacy classes of minimal parabolic subgroups. We consider the projection

$$Y = \coprod_{i=1}^{i=d} X/\Gamma \cap P_{i,\infty} \longrightarrow X/\Gamma.$$

For each $X/\Gamma \cap P_{i,\infty}$ we have the functions $h_i : X/\Gamma \cap P_{i,\infty} \to (\mathbf{R}^+)^*$ which have been introduced above. We collect these functions to a function $h: Y \longrightarrow (R^+)^*$ If $t_0 > 0$ we define

$$Y(t_0) = \{y \in Y \mid h(y) < t_0\}.$$

It is well known that we can choose $t_0 > 0$, such that the mapping $Y(t_0) \longrightarrow X/\Gamma$ is injective and locally diffeomorphic (comp. [2], Theorem 17.10). Let us identify $Y(t_0)$ with its image. The complement of $Y(t_0)$ is a compact manifold V with boundary ∂Y (comp. [2], loc. cit.,). The boundary components of V are the manifolds $X_i(t_0)/\Gamma \cap P_{i,\infty}$ where $X_i(t_0)$ is of course $\{x \in X | h_i(x) = t_0\}$.

Now we come back to the Eisenstein series. Let us consider a cohomology class $[\varphi] \in H^m(\partial V, E)$, then this class may be considered as a vector $[\varphi] = ([\varphi_1], \dots, [\varphi_d])$ where

$$[\varphi_i] \in H^m(X_i(t_0)/\Gamma \cap P_{i,\infty}, E).$$

Since we have the identification

$$\bar{m}_{t_0} : X_i(1)/\Gamma \cap P_{i,\infty} \longrightarrow X_i(t_0)/\Gamma \cap P_{i,\infty}$$

we can associate an Eisenstein series to each of the classes $[\varphi_i]$ and we put

$$E(\varphi,s) = \sum_{i=1}^{i=d} E(\varphi_i,s).$$

This is a differential form on X/Γ.

We call $[\varphi] \in H^m(\partial V, E)$ a cohomology class of weight λ if its components $[\varphi_1], \ldots, [\varphi_d]$ are of weight λ. In this case we have (Lemma 3.1)

$$\left.\begin{aligned} \Delta E(\varphi, s) &= (s^2-(\lambda+\rho)^2)E(\varphi,s) \\ \delta E(\varphi, s) &= 0. \end{aligned}\right\} \qquad (3.3)$$

For any cohomology class $[\varphi]$ on the boundary we consider the constant Fourier coefficients of $E(\varphi,s)$ along the P_i's

$$E^{P_i}(\varphi,s)=(\varphi_i)_s + \left(\sum_{j=1}^{d} c_{ij}(s)\varphi_j\right)_{-s}$$

and we collect them to a vector

$$E^P(\varphi, s) = (E^{P_1}(\varphi, s),\ldots, E^{P_d}(\varphi, s)).$$

We shall write symbolically

$$E^P(\varphi,s) = (\varphi)_s + (c(s)\varphi)_{-s}.$$

For the rest of the paper we keep this notation for the constant Fourier coefficient. This is slightly sloppy because it means that we do so as if there is only one cusp. But otherwise the notation becomes clumsy and the arguments are essentially the same.

4. Cohomology classes represented by singular values of the Eisenstein series. If φ is of weight λ, then we know that $\Delta E(\varphi,s) = (s^2 - (\rho + \lambda)^2)E(\varphi, s)$. Therefore we get a harmonic form if we evaluate $E(\varphi,s)$ at the special values $s = \pm (\rho+\lambda)$. Of course we have to be careful, since $E(\varphi,s)$ might have a pole at such a special value. Our main concern will be the case where this pole is of order one. Then we put

$$\operatorname*{Res}_{s=\rho+\lambda} E(\varphi,s) = \lim_{s\to\rho+\lambda} (s-\rho-\lambda)\, E(\varphi,s) = E'(\varphi, \rho+\lambda).$$

We call these differential forms $E(\varphi, \pm(\rho+\lambda))$ or $E'(\varphi, \pm(\rho+\lambda))$ the *singular values* of the Eisenstein series. We shall discuss the following two problems:

(A) When does a singular value represent a cohomology class, i.e., when is it closed?

(B) In case it represents a cohomology class, what is the restriction of this class to the boundary?

To attack these questions we have to consider the constant Fourier coefficient. First of all we claim that $dE(\varphi, \rho+\lambda) = 0$ if and only if $d(E^P(\varphi, \rho+\lambda)) = 0$. It is clear that the first statement implies the second one. To see the other direction we have to recall the notion of the space of cusp forms. A differential form $\omega \in \Omega^m(X/\Gamma, E)$ is called a cusp form if it is square integrable and if (comp. [8] Chap. I, § 2)

$$\omega^P(g) = \int_{U_\infty/U'_\infty \cap \Gamma} \omega(gu)\,du \equiv 0.$$

Here we interpret ω as a function on G_∞/Γ which satisfies (1.8). We know that for any cusp form $\omega \in \Omega^{m+1}(X/\Gamma, E)$, which is an eigenfunction with respect to the centre of the universal enveloping algebra, the integral

$$\langle dE(\varphi_1\rho+\lambda), \omega)\rangle = \int_{X/\Gamma} (dE(\varphi_1\rho+\lambda), *\overline{\circ \# \omega})$$

exists ([8], Chap. I, Lemma 15). The results in [8], Chap. I also imply that the value of this integral is equal to $\langle E(\varphi, \lambda+\varphi), \delta\omega\rangle$ (apply cor. to Lemma 10 and Lemma 14). But the latter integral is zero since $\delta\omega$ is a cusp form. Now our assertion follows from Theorem 4 in [8], Chap. I.

If $dE(\varphi, \rho+\lambda) = 0$ then we see from our Theorem 2.8 and the description of the boundary that $[E(\varphi, \rho+\lambda])$ and $[E^P(\varphi, \rho+\lambda)]$ represent the same cohomology class on the boundary ∂V of V.

We decompose the space $\mathrm{Hom}(\Lambda^m\mathfrak{r}_1, E)$ with respect to the action of A and write

$$\mathrm{Hom}(\Lambda^m \underline{r}_1, E) = \bigoplus_\mu \mathrm{Hom}_\mu(\Lambda^m \underline{r}_1, E)$$

where the μ's are algebraic characters on A (comp. §2). Moreover we write

$$\mathrm{Hom}(\Lambda^m \underline{r}, E) = \bigoplus_\mu \mathrm{Hom}_\mu(\Lambda^m \underline{r}_1, E) \oplus \bigoplus_\mu \frac{dt}{t} \mathbf{C} \wedge \mathrm{Hom}_\mu(\Lambda^{m-1} \underline{r}_1, E).$$

This yields a decomposition of $c(s)\,\varphi$:

$$c(s)\,\varphi = \sum_\mu c_\mu(s)\,\varphi + \sum_\mu \frac{dt}{t} \wedge n_\mu(s)\,\varphi.$$

From this we get the following rather messy formula for the exterior derivative of $E(\varphi, s)$ where φ is supposed to be of weight λ (comp. Lemma 3.1.)

$$dE^P(\varphi, s) = (-s - \rho - \lambda)\frac{dt}{t} \wedge (\varphi)_s + \sum_\mu (s - \rho - \mu)\,\frac{dt}{t} \wedge (c_\mu(s)\,\varphi)_{-s}$$

$$- \sum_\mu \left(\frac{dt}{t} \wedge dn_\mu(s)\,\varphi\right)_{-s} + \sum_\mu (dc_\mu(s)\,\varphi))_{-s}. \qquad (4.1)$$

(Actually we have to apply a slight generalization of Lemma 3.1). On the other hand we may also start from the function

$$\frac{dt}{t} \wedge \varphi : P_\infty(1)/\Gamma \cap P_\infty \longrightarrow \frac{dt}{t} \wedge \mathrm{Hom}(\Lambda^m \underline{r}_1, E) \subset \mathrm{Hom}(\Lambda^{m+1} \underline{r}, E)$$

and we may consider its associated Eisenstein series $E\left(\frac{dt}{t} \wedge \varphi, s\right)$. Then it is clear that

$$E^P\left(\frac{dt}{t} \wedge \varphi, s\right) = \frac{1}{-s - \rho - \lambda}\, dE^P(\varphi, s). \qquad (4.2)$$

The operator $* \circ \#$ induces an isomorphism

$$* \circ \# : \mathrm{Hom}(\Lambda^m \underline{r}, E) \longrightarrow \mathrm{Hom}(\Lambda^{N-m}\, \underline{r}, E^*), \quad N = \dim(X)$$

which commutes with the action of K. We apply this operator on both sides of (4.2) and since the construction of Eisenstein series commutes with the operator $* \circ \#$, we obtain

$$E^P(*_1 \circ \#_1\, \varphi, s) = \frac{1}{-s - \rho - \lambda} * \circ \#\, dE^P(\varphi, s) \qquad (4.3)$$

(comp. (3.2)).

We now assume that $\lambda > -\rho$ and that φ runs over the classes of weight λ. We shall investigate the influence of the behavior of the Eisenstein series at $\rho + \lambda > 0$ on the solution of our problems (A) and (B). We know that the Eisenstein series has at most a pole of order 1 at this point. (comp. [8], Chap. IV, §6). We put

$$C_\mu = \operatorname*{res}_{s=\rho+\lambda} c_\mu(s) \text{ and } N_\mu = \operatorname*{res}_{s=\rho+\lambda} n_\mu(s).$$

Then we get for the residue of the Eisenstein series

$$E'^{P}(\varphi, \rho+\lambda) = \sum_\mu (C_\mu \varphi)_{-\rho-\lambda} + \sum_\mu \left(\frac{dt}{t} \wedge N_\mu(\varphi)\right)_{-\rho-\lambda}$$

Moreover it follows from the scalar product formula (comp. [8], Chap. IV, §8) that $E'(\varphi, \rho+\lambda)$ is square integrable and that

$$\|E'(\varphi, \lambda+\rho)\|_2^2 = c \langle \varphi, C_\lambda \varphi \rangle = c \int_{X(1)/\Gamma \cap P_\infty} (\varphi, \overline{*_1 \circ \#_1 C_\lambda \varphi}) \qquad (4.4)$$

(for notation comp. (1. 7)) where c is a positive constant. We have to observe that $C_\mu \varphi$ for $\mu \neq \lambda$ and $\frac{dt}{t} \wedge N_\mu \varphi$ are orthogonal to φ.

A theorem of Andreotti and Vesentini (comp. [1], Prop. 7 and [5], Prop. 3. 20) tells us that $E'(\varphi, \lambda+\rho)$ is a closed form. Then it follows from formula (4.1) that the forms $C_\mu \varphi$ are closed. We write $C_\lambda \varphi = \widetilde{C}_\lambda \varphi + H$ where $\Delta_M \widetilde{C}_\lambda \varphi = 0$ and where H is a sum of eigenvectors to nonzero eigenvalues of Δ_M. Then we have

$$[C_\lambda \varphi] = [\widetilde{C}_\lambda \varphi] \text{ and } \langle \varphi, \widetilde{C}_\lambda \varphi \rangle = \langle \varphi, C_\lambda \varphi \rangle. \qquad (4.4')$$

We define $\widetilde{C}_\lambda[\varphi] = [\widetilde{C}_\lambda \varphi]$.

It follows from (4.4) and (4.4') that $\widetilde{C}_\lambda$ is a positive selfadjoint positive operator and therefore we have an orthogonal decomposition

$$H^m(\partial V, E) = \ker(\widetilde{C}_\lambda) \oplus \operatorname{im}(\widetilde{C}_\lambda).$$

Another consequence of (4.4) and (4.4') is

LEMMA 4.5. *If $\varphi \in \ker(\widetilde{C}_\lambda)$ then $E(\varphi, s)$ is holomorphic at the point $\rho + \lambda$.*

We are now ready to state and to prove the first part of our main result.

Theorem 4.6.1. *Let us assume that* $\lambda > -\rho$. *Then the form* $E'(\varphi, \rho + \lambda)$ *is closed for all* $[\varphi] \in H^m_\lambda(\partial V, E)$ *and*

$$[E'(\varphi, \rho + \lambda)]|_{\partial V} = \tilde{C}_\lambda[\varphi].$$

Proof. We have already seen that $E'(\varphi, \rho + \lambda)$ is closed. We get from (4.1)

$$0 = dE'^P(\varphi, \rho + \lambda) = \sum_\mu (\mu - \lambda)\frac{dt}{t} \wedge (C_\mu \varphi)_{-\rho-\lambda} -$$

$$- \sum_\mu \left(\frac{dt}{t} \wedge dN_\mu \varphi\right)_{-\rho-\lambda} + \sum_\mu d(C_\mu \varphi)_{-\rho-\lambda}.$$

This yields that $C_\mu \varphi$ is a boundary for $\lambda \neq \mu$, and that implies the last assertion of the theorem.

The next case which we shall consider is $\lambda < -\rho$. We have the isomorphism

$$*_1 \circ \#_1 : H^N_\lambda(\partial V, E) \longrightarrow H^{N-m-1}_{-2\rho-\lambda}(\partial V, E^*).$$

Since $-2\rho - \lambda > -\rho$ we can decompose the right hand side

$$H^{N-m-1}_{-2\rho-\lambda}(\partial V, E) = \ker(\tilde{C}_{-2\rho-\lambda}) + \operatorname{im}(\tilde{C}_{-2\rho-\lambda}).$$

We put

$$F_\lambda = *_1 \circ \#_1 \ker(\tilde{C}_{-2\rho-\lambda}).$$

Theorem 4.6.2. *If* $\lambda < -\rho$, *and if* $[\varphi] \in F_\lambda$ *then* $E(\varphi, s)$ *is holomorphic at* $-\rho - \lambda$, *Moreover the form* $E(\varphi, -\rho - \lambda)$ *is closed. The restriction of* $[E(\varphi, -\rho - \lambda)]$ *to the boundary is given by*

$$[E(\varphi, -p - \lambda)]|_{\partial V} = [\varphi] + [c_{-2\rho-\lambda}(-\rho - \lambda)\varphi].$$

Proof. If $[\varphi] \in F_\lambda$, then we have by definition $\#_1 \circ *_1 \varphi \in \ker(\tilde{C}_{-2\rho-\lambda})$. We have to apply the formula (4.3) twice. The first time we substitute $\#_1 \circ *_1\varphi$ for φ and get

$$E^P(\varphi, s) = \frac{1}{-s+p+\lambda}\cdot dE^P(\#_1 \circ *_1 \varphi, s).$$

Now we obtain from Lemma 4.5 that the right-hand side is holomorphic at $-\rho-\lambda$, and this proves that $E(\varphi, s)$ is holomorphic at $-\rho-\lambda$. The second time we apply (4.3) to φ itself; again we get from Lemma 4.5 that $E(\varphi, -\rho-\lambda)$ is closed. The rest is clear, the arguments are exactly the same as in the proof of Theorem 4.6.1.

The last case is $\lambda = -\rho$. We know from [8], Chap. IV, §7 that $E(\varphi, s)$ is holomorphic at $s = 0$. Moreover we have

$$\#_1 \circ *_1 : H^m_{-\rho}(\partial V, E) \longrightarrow H^{N-m-1}_{-\rho}(\partial V, E^*)$$

and it follows from (4.3) that $E(\varphi, 0)$ is always closed. Therefore we have to check the restriction of $[E(\varphi, 0)]$ to the boundary ∂V.

When we introduced the Eisenstein series we embedded the cohomology of the boundary into a bigger space $\mathscr{H}^{(m)}$ and we have the diagram

$$\begin{array}{ccc} c(0): H^m(\partial V, E) & \longrightarrow & \mathscr{H}^{(m)} \\ \downarrow & & \| \\ c(0): \mathscr{H}^{(m)} & \longrightarrow & \mathscr{H}^{(m)}. \end{array}$$

The functional equation for the Eisenstein series tells us that $c(0)^2 = id$ (comp. [8], Chap. IV, §6). Therefore we can decompose with respect to the eigenvalues ± 1 and get

$$\mathscr{H}^{(m)} = \mathscr{H}^{(m)}_+ \oplus \mathscr{H}^{(m)}_-.$$

It is clear that for $\psi \in \mathscr{H}^{(m)}_+$

$$E^P(\psi, 0) = 2(\psi)_0$$

and for $\psi \in \mathscr{H}^{(m)}_-$ we have $E(\psi, 0) = 0$. We claim

THEOREM 4.6.3. *We have*

$$H^m_{-\rho}(\partial V, E)'_- = H^m_{-\rho}(\partial V, E)_+ \oplus H^m_{-\rho}(\partial V, E)_-$$

where

$$H^m_{-\rho}(\partial V, E)_\pm = \mathscr{H}^{(m)}_\pm \cap H^m_{-\rho}(\partial V, E).$$

The image of the map

$$\text{proj} \circ r : H^m(X/\Gamma, E) \longrightarrow H^m_{-\rho}(\partial V, E)$$

is $H^m_{-\rho}(\partial V, E)_+$ *and for* $[\varphi] \in H^m_{-\rho}(\partial V, E)_+$ *we have*

$$[E(\varphi, 0)]|_{\partial V} = 2[\varphi].$$

PROOF. Only the first statement has to be proved. We consider the pairing

$$H^m_{-\rho}(\partial V, E) \times H^{N-m-1}_{-\rho}(\partial V, E^*) \longrightarrow \mathbf{C}$$

which is given by

$$\langle [\varphi], [\psi] \rangle \int_{\partial V} (\varphi, \psi)$$

(comp. §1). This mapping is nondegenerate and is easy to see that for classes $[\varphi]$ and $[\psi]$ which are restrictions of classes on X/Γ we have

$$\langle [\varphi], [\psi] \rangle = 0.$$

If R (resp. S) is the space of classes in $H^m_{-\rho}(\partial V, E)$ (resp. $H^{N-m-1}_{-\rho}(\partial V, E^*)$) which are restriction of classes on X/Γ then we see that R and S are orthogonal with respect to $\langle\ \rangle$ and therefore

$$\dim R + \dim S \leqslant \dim H^m_{-\rho}(\partial V, E) = \dim H^{N-m-1}_{-\rho}(\partial V, E^*).$$

On the other hand we know that for any $\varphi \in H^m_{-\rho}(\partial V, E)$ (resp. $\psi \in H^{N-m-1}_{-\rho}(\partial V, E^*)$) we have

$$[E(\varphi, 0)]|_{\partial V} \in R, \quad (\text{resp. } [E(\psi, 0)]|_{\partial V} \in S).$$

The kernel of

$$[\varphi] \longrightarrow [E(\varphi, 0)]|_{\partial V}, \quad (\text{resp. } [\psi] \longrightarrow [E(\psi, 0)]|_{\partial V})$$

is

$$\mathscr{H}^{(m)}_- \cap H^m_{-\rho}(\partial V, E) = F, \ (\text{resp. } \mathscr{H}^{(N-m-1)}_- \cap H^{N-m-1}_{-\rho}(\partial V, E^*) = H).$$

It follows from our first inequality that

$$\dim F + \dim H \geqslant \dim H^m(\partial V, E).$$

The formula (4.3) tells us that

$$\#_1 \circ *_1(F) \subset \mathscr{H}^{(N-m-1)}_+ \cap H^{(N-m-1)}_{-\rho}(\partial V, E^*)$$

$$\#_1 \circ *_1(H) \subset \mathscr{H}^{(m)}_+ \cap H^m_{-\rho}(\partial V, E).$$

Since

$$S \supset \#_1 \circ *_1(F),$$
$$R \supset \#_1 \circ *_1(H),$$

it follows that

$$R = H^m_{-\rho}(\partial V, E)_+,$$
$$S = H^m_{-\rho}(\partial V, E^*)_-,$$

and that $\dim R + \dim S = \dim H^m(\partial V, E)$. This proves the theorem.

COROLLARY 4.7. *The image of the restriction map*

$$r : H^{\cdot}_{!}(X/\Gamma, E) \longrightarrow H^{\cdot}(\partial V, E).$$

is compatible with the decomposition

$$\bigoplus_{\lambda > -\rho} (H^{\cdot}_{\lambda}(\partial V, E) \oplus H^{\cdot}_{-2\rho-\lambda}(\partial V, E)) \oplus H^{\cdot}_{-\rho}(\partial V, E).$$

If a cohomology class on the boundary is in the image of r then we can find a representative in $H^{\cdot}(X/\Gamma, E)$ which is represented by a singular value of an Eisenstein series.

PROOF. This is clear from our previous considerations.

CONCLUDING REMARKS. (1) A consequence of our results is that we have a decomposition

$$H^{\cdot}(X/\Gamma, E) = H^{\cdot}_{!}(X/\Gamma, E) \oplus H^{\cdot}_{\mathrm{inf}}(X/\Gamma, E)$$

where $H^{\cdot}_{!}$ is the image of the cohomology with compact support in the usual cohomology and where $H^{\cdot}_{\mathrm{inf}}(X/\Gamma, E)$ maps isomorphically to the image of r. We have some sort of control of this complementary space $H^{\cdot}_{\mathrm{inf}}(X/\Gamma, E)$ in terms of the Eisenstein series and we know how to compute this space if we have enough information concerning the behavior of the Eisenstein series at certain critical values of s and to get this information we have to understand the "intertwining operators" $c(s)$.

(2) We think that Theorem 4.6.2 can be sharpened since it seems to be plausible that $c_{-2\rho-\lambda}(-\rho-\lambda)=0$ or even $c_{-2\rho-\lambda}(s)=0$. If this would not be the case then we could get some rationality result. Let us assume our group G is defined over $\mathbf{Q}$ and that ρ is a representation defined over $\mathbf{Q}$. In this case the vector spaces $H^{\cdot}(X/\Gamma, E)$

and $H^{\bullet}(\partial V, E)$ have a natural **Q**-structure and the map r is also defined over **Q**. Therefore we know that the image of the restriction map is also defined over **Q** and hence it follows from Theorem 4.6.2 that $c_{-2\rho-\lambda}(-\rho-\lambda)$ is defined over **Q**. This seems to us would be a too strong consequence and therefore we tend to believe that $c_{-2\rho-\lambda}$ vanishes at $-\rho-\lambda$. This vanishing should follow from computations at infinity which unfortunately seem to be rather messy.

(3) There are of course some cases where our information is slightly better, if we are dealing with cohomology classes of weight $\lambda > 0$. In this case $\rho + \lambda > \rho$ and $E(\varphi, s)$ is certainly holomorphic as $\rho + \lambda$. This means that the map (Theorem 4.6.2)

$$H^{\bullet}(X/\Gamma, E) \longrightarrow H^{\bullet}_{\lambda}(\partial V, E) \oplus H^{\bullet}_{-2\rho-\lambda}(\partial V, E)$$

has as its image a subspace which projects isomorphically to the second summand. If our conjecture made in Remark (2) is true then the image would be exactly the second summand.

(4) The classes in $H^{\bullet}_{\mathrm{inf}}(X/\Gamma, E)$ are in general not square integrable unless they come from a pole, i.e., they are represented by a residue of an Eisenstein series (comp. [11], §5). The classes in $H^{\bullet}_{!}(X/\Gamma, E)$ are of course all square integrable and therefore we can apply Hodge theory to investigate that part of the cohomology. We have a splitting

$$H^{\bullet}_{!}(X/\Gamma, E) = H^{\bullet}_{p}(X/\Gamma, E) \oplus H^{\bullet}_{\mathrm{Eis}}(X/\Gamma, E)$$

where the first summand denotes the space which is spanned by the harmonic cusps forms and where the second space is spanned by classes which are represented by certain residues of Eisenstein series which are harmonic. Hopefully we can apply the Hodge theory verbatim to the first part; the nature of the second part seems to be unclear. We have very explicit informations if $G = SL_2/k_2$ (comp. [7], Prop. 2.3).

REFERENCES

1. A. Andreotti and E. Vesentini: Carleman estimates for the Laplace-Beltrami equation on complex manifolds, *Publ. Math., I.H.E.S., No.* 25, (1965), 313-362.

2. A. Borel: *Introduction aux groupes arithmétiques*, Hermann, Paris, (1969).

3. H. Cartan and S. Eilenberg: *Homological Algebra*, Princeton University Press, Princeton, (1956).

4. W. T. van Est: A generalization of the Cartan-Leray spectral sequence, *Nederlands Akademic van Wetenschappen, Proceedings, Series A*, vol. 61, (1958), I, 399-405, II, 406-413.

5. H. Garland: A rigidity theorem for discrete subgroups, *Trans. Amer. Math. Soc.*, 129, (1967), 1-25.

6. G. Harder: A Gauss-Bonnet formula for discrete arithmetically defined groups, *Ann. Scient. Éc. Norm. Sup.*, t. 4, (1971), 409-455.

7. G. Harder: On the cohomology of $SL_2(\mathcal{O})$ to appear in the Proceedings of the Summer School on "Representation Theory", Budapest, (1971).

8. Harish-Chandra: Automorphic forms on semisimple Lie groups, *Springer Lecture Notes*, vol. 62, (1968).

9. B. Kostant: Lie algebra cohomology and the generalized Borel-Weil theorem, *Ann. of Math.*, vol. 74, (1961), 329-387.

10. Y. Matsushima and S. Murakami: On vector bundle valued harmonic forms and automorphic forms on symmetric Riemannian manifolds, *Ann. of Math.*, vol. 78, (1963), 365-416.

11. M. S. Raghunathan: Cohomology of arithmetic subgroups of algebraic groups, *Ann. of Math.*, vol. 78, (1968), 279-304.

12. G. de Rham: *Variétés Différentiables*, Paris, Hermann, (1955).

The Institute for Advanced Study
Princeton, New Jersey

and

53 Bonn
Mathematisches Institut
Wegelerstr. 10

ON MODULAR CURVES OVER FINITE FIELDS

By YASUTAKA IHARA

INTRODUCTION. We shall prove a certain basic theorem on modular curves over finite fields. This gives a solution of a conjecture raised at the Boulder summer school in 1965 [7]. The congruence subgroup property of the modular group of degree two over $\mathbf{Z}[1/p]$ is faithfully reflected in this theorem.

Let p be a fixed prime number, F be the algebraic closure of the prime field of characteristic p, and F_{p^m} $(m \geqslant 1)$ be the finite subfield of F with p^m elements. Let j be a variable over F. For each positive integer n with $p \nmid n$, let L_n be the field of modular functions of level n over $F(j)$ (with respect to j) in the sense of Igusa [5]. Then $L_n/F(j)$ is a Galois extension with Galois group isomorphic to $SL_2(\mathbf{Z}/n)/\pm I$. We know that there is a *canonical* choice of a Galois extension $K_n/F_{p^2}(j)$ with the same Galois group as $L_n/F(j)$, satisfying $K_n \cdot F = L_n$ and $K_n \cap F = F_{p^2}$ ([7], [8]; see §1). The field K_n is obtained as the fixed field of the action of $p^{\mathbf{Z}}$ on L_n. Call a prime divisor P of K_n *supersingular* if the residue class of j at P is a supersingular "j-invariant". An important fact, which becomes apparent by lowering the field of modular functions from L_n to K_n, is that *all supersingular prime divisors of K_n are of degree one over F_{p^2}*. This proof is in [8]. For $n > 1$, the number of supersingular prime divisors of K_n is equal to $(1/12)(p-1)[K_n : K_1]$. We shall prove the following

THEOREM. *There is no non-trivial unramified extension of K_n in which all supersingular prime divisors are decomposed completely.*

This, combined with Igusa's result [5] on the ramifications of L_n/L_1 (which says that the ramifications of L_n/L_1 are almost analogous to those of the corresponding situation in characteristic 0), gives the following complete characterization of the Galois extension K_∞/K_r, where $r > 1$ and K_∞ is the composite of K_n for all n:

COROLLARY. *K_∞/K_r is the maximum Galois extension of K_r satisfying the following two properties*; (i) *it is tamely ramified, and unramified* *outside the cusps of* K_r; (ii) *the supersingular prime divisors af* K_r *are decomposed completely.*

If one wants to emphasize the arithmetic nature of the theorem, one may restate the theorem in the following way: "*the Galois group of the maximum unramified Galois extension of* K_n *is topologically generated by the* p^2*-th power Frobenius automorphisms of* (*all extensions of*) *the supersingular prime divisors of* K_n." Some direct applications of this to the distribution of supersingular prime divisors are given in §2. If one wants to look at the theorem as a theorem on the fundamental group, one may formulate it as follows. Let $\Gamma_1 = PSL_2(\mathbf{Z}^{(p)})$ be the modular group over $\mathbf{Z}^{(p)} = \mathbf{Z}[1/p]$ and Γ_n the principal congruence subgroup of Γ_1 with level $n\,(p \nmid n)$. Let X_n be a complete non-singular model of K_n. Then, for $r > 1$, "*there is a categorical equivalence between subgroups of* Γ_r *with finite indices and those finite separable irreducible coverings of* X_r *defined over* F_{p^2} *and satisfying the above two conditions* (i), (ii)." (The second condition (ii) is geometrically stated as "all points lying on the supersingular points of X_r are F_{p^2}-rational".) In this sense, the modular group of degree two over $\mathbf{Z}^{(p)}$ is *the fundamental group* defined by (i), (ii).

We shall mention here that it is essential to consider the groups over $\mathbf{Z}^{(p)}$ and the curves over F_{p^2}. It cannot be replaced by the groups over $\mathbf{Z}$ and the curves over F. Roughly speaking, considering the group over $\mathbf{Z}$ corresponds to considering the condition (i) alone. The condition (i) alone cannot characterize the system of coverings defined by the modular curves X_n, as each X_n for $n \geqslant 6$ has so many non-trivial unramified coverings and they are all from *outside* the system (see § 1.2, § 2). Such unramified coverings of X_n correspond to *non-congruence* subgroups of $PSL_2(\mathbf{Z})$. The passage from $PSL_2(\mathbf{Z})$ to $PSL_2(\mathbf{Z}^{(p)})$ kills all non-congruence subgroups. Our theorem says, as a corresponding geometric fact, that all non-trivial unramified coverings of X_n are killed by the condition (ii), i.e., by *the supersingular Frobeniuses*. The connection between the function field K_n (or the curve X_n) and Γ_n is essential, as our previous studies [7], [8], and also the proof of the present theorem show.

In §1, we shall give some preliminaries, and in §2, we shall state our theorem in several different forms, with some corollaries. The proof will be given in the rest of the paper, in §§3-6.

METHOD FOR PROOF. The modular group over $\mathbf{Z}^{(p)}$ has the congruence subgroup property (Mennicke, Serre), and our theorem is a faithful reflection of this property. The proof goes through directly by chasing away the obstacles (clouds!). It consists of the following steps (a)-(d).

(a) A geometric interpretation of the congruence subgroup property as the "simply-connectedness" of the system $\{R_n \overset{\varpi}{\leftarrow} R_n^0 \overset{\varpi'}{\rightarrow} R_n'\}$ of three compact Riemann surfaces. The Riemann surfaces R_n, R_n' and R_n^0 are defined by the fuchsian groups Δ_n, Δ_n' and Δ_n^0, where Δ_n is the principal congruence subgroup of $PSL_2(\mathbf{Z})$ of level n,

$$\Delta_n' = \begin{pmatrix} p & 0 \\ 0 & 1 \end{pmatrix}^{-1} \Delta_n \begin{pmatrix} p & 0 \\ 0 & 1 \end{pmatrix}, \text{ and } \Delta_n^0 \doteq \Delta_n \cap \Delta_n'.$$

The system is essentially the graph of the modular correspondence "$T(p)$ for level n". (However, R_n^0 is not immersed in the product $R_n \times R_n'$.) The interpretation is made possible due to the fact that Γ_n is the *free* product of Δ_n and Δ_n' with amalgamated subgroup Δ_n^0. And the congruence subgroup property is used in the following form that Γ_n is generated only by the parabolic elements (§3).

(b) A geometric interpretation of the theorem to be proved as the "simply-connectedness" of the system $\{X_n \overset{\psi}{\leftarrow} X_n^0 \overset{\psi'}{\rightarrow} X_n'\}$ of three F_{p^2}-curves. Here, X_n is as above, X_n' is its conjugate over F_p, and X_n^0 is the union of the graphs Π, Π' of two p-th power morphisms $X_n \to X_n'$, $X_n' \to X_n$ (both graphs being taken on $X_n \times X_n'$. Actually, X_n^0 is partially normalized so that Π and Π' are *crossing only at the supersingular points* (§4).

(c) The bridge connecting these two systems, i.e., the congruence relation "$R_n^0 \equiv \Pi + \Pi'$ (mod p)". This is presented in §5.1. Due to our canonical choice of the curves over F_{p^2} and their p-adic liftings, where the action of $p^{\mathbf{Z}}$ is trivialized, the congruence relation is in its *symmetric* form. Also, we use the non-singular curve R_n^0 instead of its image in $R_n \times R'_n$, and this corresponds to that Π, Π'

are crossing only at the supersingular points.† Besides the general formulations of congruence relations by Shimura [18], it is also meaningful to draw attention to this canonical and natural form of the congruence relation.

(d) Passing the bridge, using a celebrated theorem of Grothendieck on the unique p-adic liftability of étale coverings (§6).

I am very grateful to Professor P. Deligne who read my first version of the proof and kindly pointed out to me a simplification by an extended use of a theorem of Grothendieck (see §6.2).

1. Preliminaries.

We shall review the definitions and some basic facts related to the fields of modular functions over finite fields.

1.1. Let p be the fixed prime number, and n be a positive integer not divisible by p. In §1.1, F_p will denote any field of characteristic either 0 or p, satisfying the following condition. Let W_n be the group of n-th roots of unity in the fixed separable closure of F_p, let W_∞ be the inductive union of W_n for all n (with $p \nmid n$), and consider the cyclotomic extension $F = F_p(W_\infty)$. Then our condition on F_p is that the Galois group $G(F/F_p)$ contains, and is topologically generated by, a special element σ that acts on W_∞ as $\zeta \to \zeta^p$. This being assumed, put $F_{p^m} = F_p(W_{p^m-1})$, for each positive integer m. Then F_{p^m} is a cyclic extension of F_p with degree m, and it is the fixed field of σ^m in F. The Krull topology of $G(F/F_p)$ induces, *via* $\sigma^{\mathbf{Z}}$, the product of l-adic topologies of $\mathbf{Z}$ over all prime numbers l (including p), and $G(F/F_p)$ is canonically isomorphic to the completion $\hat{\mathbf{Z}}$ of $\mathbf{Z}$ by this topology. For each $a \in \hat{\mathbf{Z}}$, the corresponding element of $G(F/F_p)$ will be denoted by σ^a, and the automorphism of W_∞ induced by σ^a will be denoted simply as "p^a". Examples of F_p are the prime field of characteristic p, the p-adic field $\mathbf{Q}_p$, the decomposition field of p in the cyclotomic field $\mathbf{Q}(W_\infty)$ over the rational number field $\mathbf{Q}$, etc.

Now let j be a variable over F_p, and E be any elliptic curve over the rational function field $F_p(j)$, having j as its absolute invariant. The equation of Tate,

† We learnt this from Igusa [5], p.472 (footnote); cf. also Deligne [1], No. 4

$$Y^2+XY=X^3-\frac{36}{j-1728}X-\frac{1}{j-1728}, \qquad (\mathrm{T})_j$$

gives an example of E. For each n, let E_n be the group of n-th division points of E, and put $E_\infty=\bigcup_{p\nmid n} E_n$. Then, since $E_n\simeq(\mathbf{Z}/n)^2$, the determinant gives a homomorphism Aut $E_n\to(\mathbf{Z}/n)^\times$. Identifying the two groups $(\mathbf{Z}/n)^\times$and Aut W_n in the canonical way, we shall consider the determinant as giving a homomorphism Aut $E_n\to$Aut W_n, which will be filtrated to the homomorphism at infinity,

$$\det: \mathrm{Aut}\ E_\infty\to\mathrm{Aut}\ W_\infty. \qquad (1.1.1)$$

Let $\mathfrak{g}$ be the Galois group over $F_p(j)$ of its separable closure. Then $\mathfrak{g}$ acts on E_∞ and W_∞, and the two actions are compatible with the homomorphism (1.1.1). In particular, by our assumption on F_p, the determinant of the action of each element of $\mathfrak{g}$ on E_∞ must belong to $p^{\hat{\mathbf{Z}}}$.

PROPOSITION 1.1.2 (Igusa). *The subgroup of* Aut E_∞ *generated by the change of sign* -1_∞ (1_∞: *the identity map of* E_∞) *and the actions of all elements of* $\mathfrak{g}$ *on* E_∞ *consists of all those elements of* Aut E_∞ *whose determinants belong to* $p^{\hat{\mathbf{Z}}}$.

PROOF. By Igusa [5], this subgroup contains all elements with determinant unity, as j is a variable over F. The Proposition follows immediately from this and our assumption on F_p. (For characteristic 0, see also Shimura [17], [18].)

COROLLARY. *For at least one choice of the sign* $\pm$, *there exists an element of* $\mathfrak{g}$ *acting on* E_∞ *as* $\pm p\cdot 1_\infty$.

DEFINITIONS. The fields on the left are by definition the fixed fields of the groups on the right. The groups are closed subgroups of $\mathfrak{g}$ defined by their actions on E_∞ and W_∞ (the two actions being connected by the determinant). The symbol "1" denotes the identity map of the indicated group.

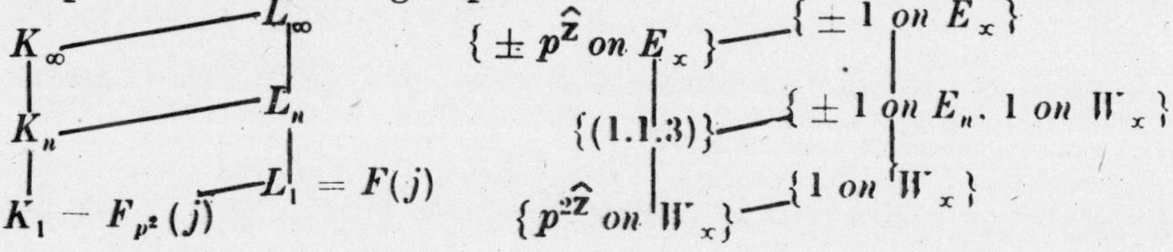

The group corresponding to K_n is defined by

$$\{\pm p^a \text{ on } E_n \text{ and } p^{2a} \text{ on } W_\infty \text{ for some common } a \in \hat{\mathbf{Z}}\}. \qquad (1.1.3)$$

It is the composite of the groups corresponding to K_∞ and L_n.

Thus, K_n is a finite Galois extension of $F_p(j)$, and it is an algebraic function field of one variable with exact constant field F_{p^2}. The field L_n is the field of modular functions of level n over $F(j)$ in the sense of [5], and the relations between L_n and K_n are $L_n = K_n \cdot F$, $K_n = L_n \cap K_\infty$. The fields L_∞, K_∞ are the composites of L_n, K_n for all n, and $L_\infty = K_\infty \cdot F$.

The Galois group $G(K_n/K_1)$ is canonically isomorphic to $G(L_n/L_1)$, and hence by Prop. 1.1.2., canonically isomorphic to $\mathrm{Aut}^1 E_n/\pm I$, where $\mathrm{Aut}^1 E_n$ is the group of all automorphisms of E_n with determinant unity. So, each choice of an isomorphism $\epsilon_n: E_n \xrightarrow{\sim} (\mathbf{Z}/n)^2$ defines an isomorphism $\alpha_n: G(K_n/K_1) \xrightarrow{\sim} SL_2(\mathbf{Z}/n)/\pm I$. If we fix an isomorphism $\omega_n: W_n \xrightarrow{\sim} \mathbf{Z}/n$, we can impose on ϵ_n the following condition that

$$\omega_n(e(u, v)) = |\epsilon_n u, \epsilon_n v|, \ (u, v \in E_n), \qquad (1.1.4)$$

where $e: E_n \times E_n \to W_n$ is the Weil's Riemannian form with respect to the divisor ($\equiv -$) class of *degree one* (Weil [19] Ch. IX), and is the matricial determinant. This condition (1.1.4) defines a unique $SL_2(\mathbf{Z}/n)$-class of ϵ_n, and hence a unique class of α_n modulo inner automorphisms. We shall fix an isomorphism

$$\omega : W_\infty \xrightarrow{\sim} \varinjlim \mathbf{Z}/n, \qquad (1.1.5)$$

which defines a unique class of isomorphisms

$$\alpha: G(K_\infty/K_1) \xrightarrow{\sim} \varprojlim (SL_2(\mathbf{Z}/n)/\pm I) \qquad (1.1.6)$$

modulo inner automorphisms.

Let $\Gamma_1 = PSL_2(\mathbf{Z}^{(p)})$ be the modular group over $\mathbf{Z}^{(p)} = \mathbf{Z}[1/p]$, and Γ_n be the principal congruence subgroup of level n. Then Γ_1/Γ_n is canonically isomorphic to $SL_2(\mathbf{Z}/n)/\pm I$; hence the right side of (1.1.6) is canonically isomorphic to $\varprojlim (\Gamma_1/\Gamma_n)$. Therefore, (1.1.6) induces an injection

$$\iota : \Gamma_1 \hookrightarrow G(K_\infty/K_1). \qquad (1.1.7)$$

which is intrinsic up to inner automorphisms of $G(K_\infty/K_1)$, once ω is fixed. The congruence subgroups of Γ_1 are in one-to-one correspondence with finite extensions of K_1 in K_∞. In particular, Γ_n corresponds to K_n (for any choice of ω).

It is easy to check that the fields K_n, L_n are independent of the special choice of E. Also, the classes of α and ι, considered up to inner automorphisms of $G(K_\infty/K_1)$, depend only on ω and not on the special choice of E.

We shall call K_n *the field of modular functions of level n over* F_{p^2}.

1.2. Now we shall specify F_p as the prime field of characteristic p, so that F is the algebraic closure of F_p and $W_\infty = F^\times$. For each $a \in F \cup (\infty)$, P_a will denote the prime divisor of the function field $K_1 = F_{p^2}(j)$ defined by $j \equiv a \pmod{P_a}$. So, for $a, b \in F$, $P_a = P_b$ holds if and only if a, b are conjugate over F_{p^2}. The prime divisor P_∞ is called cuspidal, and P_a is called supersingular if a is a supersingular "j-invariant" In the latter case, we have $a \in F_{p^2}$, so that all supersingular prime divisors of K_1 are of degree one over F_{p^2}. Other special prime divisors of K_1 are P_{1728} and P_0. When $p = 2$ or 3, $P_{1728} = P_0$ is the unique supersingular prime divisor of K_1. In all other cases, P_{1728} (resp. P_0) is supersingular if and only if $p \equiv -1 \pmod 4$ (resp. $p \equiv -1 \pmod 3$). For any intermediate field K of K_∞/K_1 and a prime divisor P of K, we call P cuspidal (resp. supersingular) when its restriction to K_1 is cuspidal (resp. supersingular). We shall review some basic facts about the Galois extension K_∞/K_1.

THE RAMIFICATION. It is independent of the constant field. Hence the ramification of K_∞/K_1 is described by that of L_∞/L_1, which was determined by Igusa [5] as follows.

(IG.1) $P = P_\infty$. It is tamely ramified in K_∞/K_1, and has an extension to K_∞ whose inertia group is topologically generated by $\iota\left(\begin{pmatrix} 1 & 1 \\ 0 & 1 \end{pmatrix}\right)$.†

† This last point is not explicitly stated in [5], but follows easily from the arguments used there. See also [8], II, Ch. 5, §28 for an alternative proof.

In particular, the ramification index of P_∞ in K_n/K_1 is equal to n, and moreover, K_n is the maximum extension of K_1 *in* K_∞ with this property. The second point follows immediately from the *local* congruence subgroup property of PSL_2, that the principal congruence subgroup of level n of the group $\varprojlim(SL_2(\mathbf{Z}/n'))$ is normally generated by $\begin{pmatrix} 1 & n \\ 0 & 1 \end{pmatrix}$ in the topological sense. Thus:

PROPOSITION 1.2.1. *There is no non-trivial unramified extension of K_n contained in K_∞.*

This implies that all non-trivial unramified extensions of K_n, for each n, are *outside* K_∞.

(IG.2). $P = P_{1728}, P_0$. If $p \neq 2, 3$, these prime divisors are tamely ramified in K_∞ with indices 2, 3, respectively. If $p = 2$ or $p=3$, $P_{1728}=P_0$ is wildly ramified with indices 12 $(p=2)$ or $6 (p=3)$. In all cases, the ramifications of P_{1728}, P_0 in K_∞/K_1 are accomplished in K_n/K_1 for any $n > 1$, so that each extension of P_{1728} or P_0 is unramified in K_∞/K_n for $n>1$.

(IG.3) $P \neq P_\infty, P_{1728}, P_0$. These prime divisors P are unramified in K_∞/K_1.

1.3. *The decomposition.* The following assertion is an immediate consequence of Theorems 3, 4, 5 of our previous note [8], II, Ch. 5 (§§ 24,26,28). It can also be proved directly without any essential difficulty.

PROPOSITION 1.3.1. *A prime divisor of K_∞ is of degree one over F_{p^2} if and only if it is supersingular.*

(This includes in particular that the cuspidal prime divisors of K_∞ are *not* of degree one over F_{p^2}.)

As a direct formal consequence of Prop. 1.3.1., we have:

PROPOSITION 1.3.1′. *There exists a non-zero ideal $n_0\mathbf{Z}$ of $\mathbf{Z}$ such that for each $n \in n_0\mathbf{Z}$, the prime divisors of K_n of degree one over F_{p^2} are precisely the supersingular prime divisors of K_n.*

Let h_n denote the number of supersingular prime divisors of K_n. It is well known that

$$h_1 = \frac{1}{12}(p+13) - \frac{1}{4}\left(1+\left(\frac{-1}{p}\right)\right) - \frac{1}{3}\left(1+\left(\frac{-3}{p}\right)\right), \quad (1.3.2)_1$$

$$h_n = \frac{1}{12}(p-1)\,[\,K_n : K_1], \quad (n > 1). \quad (1.3.2)_n$$

The second formula follows from the first by using (IG2), (IG3).

In [8], Ch. 5, more intimate arithmetic relations between the group Γ_n and the field K_n are presented, but they will not be directly used here.

2. The main theorem.

2.1. For each positive integer n with $p \nmid n$, let K_n be the modular function field of level n over the finite field F_{p^2} defined in §1. Then, K_n has a special finite set of prime divisors $s_1, \dots, s_{h_n}$ of degree one, the supersingular prime divisors, where $h_n = \frac{1}{12}(p-1)[K_n : K_1]$ for $n > 1$ (see § 1.3).

Main Theorem (First formulation). *Let M_n be the maximum unramified Galois extension of K_n, and $s_1, \dots, s_{h_n}$ be the set of all supersingular prime divisors of K_n. Then the Frobenius conjugacy classes $\left\{\frac{M_n/K_n}{s_i}\right\}$ of s_i $(i = 1, \dots, h_n)$ generate, in the topological sense, the Galois group of M_n/K_n.*

(Since M_n/K_n is generally non-abelian, the Frobenius automorphism associates to each prime divisor of K_n a conjugacy class of the Galois group, which is called the Frobenius conjugacy class.)

In a more suggestive way, one may state it as "*the non-abelian divisor class group of K_n is generated by the supersingular prime divisors $s_1, \dots, s_{h_n}$.*"

The restriction of the theorem to abelian extensions gives:

Corollary 1. *The divisor class group of K_n is generated by the classes of supersingular prime divisors $s_1, \dots, s_{h_n}$.*

REMARK 1. According to the Grothendieck theory of fundamental groups, the Galois group of the maximum unramified Galois extension of an algebraic function field of genus $g > 0$ (over any algebraically closed constant field) is isomorphic to the projective limit of *some subsystem* of the system of all finite factor groups of the abstract group defined by $2g$ generators $\alpha_1, \dots, \alpha_g, \beta_1, \dots, \beta_g$ with a single relation

$$\alpha_1 \beta_1 \alpha_1^{-1} \beta_1^{-1} \dots \alpha_g \beta_g \alpha_g^{-1} \beta_g^{-1} = 1.$$

For characteristic 0, it is the projective limit of *all* finite factor groups (as is classically well known), and for characteristic $p > 0$, the best (general type) information on this subsystem is that it contains all those finite factor groups whose orders are not divisible by p. This gives a general idea on the structure of the Galois group. Note that *our result is of a completely different type.* It also gives some topological generators of the Galois group, but its aim is not to describe the structure of the Galois group, but *to describe the arithmetic distribution of supersingular prime divisors and to characterize the field* K_n *by ramifications and supersingular complete decompositions* (see [MT 3] below). While the structure of the Galois group of M_n/K_n itself is similar to the case of characteristic 0, our assertion is what properly belongs to characteristic $p > 0$.

It is necessary to look at our theorem from several different angles, and so we shall present various different (but equivalent) versions of the theorem under the names [MT i]. The first formulation will hereafter be cited as $[\text{MT } 1]_n$. For example, an immediate variation is the following:

$[\text{MT } 2]_n$. *There is no non-trivial unramified extension of* K_n *in which all supersingular prime divisors* $s_1, \dots, s_{h_n}$ *of* K_n *are decomposed completely.*

REMARK 2. If K' is a finite unramified extension of K_n in which $s_1, \dots, s_{h_n}$ are decomposed completely, the curve for K' has at least $[K':K_n].h_n$ number of F_{p^2}-rational points. It should be mentioned here that $[\text{MT } 2]_n$ *cannot* be proved only by the estimation of the number of rational points.

The above two formulations of the theorem are convenient for the presentations of the *new* part of its content. But to exhibit its natural form, we must let $n \to \infty$. As noted in §1, the constant field extension $L_n = K_n \cdot F$ is nothing but the modular function field of level n over $F(j)$ in the sense of [5]. Let M_n be, as in $[\mathrm{MT}\ 1]_n$, the maximum unramified Galois extension of K_n (or equivalently, of L_n), and take the composites M_∞, L_∞, K_∞ of M_n, L_n, K_n for all n (with $p \nmid n$):

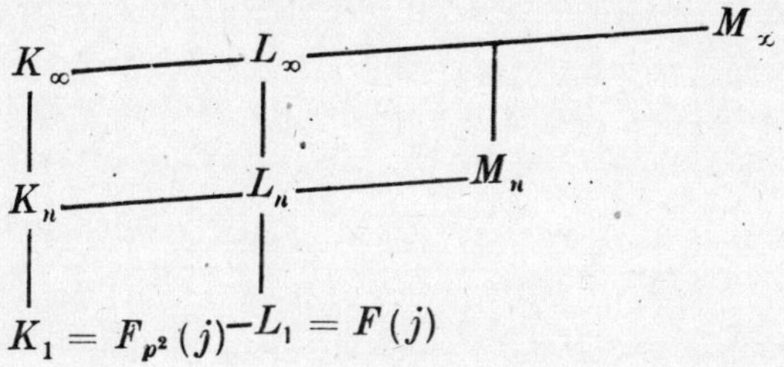

By Prop. 1.2.1, we have $M_n \cap K_\infty = K_n$ and (equivalently) $M_n \cap L_\infty = L_n$. Now, $[\mathrm{MT}\ 1]_n$ gives immediately:

$[\mathrm{MT}\ 1]_\infty$ (resp. $[\mathrm{MT}\ 2]_\infty$). *One can replace n by ∞ in* $[\mathrm{MT}\ 1]_n$ (resp. $[\mathrm{MT}\ 2]_n$).

Here, by definition, the unramified extensions of K_∞ are those extensions of K_∞ contained in M_∞.

Fix any positive integer r with $r > 1$ and $p \nmid r$. Let n be any positive integer with $p \nmid n$. Then, as reviewed in §1.2, L_{nr}/L_r is unramified outside the cusps, and the ramification index at each cusp is precisely n. In particular, the cusps are tamely ramified. Therefore, M_{nr} (resp. M_∞) is the maximum Galois extension of K_r satisfying the following condition (i 1)(i 2)$_n$ (resp. (i 1) (i 2)$_\infty$);

(i 1) *it is unramified outside the cusps of K_r,*

(i 2)$_n$ *the ramification index of each cusp is a factor of n, resp.*

(i 2)$_\infty$ *each cusp is at most tamely ramified.*

Therefore, we obtain from [MT 2] the following simple characterization of K_{nr} (resp. K_∞).

[MT 3] K_{nr} (resp. K_{∞}) *is the maximum Galois extension of* K_r *satisfying* (i 1)(i 2)$_n$ (*resp.* (i 1)(i 2)$_{\infty}$) *and the following property* (ii):

(ii) *the supersingular prime divisors of* K_r *are decomposed completely.*

Of course, to be able to say that "[MT 3] characterizes K_{nr} (or K_{∞})", it is necessary that the base field K_r, its cusps, and its supersingular prime divisors are all explicitly presentable. A well-known case when they are explicitly presented is where $r=2$ ($p \neq 2$). Then $K_2 = F_{p^2}(\lambda)$, the rational function field, with cusps at $0, 1, \infty$, and supersingular prime divisors at the zeros of the polynomial

$$P(\lambda) = \sum_{i=0}^{b} \binom{b}{i}^2 \lambda^i, \qquad (b = \tfrac{1}{2}(p-1)),$$

the prime divisors of K_2 being expressed by the residue classes of λ. Therefore, this special case gives the following.

Let p be an odd prime and n be a positive integer with $p \nmid n$. Then K_{∞}(resp. K_{2n}) is the maximum Galois extension of $F_{p^2}(\lambda)$ satisfying:

(i 1) *it is unramified outside* $0, 1, \infty$;

(i 2) (*resp.* (i 2)$_n$) $0,1,\infty$ *are tamely ramified* (*resp.* $0,1,\infty$ *are tamely ramified with ramification index dividing* n);

(iii) *the zeros of* $P(\lambda)$ *are decomposed completely.*

One may restate [MT 3] in a more suggestive way as "*the subgroup of the non-abelian divisor class group of* K_r *with conductor* $\prod_{i=1}^{\nu} c_i$, *the product of all distinct cusps of* K_r, *and generated by the supersingular prime divisors* $s_1, \ldots, s_{h_r}$ *of* K_r, *is precisely that group corresponding to the Galois extension* K_{∞} *of* K_r".

The restriction to abelian extensions gives:

COROLLARY 2. *The subgroup of the divisor class group of* K_r *with cuspidal conductor* $\prod_{i=1}^{\nu} c_i$ *and generated by the supersingular prime divisors* $s_1, \ldots, s_{h_r}$ *is precisely that group corresponding to the maximum abelian extension of* K_r *contained in* K_{∞}.

For example, if $r = 2$, the maximum abelian extension of K_2 in K_∞ is given by

$$F_{p^2}(\lambda; (-\lambda)^{1/8}, (-\lambda')^{1/8}, (\tfrac{1}{2}\lambda\lambda')^{1/3}) \qquad (p \neq 2,3),$$

$$F_{p^2}(\lambda; (-\lambda)^{1/8}, (-\lambda')^{1/8}) \qquad (p = 3),$$

where $\lambda' = 1 - \lambda$. Therefore, we obtain easily from Corollary 2 the following multiplicative property of the set of supersingular "λ-invariants".

Corollary 3 ($p \neq 2, 3$). *Consider the group* $\tilde{G} = (F_{p^2}^\times)^8 \times (F_{p^2}^\times)^8$ *and the subgroup* $H = \{(x, y) \in \tilde{G} \mid xy \in (F_{p^2}^\times)^{24}\}$ *of* $\tilde{G}$ *with index* 3. *Define an H-coset H' by* $H' = \{(x, y) \in \tilde{G} \mid \tfrac{1}{2}xy \in (F_{p^2}^\times)^{24}\}$, *and let G be the subgroup of $\tilde{G}$ generated by H'. Let S be the set of all zeros of $P(\lambda)$. For each $s \in S$ put $g_s = (-s, -s')$, where $s' = 1 - s$. Then*

(a) $g_s \in H'$ *for any* $s \in S$,

(b) $g_s \cdot g_t^{-1} (s, t \in S)$ *generate* H;

hence (c) $g_s\, (s \in S)$ *generate* G.

Finally, let $\Gamma_1 = PSL_2(\mathbf{Z}^{(p)})$ and Γ_n be the principal congruence subgroup of level n. Then, in view of the isomorphism $G(K_\infty/K_1) \cong \varprojlim \Gamma_1/\Gamma_n$ (§1.1) and the congruence subgroup property of Γ_1, [MT 3] can also be formulated as follows:

[MT 4]. *There is a categorical equivalence between subgroups with finite indices of* Γ_r *and finite extensions of* K_r *satisfying* (i 1), (i 2) *and* (ii).

Or in short, "Γ_r *is the strict fundamental group of* K_r *defined by* (i 1), (i 2) *and* (ii)".

In §4, a more geometric version of [MT]'s will be given.

Remark 3. [MT 3] characterizes the extensions K_{nr}/K_r, *for* $r > 1$. If one wants to take K_1 as the base field, one must first observe that not only the cusp P_∞ but also P_{1728} and P_0 are ramified in K_∞/K_1. If $p \neq 2, 3$ and $n > 1$, the ramification indices of these three prime divisors of K_1 in K_n/K_1 are n, 2, 3, respectively (and all other prime divisors of K_1 are unramified). So, by [MT 2]$_n$, K_n is a *maximal* extension of K_1 having (i) this ramification property, and (ii) the property that all prime divisors lying on the supersingular prime divisors of K_1 are of degree one over F_{p^2}. But here, we cannot replace "*maximal*" by "*maximum*". In fact, due to the situation that the ramifying prime divisors P_{1728} or P_0 can be supersingular (for $p \equiv -1 \pmod 4$ or $-1 \pmod 3$ respectively), the composite of two extensions of K_1 satisfying above two properties may *not* satisfy them. This is why we did not formulate the characterization of the extension K_{nr}/K_r for $r = 1$. Our theorem was formulated as a conjecture in [7], [8] and [9]. Among them, the formulation in [8], II, Ch. 5, " $\hat{\mathfrak{K}} = \mathfrak{K}$?", contains an error in the definition of $\hat{\mathfrak{K}}$ arising from the misobservation of this subtle situation which, of course, can be easily corrected. †

3. The complex system $\{ R_n \xleftarrow{\varphi} R_n^0 \xrightarrow{\varphi'} R_n' \}$. In this section, we shall prove a lemma in characteristic 0, which is *the complex version* of the theorem to be proved. It is a reflection of the congruence subgroup property of the modular group over $\mathbf{Z}^{(p)}$. This property is in fact faithfully reflected due to the fact that the modular group over $\mathbf{Z}^{(p)}$ is a free product of two modular groups over $\mathbf{Z}$ with amalgamation. Our theorem will finally be reduced to this lemma.

3.1. Consider a system $\{R \xleftarrow{\varphi} R^0 \xrightarrow{\varphi'} R'\}$ of compact Riemann surfaces R, R', R^0 and surjective morphisms φ, φ'. *Unramified covering* of such a system $\{R \xleftarrow{\varphi} R^0 \xrightarrow{\varphi'} R'\}$ is defined by a commutative diagram

$$\begin{array}{ccccc} R^* & \xleftarrow{\varphi^*} & R^{*0} & \xrightarrow{\varphi^{*\prime}} & R^{*\prime} \\ {\scriptstyle f}\downarrow & & \downarrow{\scriptstyle f^0} & & \downarrow{\scriptstyle f'} \\ R & \xleftarrow[\varphi]{} & R^0 & \xrightarrow[\varphi']{} & R' \end{array} \qquad (3.1.1)$$

† The second definition on p.179 ([8] II) is the correct one.

of surjective morphisms beween compact Riemann surfaces satisfying the following conditions (a), (b), (c):

(a) *The degrees of morphisms indicated by parallel arrows are equal.*

(b) *Let* F, F' *be the smallest Galois coverings of* R, R' *containing* f, f' *as subcoverings, respectively. Then* F *and* φ, F' *and* φ', *are both linearly disjoint.*

(c) *The vertical coverings* f, f' *and* f^0 *are unramified.*

The degree of the vertical coverings is called *the degree of the covering* (3.1.1). Equivalence of two such coverings is defined by a commutative diagram connecting the two covering systems by three isomorphisms. For example, if ϵ^0 is an automorphism of R^{*0} commuting with f^0, then replacing φ^* by $\varphi^* \circ \epsilon^0$ (and leaving all others fixed) only gives an equivalent covering. In fact, by (b), there is some automorphism ϵ of R^* commuting with f, such that $\varphi^* \circ \epsilon^0 = \epsilon \circ \varphi^*$. But then, it is clear that the new covering is equivalent to the old one. *Up to equivalence, giving an unramified covering* (3.1.1) *is the same thing as giving a pair* $\{f, f'\}$ *of unramified coverings* $f: R^* \longrightarrow R$, $f': R^{*\prime} \longrightarrow R'$ *satisfying* (b), *for which the fiber products* $R^* \underset{R}{\times} R^0$ *and* $R^0 \underset{R'}{\times} R^{*\prime}$ *are isomorphic as coverings of* R^0. (These products are *connected* by the linear disjointness, and *non-singular* by the unramifiedness of f, f' ; so that they are also compact Riemann surfaces.)

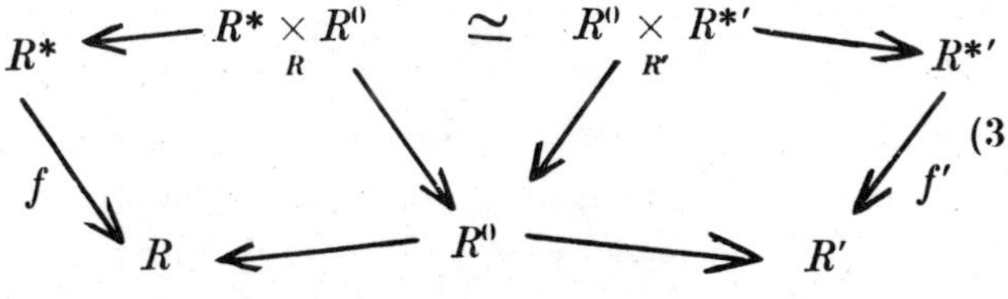

(3.1.1′)

By the above remark, the choice of $\simeq$ in (3.1.1′) does not affect the equivalence class.

DEFINITION. *The system* $\{R \overset{\varphi}{\leftarrow} R^0 \overset{\varphi'}{\rightarrow} R'\}$ *will be called 'simply-connected, if it has no other unramified covering than that of degree one.*

3.2. Now let n be any positive integer with $p \nmid n$ and Δ_n be the principal congruence subgroup of level n of the modular group $\Delta_1 = PSL_2(\mathbf{Z})$ over $\mathbf{Z}$. Put

$$\Delta'_n = \begin{pmatrix} p & 0 \\ 0 & 1 \end{pmatrix}^{-1} \Delta_n \begin{pmatrix} p & 0 \\ 0 & 1 \end{pmatrix} = \left\{ \begin{pmatrix} a & p^{-1}b \\ pc & d \end{pmatrix} \middle| \begin{pmatrix} a & b \\ c & d \end{pmatrix} \in \Delta_n \right\},$$

$$\Delta_n^0 = \Delta_n \cap \Delta'_n = \left\{ \begin{pmatrix} a & b \\ c & d \end{pmatrix} \in \Delta_n \middle| c \equiv 0 \pmod{p} \right\}.$$

Then they define a system $\{\Delta_n \leftarrow \Delta_n^0 \rightarrow \Delta'_n\}$ of fuchsian groups, the arrows indicating the inclusions. Note that Δ_n^0 is of index $p+1$ in Δ_n and Δ'_n. Let $\{R_n \overset{\varphi}{\leftarrow} R_n^0 \overset{\varphi'}{\rightarrow} R'_n\}$ be the corresponding system of compact Riemann surfaces, i.e., R_n, R'_n, R_n^0 are the compactified quotients of the complex upper half plane by Δ_n, Δ'_n, Δ_n^0; and φ, φ' are the canonical coverings defined by the inclusions $\Delta_n^0 \hookrightarrow \Delta_n$, $\Delta_n^0 \hookrightarrow \Delta'_n$.

LEMMA 3.2. *For each positive integer n with $p \nmid n$, the system*

$$\left\{ R_n \overset{\varphi}{\longleftarrow} R_n^0 \overset{\varphi'}{\longrightarrow} R'_n \right\} \text{ is simply-connected.}$$

Let $\Gamma_1 = PSL_2(\mathbf{Z}^{(p)})$ be the modular group over $\mathbf{Z}^{(p)}$ and Γ_n be the principal congruence subgroup of level n. The proof is based on the following properties (A), (B) of Γ_n.

(A) *Γ_n is the free product of Δ_n and Δ'_n with amalgamated subgroup Δ_n^0.*

(B) *Γ_n is normally generated by $\begin{pmatrix} 1 & n \\ 0 & 1 \end{pmatrix}$ in Γ_1; in particular, it is generated only by parabolic elements.*

The property (A) was proved in our previous work. It is a corollary of the corresponding property of the local groups ([8], I, Ch. 2, §28, p. 111, or a more full-fledged geometric exposition in [15]). The property (B) was proved by Mennicke [12] and Serre [16]. It is equivalent to the congruence subgroup property of Γ_1, modulo the local congruence subgroup property.

3.3. PROOF OF LEMMA 3.2. Write

$$\left\{ R \overset{\varphi}{\longleftarrow} R^0 \overset{\varphi'}{\longrightarrow} R' \right\} \quad \text{and} \quad \left\{ \Delta \longleftarrow \Delta^0 \longrightarrow \Delta' \right\}$$

instead of writing with suffix n. Suppose that there is an unramified covering (3.1.1) of degree m. Then there is a corresponding commutative diagram of inclusions of fuchsian groups

$$\begin{array}{ccccc} \Delta^* & \longleftarrow & \Delta^{*0} & \longrightarrow & \Delta^{*\prime} \\ \downarrow & & \downarrow & & \downarrow \\ \Delta & \longleftarrow & \Delta^0 & \longrightarrow & \Delta' \end{array} \qquad (3.3.1)$$

with

$$(\Delta : \Delta^*) = (\Delta' : \Delta^{*\prime}) = (\Delta^0 : \Delta^{*0}) = m, \qquad (3.3.2)$$

$$\Delta^* \cdot \Delta^0 = \Delta, \quad \Delta^{*\prime} \cdot \Delta^0 = \Delta',$$

$$\Delta^* \cap \Delta^0 = \Delta^{*\prime} \cap \Delta^0 = \Delta^{*0}, \qquad (3.3.3)$$

$$Pb(\Delta) = Pb(\Delta^*),\ Pb(\Delta') = Pb(\Delta^{*\prime}),\ Pb(\Delta^0) = Pb(\Delta^{*0}), \qquad (3.3.4)$$

where Pb ( ) is the set of all parabolic elements of the group inside the parenthesis. The first two equalities of (3.3.3) are the consequences of the condition (b), the rest of (3.3.3) is the consequence of the first two equalities and the condition (a), and (3.3.4) is a consequence of (c).

Now let Γ^* be the subgroup of $\Gamma = \Gamma_n$ generated by Δ^* and $\Delta^{*\prime}$. We shall deduce from (3.3.2), (3.3.3) and the property (A) of Γ that

$$(\Gamma : \Gamma^*) = m, \qquad (3.3.5)$$

and from (3.3.4) that

$$Pb(\Gamma) = Pb(\Gamma^*). \qquad (3.3.6)$$

But then, by the property (B) of Γ, (3.3.6) would imply $\Gamma = \Gamma^*$, and hence $m=1$ by (3.3.5). So it remains to prove (3.3.5) and (3.3.6).

Let $1 = M_0, M_1, \ldots, M_p$ (resp. $1 = M'_0, M'_1, \ldots, M'_p$) be representatives of the coset spaces $\Delta^{*0}\backslash\Delta^*$ (resp. $\Delta^{*0}\backslash\Delta^{*\prime}$). Then by (3.3.3), they are also representatives of $\Delta^0\backslash\Delta$ (resp. $\Delta^0\backslash\Delta'$). Since Γ is generated by Δ and Δ', each element γ of Γ can be expressed in the form

$$\gamma = \delta_0 M_{i_1} M'_{j_1} \cdots M_{i_r} M'_{j_r} \quad (\delta_0 \in \Delta^0, j_1, \ldots, i_r \neq 0). \qquad (3.3.7)$$

The set of all those $\gamma\in\Gamma$ having an expression (3.3.7) with $\delta_0\in\Delta^{*0}$ forms a subgroup of Γ, since M_i and M_j' belong to Δ^* and $\Delta^{*\prime}$ respectively. It is clear that this subgroup is Γ^*. Therefore, $\Delta^0\cdot\Gamma^*=\Gamma$; hence $(\Gamma:\Gamma^*)=(\Delta^0:\Delta^0\cap\Gamma^*)$. But *by the property* (A) *of* Γ, *the expression* (3.3.7) *is unique* (see Kurosh [10] II, for the general theorems on free products with amalgamations). Therefore, we have $\Delta^0\cap\Gamma^*=\Delta^{*0}$, which gives $(\Gamma:\Gamma^*)=(\Delta^0:\Delta^{*0})=m$. Thus, (3.3.5) is settled.

To check (3.3.6), let $\gamma\in\Gamma$ be parabolic, and put $\gamma=\pm I+p^{-k}A$ ($k\geqslant 1$), where I is the identity matrix and A is a $\mathbf{Z}$-integral matrix with $A^2=0$. Then $\gamma^{p^k}=\pm I+A\in PSL_2(\mathbf{Z})$. Hence γ^{p^k} belongs to Δ and hence to Δ^* by (3.3.4). Therefore, $\gamma^{p^k}\in\Gamma^*$. But for any parabolic element γ of Γ_1, its order relative to any subgroup with finite index of Γ_1 is not divisible by p. In fact, by the above argument, $\gamma^{p^k}\in PSL_2(\mathbf{Z})$, and hence γ^{p^k} is conjugate to an integral power of $\begin{pmatrix}1&1\\0&1\end{pmatrix}$. Therefore, γ is conjugate to $\begin{pmatrix}1&b\\0&1\end{pmatrix}$ with some $b\in\mathbf{Z}^{(p)}$. But $\begin{pmatrix}1&b\\0&1\end{pmatrix}$ is conjugate to its p^2-th power $\left(\text{by }\begin{pmatrix}p^{-1}&0\\0&p\end{pmatrix}\right)$; hence γ is also conjugate to its p^2-th power γ^{p^2}. This gives the above assertion, and hence also that $\gamma\in\Gamma^*$. Therefore, (3.3.6) is also settled. The lemma follows.

3.4. As for the covering (3.1.1) of those systems obtained from the system of fuchsian groups $\{\Delta\leftarrow\Delta^0\rightarrow\Delta'\}=\{\Delta_n\leftarrow\Delta_n^0\rightarrow\Delta_n'\}$, *the linear disjointness of* F *and* φ (*resp.* F' *and* φ') *follows automatically from the unramifiedness of* f (*resp.* f'). In fact, let Δ^* be the fuchsian group corresponding to R^*, and Δ^{**} be the intersection of all conjugates of Δ^* in Δ. Then the unramifiedness of f gives $Pb(\Delta)=Pb(\Delta^*)$; hence also $Pb(\Delta)=Pb(\Delta^{**})$. But $Pb(\Delta^0)\subsetneq Pb(\Delta)$, as $\begin{pmatrix}1&0\\n&1\end{pmatrix}\notin\Delta^0$. In particular, Δ^{**} is not contained in Δ^0. But as is well known, there is no proper intermediate group between Δ and Δ^0. (This is clear since $\Delta^0\backslash\Delta$ can be identified with the projective line over F_p by $\begin{pmatrix}a&b\\c&d\end{pmatrix}\rightsquigarrow dc^{-1}\pmod p$, and Δ acts doubly transitively on this line as a group of linear fractional transforms.) Therefore, $\Delta^0\cdot\Delta^{**}=\Delta$, which implies the linear disjointness of F and φ.

3.5. *The general picture.* To clarify the general situation, we note the following. As in § 3.2, let $\Gamma_1 = PSL_2(\mathbf{Z}^{(p)})$, and put $\Delta_1 = PSL_2(\mathbf{Z})$, $\Delta_1' = \begin{pmatrix} p & 0 \\ 0 & 1 \end{pmatrix}^{-1} \Delta_1 \begin{pmatrix} p & 0 \\ 0 & 1 \end{pmatrix}$ and $\Delta_1^0 = \Delta_1 \cap \Delta_1'$. Then, (i) Γ_1 is a free product of Δ_1 and Δ_1' with amalgamated subgroup Δ_1^0, and (ii) for any subgroup Γ of Γ_1 with finite index, it holds that $\Delta_1^0 \cdot \Gamma = \Gamma_1$. (Since Γ_1 is dense in $PSL_2(\mathbf{Q}_p)$, $\mathbf{Q}_p$ being the p-adic field, the topological closure of Γ in $PSL_2(\mathbf{Q}_p)$ is of finite index in $PSL_2(\mathbf{Q}_p)$. But since $PSL_2(\mathbf{Q}_p)$ is an infinite simple group, it has no non-trivial subgroups with finite indices. Therefore, Γ is *dense* in $PSL_2(\mathbf{Q}_p)$. This gives $\Delta_1^0 \cdot \Gamma = \Gamma_1$.) From these two properties (i), (ii), it follows directly that the association

$$\Gamma \rightsquigarrow \{\Delta, \Delta'\}; \quad (\Delta = \Gamma \cap \Delta_1, \ \Delta' = \Gamma \cap \Delta_1') \tag{3.5.1}$$

gives a one-to-one correspondence between subgroups Γ of Γ_1 with finite indices and the pairs $\{\Delta, \Delta'\}$ of subgroups $\Delta \subset \Delta_1$, $\Delta' \subset \Delta_1'$ with finite indices satisfying $\Delta_1^0 \cdot \Delta = \Delta_1$, $\Delta_1^0 \cdot \Delta' = \Delta_1'$ and $\Delta_1^0 \cap \Delta = \Delta_1^0 \cap \Delta'$. It is also easy to check, by the argument similar to that used in the proof of Lemma 3.1, that Γ is a free product of Δ and Δ' with amalgamated subgroup $\Delta^0 = \Delta \cap \Delta'$, and that $(\Gamma_1 : \Gamma) = (\Delta_1 : \Delta) = (\Delta_1' : \Delta') = (\Delta_1^0 : \Delta^0)$. We may call $\{\Delta, \Delta'\}$ *the canonical generating pair of fuchsian groups for* Γ. For $\Gamma = \Gamma_n$, it is nothing but $\{\Delta_n, \Delta_n'\}$ defined in § 3.2.

Now let $\{ R \overset{\varphi}{\leftarrow} R^0 \overset{\varphi'}{\rightarrow} R' \}$ be the system of compact Riemann surfaces corresponding to $\{\Delta \leftarrow \Delta^0 \rightarrow \Delta'\}$. Then there is a canonical covering

$$\begin{array}{ccccc} R & \overset{\varphi}{\longleftarrow} & R^0 & \overset{\varphi'}{\longrightarrow} & R' \\ {\scriptstyle f}\downarrow & & {\scriptstyle f^0}\downarrow & & {\scriptstyle f'}\downarrow \\ R_1 & \underset{\varphi_1}{\longleftarrow} & R_1^0 & \underset{\varphi_1'}{\longrightarrow} & R_1' \end{array} \quad \textit{corresponding to} \quad \begin{array}{ccccc} \Delta & \longleftarrow & \Delta^0 & \longrightarrow & \Delta' \\ \downarrow & & \downarrow & & \downarrow \\ \Delta_1 & \longleftarrow & \Delta_1^0 & \longrightarrow & \Delta_1' \end{array} \tag{3.5.2}$$

The covering (3.5.2) satisfies the conditions (a), (b) of §3.1, and instead of (c), the following weaker condition:

(c′) *The ramification index of* f^0 *at each non-cuspidal point* P *of* R_1^0 *is a factor of the ramification index at* P *of the covering* $\mathscr{H} \to \Delta_1^0\backslash\mathscr{H} \hookrightarrow R_1^0$ *of* R_1^0 *by the upper half plane* $\mathscr{H}$.

It follows immediately from the above remarks that *the functor*

$$\Gamma \rightsquigarrow \{R \overset{\varphi}{\leftarrow} R^0 \overset{\varphi'}{\rightarrow} R'\} \tag{3.5.3}$$

gives a categorical equivalence between subgroups with finite indices of Γ_1 *and those coverings of the system* $\left\{R_1 \xleftarrow{\varphi_1} R_1^0 \xrightarrow{\varphi_1'} R_1'\right\}$ *satisfying* (a), (b), (c′).

Finally, let Γ^* denote the subgroup of Γ generated by $Pb(\Gamma)$. It is of finite index, as Γ contains some Γ_n and $\Gamma_n^* = \Gamma_n$. If Γ is torsion-free, Γ^* corresponds to the maximum unramified covering of $\{R \overset{\varphi}{\leftarrow} R^0 \overset{\varphi'}{\rightarrow} R'\}$ so that Lemma 3.2 *is valid for* Γ *if and only if* $\Gamma^* = \Gamma$. As an example, let $n>1$ and $\Gamma \supset \Gamma_n$ be that group with which Γ/Γ_n is the center of Γ_1/Γ_n. Then Γ is torsion-free, $\Gamma^* = \Gamma_n$, and Γ/Γ_n is an elementary 2-group of rank r, where r is the number of distinct prime factors of the numerator of $n/2$, so that $\left\{R_n \xleftarrow{\varphi_n} R_n^0 \xrightarrow{\varphi_n'} R_n'\right\}$ is the maximum unramified covering of $\left\{R \xleftarrow{\varphi} R^0 \xrightarrow{\varphi'} R'\right\}$ with the elementary 2-group of rank r as the Galois group.

4. The characteristic p **system** $\left\{X_n \xleftarrow{\psi} X_n^0 \xrightarrow{\psi'} X_n'\right\}$. In §3, we gave a geometric interpretation of the congruence subgroup property of $PSL_2(\mathbf{Z}^{(p)})$ as the simply-connectedness of the system $\left\{R_n \xleftarrow{\varphi} R_n^0 \xrightarrow{\varphi'} R_n'\right\}$ of three compact Riemann surfaces (for all n with $p \nmid n$). In this section, we shall give another geometric interpretation, an interpretation of some arithmetic condition in characteristic p. It is based on a simple principle, which makes it possible to reformulate our theorem in geometric terms.

4.1. Let q be any positive power of p, F_{q^2} be the finite field with q^2 elements, and F be its algebraic closure. *Suppose given an algebraic curve X over F_{q^2} and a specified set S of F_{q^2}-rational points of X; assumed that X is complete, non-singular and absolutely irreducible, and that S is non-empty.* Then, starting from such a pair of X and S we can construct a system

$$\left\{ X \xleftarrow{\psi} X^0 \xrightarrow{\psi'} X' \right\} \tag{4.1.1}$$

of three algebraic curves X, X', X^0 and surjective morphisms ψ, ψ' of degree $q+1$, as follows. First, X' is the conjugate of X over F_q. Then, there are two q-th power morphisms $\pi : X \to X'$ and $\pi' : X' \to X$. Let Π, Π' be their graphs, both taken on $X \times X'$;

$$\Pi = \{(x, x^q) \mid x \in X\},\ \Pi' = \{(x'^q, x') \mid x' \in X'\}. \tag{4.1.2}$$

Observe that the intersection $\Pi \cap \Pi'$ consists of all points of $X \times X'$ of the form (x, x'), where x, x' are F_{q^2}-rational points of X, X' that are mutually conjugate over F_q. Put

$$S^0 = \{(x, x') \in \Pi \cap \Pi' \mid x \in S\}. \tag{4.1.3}$$

Each of the two curves Π, Π' is a complete non-singular absolutely irreducible curve lying on $X \times X'$, and they intersect *transversally* at each point of $\Pi \cap \Pi'$. The union $\Pi \cup \Pi'$ can be regarded as a reducible curve having ordinary double points arising from these intersections. Denote by $\Pi + \Pi'$ the *disjoint* sum, and by $\iota : \Pi + \Pi' \to \Pi \cup \Pi'$ the canonical covering. Then ι separates all double points of $\Pi \cup \Pi'$. Subcoverings of ι are those which separate *some parts* of them. Let

$$\iota(S^0) : \Pi \underset{S^0}{+} \Pi' \longrightarrow \Pi \cup \Pi' \tag{4.1.4}$$

be that subcovering of ι defined by the condition that the double points of $\Pi \underset{S^0}{+} \Pi'$ are precisely the points lying on S^0. It is obtained by separating precisely those double points of $\Pi \cup \Pi'$ belonging to the *complement of* S^0 in $\Pi \cap \Pi'$. Put

$$X^0 = \Pi \underset{S^0}{+} \Pi', \quad \left\{ \begin{array}{l} \psi = pr_1 \circ \iota(S^0), \\ \psi' = pr_2 \circ \iota(S^0), \end{array} \right. \tag{4.1.5}$$

where $pr_i(i = 1, 2)$ is the projection to the i-th factor of $X \times X'$. Then $\psi : X^0 \to X$, $\psi' : X^0 \to X'$ are surjective morphisms of degree $q + 1$. Note that $\psi|\Pi$ $\psi'|\Pi'$ are isomorphisms and that $\psi|\Pi', \psi'|\Pi$ are purely inseparable with degree q. The system (4.1.1) is thus constructed.

4.2. Now consider an *unramified* covering $f : Y \to X$ of X. Here, Y is an algebraic curve over F_{q^2} assumed to be complete non-singular and irreducible over F_{q^2} (but may not be absolutely irreducible), and f is assumed to be defined over F_{q^2}. We shall be concerned with the following conditions (A) and (A′) on X, S, Y and f.

(A) *All points of Y lying on S are F_{q^2}-rational points.*

If (A) is satisfied, Y has at least one F_{q^2}-rational point and such a point of Y must lie on all absolute irreducible components of Y. But since Y is non-singular, this implies that Y *must be absolutely irreducible.* A slightly weaker condition is the following :

(A′) *Each absolute irreducible component of Y has an F_{q^2}-structure satisfying* (A).

More precisely, let Y_1 be an absolute irreducible component of Y. Then (A′) imposes that there is an algebraic curve Y_2 over F_{q^2} and an isomorphism $\alpha : Y_1 \xrightarrow{\sim} Y_2$ over F such that $f \circ \alpha^{-1}: Y_2 \to X$ satisfies (A). Since all absolute irreducible components of Y are mutually conjugate over F_{q^2}, it is satisfied for *all* components if so for *one* component. If (A′) is satisfied, all absolute irreducible components of Y must be isomorphic (over F). A covering f is called a *constant field extension* if its restriction to each absolute irreducible component is an isomorphism (over F). Such a covering satisfies (A′), and it is essentially the covering of X by $R_{F_{q^{2m}}/F_{q^2}}(X)$, m being the number of absolute irreducible components of Y. The following proposition is an elementary exercise.

PROPOSITION 4.2.1. *The condition* (A′) *is equivalent to the existence of an automorphism ϵ of Y defined over F_{q^2} satisfying the following conditions* : (i) $f \circ \epsilon = f$, (ii) $\epsilon y = y^{q^2}$ *for all* $y \in Y$ *with* $f(y) \in S$.

It is clear that such ϵ is unique and that (A) is equivalent to $\epsilon = 1$.

We shall give some geometric interpretations of the conditions (A), (A′). For this purpose, let S_Y be the set of all F_{q^2}-rational points of Y lying on S;

$$S_Y = \{y \in Y \mid y^{q^2} = y,\ f(y) \in S\}.$$

If S_Y is non-empty, Y must be absolutely irreducible. Define π_Y, π'_Y, Π_Y, Π'_Y, S_Y^0 and the system

$$\left\{ Y \xleftarrow{\psi_Y} Y^0 \xrightarrow{\psi'_Y} Y' \right\} \tag{4.2.2}$$

in the same manner as in §4.1 starting from Y and S_Y. (If Y is not absolutely irreducible and hence without F_{q^2}-rational points, then Π_Y and Π'_Y do not intersect on $Y \times Y'$, and Y^0 is the disjoint sum of these two graphs.) Let $f' : Y' \longrightarrow X'$ be the conjugate of f over F_q. Then $f' \circ \pi_Y = \pi \circ f$ and $f \circ \pi'_Y = \pi' \circ f'$. Therefore, $(f \times f')$: $Y \times Y' \to X \times X'$ maps Π_Y into Π, and Π'_Y into Π', thus inducing the morphisms

$$\left.\begin{array}{ll} f^{01}\colon \Pi_Y \longrightarrow \Pi \quad , & f^{02}\colon \Pi'_Y \longrightarrow \Pi', \\ (y, y^q) \rightsquigarrow (x, x^q) & (y'^q, y') \rightsquigarrow (x'^q, x') \end{array}\right\} \tag{4.2.3}$$

where $x = f(y)$ and $x' = f'(y')$. They are unramified coverings, since f^{01} corresponds to the unramified covering $f : Y \longrightarrow X$ *via* the two isomorphisms $pr_1 : \Pi_Y \xrightarrow{\sim} Y$, $pr_1 : \Pi \xrightarrow{\sim} X$.(Similar for f^{02}.) By the definition of S_Y, the points $(y, y^q) \in \Pi_Y$ and $(y'^q, y') \in \Pi'_Y$ intersect on Y^0 if and only if $y' = y^q$, $y = y'^q$ and $f(y) \in S$. But in this case, $f^{01}((y, y^q))$ and $f^{02}((y'^q, y'))$ intersect on X^0. Therefore, there is a unique morphism, $f^0 : Y^0 \longrightarrow X^0$ such that $f^0|\Pi_Y = f^{01}$ and $f^0|\Pi'_Y = f^{02}$. The diagram

$$\begin{array}{ccccc} & \psi_Y & Y^0 & \psi'_Y & \\ Y & \longleftarrow & & \longrightarrow & Y' \\ f \big\downarrow & & \big\downarrow f^0 & & \big\downarrow f' \\ & & X^0 & & \\ X & \longleftarrow & & \longrightarrow & X' \\ & \psi & & \psi' & \end{array} \tag{4.2.4}$$

is commutative.

Proposition 4.2.5. *The condition* (A) *is equivalent to that all points of Y^0 lying on the double points of X^0 are double points, and equivalent also to that f^0 is étale.*

Proof. The double points of X^0 correspond to the points of S^0. Take $(x, x') \in S^0$ and let $y_1, \ldots, y_m (m = \deg f)$ be the points of Y lying on x. We can assume that $y_1, \ldots, y_r$ are F_{q^2}-rational points and $y_{r+1}, \ldots, y_m$ are not $(0 \leqslant r \leqslant m)$. Then the points of Π_Y (resp. Π'_Y) lying on (x, x') by f^{01} (resp. f^{02}) are (y_i, y_i^q) $(1 \leqslant i \leqslant m)$ (resp. $(y_i, y_i^{q^{-1}})$ $(1 \leqslant i \leqslant m)$). Therefore, the points of Y^0 lying on (x, x') by f are:

(a) *the double points* $(y_i, y_i^q) = (y_i, y_i^{q^{-1}})$ $(1 \leqslant i \leqslant r)$,

(b) *the simple points* (y_j, y_j^q) $(r+1 \leqslant j \leqslant m)$,

(b′) *the simple points* $(y_j, y_j^{q^{-1}})$ $(r+1 \leqslant j \leqslant m)$.

This proves the first equivalence. On the other hand, f^0 is always unramified, and the condition $r = m$ is equivalent to the flatness of f^0; whence the second equivalence. q.e.d.

Lemma 4.2.6. *Let $f: Y \to X$ and $g: Z \to X$ be two unramified coverings of X over F_{q^2} by irreducible complete non-singular curves over F_{q^2}, and $g': Z' \to X'$ be the conjugate of g over F_q. Then*

$$Y \underset{X}{\times} X^0 \simeq X^0 \underset{X'}{\times} Z' \quad \text{over } X^0$$

holds if and only if (i) $Y \simeq Z$ *over* X *and* (ii) f *satisfies* (A′).

Proof. Look at the left half of the diagram (4.2.4). Let $\alpha: Y^0 \to Y \underset{X}{\times} X^0$ be the canonical morphism, i.e., the morphism defined by $pr_1 \circ \alpha = \psi_Y$ and $pr_2 \circ \alpha = f^0$. Then α is an injective isomorphism on each irreducible component Π_Y, Π'_Y of Y^0 and produces *new double points* on the image, by $\alpha((y_j, y_j^q)) = \alpha((y_j, y_j^{q^{-1}}))$ $(r+1 \leqslant j \leqslant m)$ for each $(x, x') \in S^0$. Therefore $Y \underset{X}{\times} X^0$ is the sum of Π_Y and Π'_Y where $(y, y^q) \in \Pi_Y$ and $(y, y^{q^{-1}}) \in \Pi'_Y$ intersect transversally for each $y \in Y$ with $f(y) \in S$. Similarly, $X^0 \underset{X'}{\times} Z'$ is the sum of Π_Z and Π'_Z where $(z, z^q) \in \Pi_Z$ and $(z^{q^2}, z^q) \in \Pi_Z$ intersect transversally for each $z \in Z$ with $g(z) \in S$. Suppose that there is an isomorphism $\delta : Y \underset{X}{\times} X^0 \xrightarrow{\sim} X^0 \underset{X'}{\times} Z'$ over X^0. Then δ induces an isomorphism $\Pi_Y \xrightarrow{\sim} \Pi_Z$ over Π and $\Pi'_{Y'} \xrightarrow{\sim} \Pi'_Z$ over Π'; consequently an isomorphism $\epsilon_1 : Y \xrightarrow{\sim} Z$ over X and $\epsilon'_2 : Y' \xrightarrow{\sim} Z'$ over X'. Let ϵ_2 be the conjugate of ϵ'_2 over F_q. Now take any $y \in Y$ with $f(y) \in S$. Then δ maps (y, y^q) to $(y^{\epsilon_1}, (y^{\epsilon_1})^q)$ and $(y, y^{q^{-1}})$ to $(y^{\epsilon_2}, (y^{\epsilon_2})^{q^{-1}})$. Therefore, $(y^{\epsilon_1}, (y^{\epsilon_1})^q)$ and $(y^{\epsilon_2}, (y^{\epsilon_2})^{q^{-1}})$ must coincide on $X^0 \underset{X'}{\times} Z'$. We obtain $y^{\epsilon_2} = (y^{\epsilon_1})^{q^2}$. Therefore, $\epsilon = \epsilon_2 \epsilon_1^{-1}$ satisfies the conditions (i), (ii) of Prop. 4.2.1; hence f satisfies (A′). Conversely, if there are isomorphisms $\epsilon_2 : Y \xrightarrow{\sim} Z$ over X and $\epsilon : Y \xrightarrow{\sim} Y$ over X, satisfying the conditions (i), (ii) of Prop. 4.2.1, then we can define an isomorphism $\delta : Y \underset{X}{\times} X^0 \xrightarrow{\sim} X^0 \underset{X'}{\times} Z'$ over X^0 by combining $\epsilon_1 = \epsilon^{-1}\epsilon_2$ and ϵ_2 in the above way. (Since the intersections are transversal, the morphisms of two irreducible components give rise to a morphism of the sum if and only if they coincide on the interesecting points.) q.e.d.

COROLLARY. *The condition* (A) *is equivalent to*

$$Y^0 \simeq Y \underset{X}{\times} X^0 \simeq X^0 \underset{X'}{\times} Y' \qquad \textit{over } X^0, \tag{4.2.7}$$

and (A′) *is equivalent to*

$$Y \underset{X}{\times} X^0 \simeq X^0 \underset{X'}{\times} Y' \qquad \textit{over } X^0. \tag{4.2.7′}$$

PROOF. The second assertion follows immediately from the lemma. That (A) implies (4.2.7) is obvious by the proof of the lemma (the explicit constructions of the fiber products). Conversely, since $Y \underset{X}{\times} X^0$ is étale over X^0 (being a base change of an étale X-scheme Y, with X, Y noetherian; or as is clear by its explicit presentation given above), (4.2.7) implies that Y^0 is étale over X^0, which implies (A) by Prop. 4.2.5. q.e.d.

4.3. Now, let K be any finite extension of $K_1 = F_{p^2}(j)$ contained in K_∞ (§1.1), and X be a complete non-singular model of K over F_{p^2}. *Let S be the set of all supersingular prime divisors of K.*

Since the elements of S are of degree one over F_{p^2} (§1.3), they can be regarded as F_{p^2}-rational points of X. Define the system $\left\{X \xleftarrow{\psi} X^0 \xrightarrow{\psi'} X'\right\}$ as in §4.1, starting from these X and S. When $K = K_n$, we shall denote the system by

$$\left\{X_n \xleftarrow{\psi} X_n^0 \xrightarrow{\psi'} X_n'\right\}. \tag{4.3.1}$$

Then by §4.2 and by the Corollary of Lemma 4.2.6, our Main Theorem $[\mathrm{MT}\ 2]_n$ is equivalent to the following:

$(\mathrm{MT}\ 5)_n$. *Let $f: Y \to X_n$ be an unramified covering of X_n over F_{p^2} by a complete non-singular irreducible curve Y over F_{p^2}. Let f' : $Y' \to X_n'$ be the conjugate of f over F_p, and suppose that there is an isomorphism*

$$Y \underset{X_n}{\times} X_n^0 \simeq X_n^0 \underset{X'_n}{\times} Y' \quad \text{over } X_n^0. \tag{4.3.2}$$

Then f must be a constant field extension (i.e., an isomorphism on each absolute irreducible component).

We shall prove our theorem in this form.

Note that if n belongs to the ideal $n_0\mathbf{Z}$ of Prop 1.3.1′, so that all F_{p^2}-rational points of X_n are supersingular, then we have $X_n^0 = \Pi \cup \Pi'$ for the ordinary union taken on $X_n \times X_n'$.

4.4. Supplement. *Étale coverings and simply-connectedness of* $\left\{X \xleftarrow{\psi} X^0 \xrightarrow{\psi'} X'\right\}$.

In general, let $\left\{X \xleftarrow{\psi} X^0 \xrightarrow{\psi'} X'\right\}$ be as in §4.1, and

$$\begin{array}{ccccc} Y & \longleftarrow & W & \longrightarrow & Y' \\ f\big\downarrow & & h\big\downarrow & & f'\big\downarrow \\ X & \underset{\psi}{\longleftarrow} & X^0 & \underset{\psi'}{\longrightarrow} & X' \end{array} \quad ; \ Y, Y' \text{ irreducible over } F_{q^2}; \tag{4.4.1}$$

be a commutative diagram of surjective morphisms of complete algebraic curves, all defined over F_{q^2}. It will be called *an étale covering of degree m of* $\left\{X \overset{\psi}{\longleftarrow} X^0 \overset{\psi'}{\longrightarrow} X'\right\}$, if the three vertical morphisms f, f', h are étale coverings (finite étale morphisms) of the same well-defined degree m. Since X, X' are non-singular, f, f' are étale if and only if they are unramified. If (4.4.1) is étale, the canonical morphisms $W \to Y \underset{X}{\times} X^0$ and $W \to X^0 \underset{X'}{\times} Y'$ must be isomorphisms over X^0, so that

$$Y \underset{X}{\times} X^0 \simeq X^0 \underset{X'}{\times} Y' \quad \text{over } X^0. \tag{4.4.2}$$

Conversely, if f, f' are unramified coverings of X, X' with which (4.4.2) holds, then we can construct an étale covering (4.4.1) by putting $W = Y \underset{X}{\times} X^0$. (Equivalence of two étale coverings of $\{X \overset{\psi}{\longleftarrow} X^0 \overset{\psi'}{\longrightarrow} X'\}$ is defined in the natural way, so that the choice of isomorphisms of (4.4.2) does not affect the equivalence class; see §3.1 for a similar argument. Note also that the linear disjointness condition, the condition (b) in §3.1, is automatically satisfied for the étale coverings of $\{X \overset{\psi}{\longleftarrow} X^0 \overset{\psi'}{\longrightarrow} X'\}$). Thus, up to equivalence, giving an étale covering of $\{X \overset{\psi}{\longleftarrow} X^0 \overset{\psi'}{\longrightarrow} X'\}$ is the same thing as giving a pair of unramified coverings $\{f, f'\}$ satisfying (4.4.2). In view of Lemma 4.2.6, we can assume that f': $Y' \to X'$ is the conjugate of f over F_q. Therefore, *étale coverings of* $\{X \overset{\psi}{\longleftarrow} X^0 \overset{\psi'}{\longrightarrow} X'\}$ *and unramified coverings of X satisfying* (A′) *are equivalent notions* (there is a categorical equivalence between them). Each étale covering (4.4.1) defines the automorphism ϵ of Y over X, which may be called *the twist of* this covering. Then, *étale coverings of* $\{X \overset{\psi}{\longleftarrow} X^0 \overset{\psi'}{\longrightarrow} X'\}$ *without twist and unramified coverings of X satisfying* (A) *are also the equivalent notions.*

One may define the *simply-connectedness of* $\{X \overset{\psi}{\longleftarrow} X^0 \overset{\psi'}{\longrightarrow} X'\}$ as follows, that it is simply-connected if the only étale coverings are those with which the vertical coverings are isomorphisms on each absolute irreducible component. In these terms, our Main Theorem can also be stated as:

[MT 6]$_n$. *The system* $\{X_n \xleftarrow{\psi} X_n^0 \xrightarrow{\psi'} X_n'\}$ *is simply-connected.*

5. The congruence relation. In this section we shall present and prove one natural version of the congruence relation.†

5.1. Let p be the fixed prime number, and n be a positive integer not divisible by p. Let $\mathscr{W}_n$ be the group of complex n-th roots of unity, put $\mathscr{W}_\infty = \bigcup_{p \nmid n} \mathscr{W}_n$, and consider the cyclotomic extension $\mathscr{F} = \mathbf{Q}(\mathscr{W}_\infty)$ over the rational number field $\mathbf{Q}$. Let $\mathscr{F}_p$ be the decomposition field of p in $\mathscr{F}$. Then $\mathscr{F} = \mathscr{F}_p(\mathscr{W}_\infty)$, and the field $\mathscr{F}_p$ satisfies the condition of § 1.1 for F_p. Let $\mathscr{F}_{pm}$ $(m \geqslant 1)$ be the unique extension of $\mathscr{F}_p$ of degree m in $\mathscr{F}$. Let

$$\omega^{\mathbf{C}} : \mathscr{W}_\infty \xrightarrow{\sim} \varinjlim \mathbf{Z}/n \tag{5.1.1}$$

be *the standard isomorphism* that maps $\exp(2\pi ia/n)$ to $a(\mathrm{mod}\ n)$ for each a, n.

On the other hand, let F_p be the prime field of characteristic p, F be its algebraic closure, W_n be the group of n-th roots of unity in F, and $W_\infty = \bigcup_{p \nmid n} W_n$, so that $F = F_p(W_\infty)$ (and in fact $W_\infty = F^\times$). Then each place $\mathscr{F} \to F$ maps $\mathscr{W}_\infty$ isomorphically onto W_∞, and there are precisely as many places $\mathscr{F} \to F$ as elements of the Galois group $G(\mathscr{F}/\mathbf{Q})$ (and as many equivalence classes of places as elements of $G(\mathscr{F}_p/\mathbf{Q})$). Since the canonical homomorphism $G(\mathscr{F}/\mathbf{Q}) \to \mathrm{Aut}\ \mathscr{W}_\infty$ is a surjective isomorphism, we see that there are precisely as many places $\mathscr{F} \to F$ as isomorphisms $\mathscr{W}_\infty \simeq W_\infty$. Each isomorphism $\mathscr{W}_\infty \xrightarrow{\sim} W_\infty$ determines *via* $\omega_{\mathbf{C}}$ an isomorphism $W_\infty \xrightarrow{\sim} \varinjlim \mathbf{Z}/n$, so that places $\mathscr{F} \to F$ and isomorphisms $W_\infty \xrightarrow{\sim} \varinjlim \mathbf{Z}/n$ are in a canonical one-to-one correspondence. We shall fix

$$\omega : W_\infty \xrightarrow{\sim} \varinjlim \mathbf{Z}/n, \tag{5.1.2}$$

† cf. the previous versions [3], [17], [6], [1], [18].

and the corresponding place

$$pl : \mathscr{F} \to F. \tag{5.1.3}$$

Let $\mathfrak{z}$ be the valuation ring of (5.1.3) and put $\mathfrak{z}_{p^m} = \mathfrak{z} \cap \mathscr{F}_{p^m}$.

Now let Γ be any subgroup of $\Gamma_1 = PSL_2(\mathbf{Z}^{(p)})$ with finite index, and $\{\Delta, \Delta'\}$ be the canonical generating pair of fuchsian groups for Γ (§ 3.5), i.e.,

$$\Delta = \Gamma \cap PSL_2(\mathbf{Z}), \quad \Delta' = \Gamma \cap \begin{pmatrix} p & 0 \\ 0 & 1 \end{pmatrix}^{-1} PSL_2(\mathbf{Z}) \begin{pmatrix} p & 0 \\ 0 & 1 \end{pmatrix} \tag{5.1.4}$$

Put $\Delta^0 = \Delta \cap \Delta'$, and let $\{R \xleftarrow{\varphi} R^0 \xrightarrow{\varphi'} R'\}$ be the system of compact Riemann surfaces corresponding to $\{\Delta \leftarrow \Delta^0 \to \Delta'\}$. On the other hand, let K be the finite extension of $K_1 = F_{p^2}(j)$ contained in K_∞ that corresponds to Γ by an injective isomorphism $\iota : \Gamma_1 \hookrightarrow G(K_\infty/K_1)$ belonging to ω (see § 1.1; recall that ι is determined by ω up to inner automorphisms of $G(K_\infty/K_1)$, so that K is determined up to isomorphisms over K_1). Take a complete non-singular model X of K, let X' be its conjugate over F_p, and let X^0 be the sum of Π and Π' *with precisely the supersingular crossings*, where Π, Π' are the graphs on $X \times X'$ of the p-th power morphisms $X \to X'$, $X' \to X$, respectively (see § 4.1, § 4.3). Let $\{X \xleftarrow{\psi} X^0 \xrightarrow{\psi'} X'\}$ be the system of F_{p^2}-curves thus defined. If $\Gamma = \Gamma_n$, the principal congruence subgroup of level n, then the systems $\{R \xleftarrow{\varphi} R^0 \xrightarrow{\varphi'} R'\}$ and $\{X \xleftarrow{\psi} X^0 \xrightarrow{\psi'} X'\}$ are nothing but $\{R_n \xleftarrow{\varphi} R_n^0 \xrightarrow{\varphi'} R_n'\}$ and $\{X_n \xleftarrow{\psi} X_n^0 \xrightarrow{\psi'} X_n'\}$ defined in § 3.2 and § 4.3, respectively.

THEOREM (CONGRUENCE RELATION). *There exists a system* $\left\{\mathscr{X} \xleftarrow{\Psi} \mathscr{X}^0 \xrightarrow{\Psi'} \mathscr{X}'\right\}$ *of* $\mathfrak{z}_{p^2}$*-schemes which are proper over* $\mathfrak{z}_{p^2}$, *and* $\mathfrak{z}_{p^2}$*-morphisms* Ψ, Ψ', *such that*

$$\left\{\mathscr{X} \xleftarrow{\Psi} \mathscr{X}^0 \xrightarrow{\Psi'} \mathscr{X}'\right\} \underset{\mathfrak{z}_{p^2}}{\otimes} \mathbf{C} \simeq \left\{R \xleftarrow{\varphi} R^0 \xrightarrow{\varphi'} R'\right\}, \tag{5.1.5 C}$$

(*as systems of complex algebraic curves*),

and

$$\left\{\mathscr{X} \xleftarrow{\Psi} \mathscr{X}^0 \xrightarrow{\Psi'} \mathscr{X}'\right\} \underset{\mathfrak{z}_{p^2}}{\otimes} F_{p^2} \simeq \left\{X \xleftarrow{\psi} X^0 \xrightarrow{\psi'} X'\right\}, \qquad (5.1.5\ p)$$

where the first tensoring is with respect to the identity embedding $\mathfrak{z}_{p^2} \hookrightarrow \mathscr{F} \hookrightarrow \mathbf{C}$, *and the second tensoring is with respect to the place* (5.1.3) *that corresponds to the isomorphism* (5.1.2) *used to define* X.

Here, we listed only those properties of the schemes that are necessary for our purpose. By our construction, the schemes $\mathscr{X}, \mathscr{X}^0$, $\mathscr{X}'$ are also *normal*, and $\mathscr{X}$, $\mathscr{X}'$ are mutually conjugate over $\mathfrak{z}_p$.

An isomorphism between two systems is by definition a triple of isomorphisms between curves commuting with the four given morphisms.

We shall write down the proof for the principal congruence subgroups Γ_n. Then, the other cases follow directly, only by checking a number of compatibilities. (And we need use the theorem only for $\Gamma = \Gamma_n$.)

5.2. In this section, we shall construct the function fields with p-adic valuations and verify the "generic" congruence relation. The proof is obtained by Shimura's method [17]. Owing to a result of Igusa [6], it works for *all* p with $p \nmid n$. The only new point to be noted here is that the action of $p^{\mathbf{Z}}$ is trivialized. In any case, we shall present a full proof for the sake of completeness.

Let j be a variable over F_p, the prime field of characteristic p, and J be a variable over $\mathscr{F}_p$. Let K_n, L_n be the fields defined in §1.1 with respect to F_p and j; and $\mathscr{K}_n$, $\mathscr{L}_n$ be the corresponding fields defined with respect to $\mathscr{F}_p$ and J.

(A) The relation between the two characteristics is as follows. Extend the place (5.1.3) to the place of $\mathscr{F}(J)$ onto $F(j)$ by $J \rightsquigarrow j$, and call it also as pl.

$$pl: \mathscr{F}(J) \longrightarrow F(J) \qquad (J \rightsquigarrow j). \qquad (5.2.1)$$

Let pl_n be an extension of pl to $\mathscr{L}_n$. *Then* pl_n *maps* $\mathscr{L}_n$ *onto* L_n. This is due to Igusa [5], [6]. In fact, this follows directly from the following two facts, that $[\mathscr{L}_n : \mathscr{F}(J)] = [L_n : F(j)]$ (due to Prop. 1.1.2) and that there is an elliptic curve over $\mathscr{F}_p(J)$ with absolute invariant J having a good reduction at (5.2.1) (e.g., the curve defined by $(T)_J$, §1.1). Therefore, the discrete valuation of $\mathscr{F}(J)$ corresponding to (5.2.1) has a unique and unramified extension to $\mathscr{L}_n$, and pl_n is unique up to automorphisms of L_n over $F_p(J)$. By [5], [6], $\mathscr{L}_n$ and L_n have the same genus. It is easy to check that *pl_n maps $\mathscr{K}_n$ onto K_n.* (Just go back to the definitions of $\mathscr{K}_n$, K_n, and use the curves $(T)_J$, $(T)_j$ to compare the two characteristics.)

(B) Fix an algebraic closure Ω of $\mathscr{F}(J)$. Let $J' \in \Omega$ be a modular transform of degree p of J, i.e., a root of the absolute modular equation $\Phi_p(X, J) = 0$. Then J' is of degree $p+1$ over $\mathbf{Q}(J)$ and also over $\mathscr{L}_\infty$. Let $\mathscr{K}'_n$ be the transform of $\mathscr{K}_n$ by an isomorphism $J \rightsquigarrow J'$ over $\mathscr{F}_p$. Then

$$\mathscr{K}_n(J') = \mathscr{K}'_n(J), \tag{5.2.2}$$

so that we have a diagram

$$\begin{array}{ccccc} \mathscr{K}_n & \text{———} & \mathscr{K}^0_n & \text{———} & \mathscr{K}'_n \\ | & & | & & | \\ \mathscr{F}_p(J) & \text{———} & \mathscr{F}_p(J, J') & \text{———} & \mathscr{F}_p(J') \end{array} \qquad \mathscr{K}^0_n = \mathscr{K}_n(J') = \mathscr{K}'_n(J). \tag{5.2.3}$$

The proof of (5.2.2) proceeds as follows. Let E be an elliptic curve over $\mathscr{F}_p(J)$ with absolute invariant J, E' be its transform by $J \rightsquigarrow J'$ over $\mathscr{F}_p$, and $\lambda : E \to E'$ be an isogeny of degree p. Then λ is unique up to the sign. Hence an automorphism τ of Ω over $\mathscr{F}_p(J, J')$ can change λ only up to the sign. Therefore, $\tau(\lambda(u)) = \pm\lambda(\tau(u))$ $(u \in E_\infty)$ holds for such τ, and the equality (5.2.2) is reduced to the definitions of $\mathscr{K}_n$, $\mathscr{K}'_n$. We denote by $\mathscr{L}'_n$ the transform of $\mathscr{L}_n$ by $J \rightsquigarrow J'$ over $\mathscr{F}_p$, so that $\mathscr{L}'_n = \mathscr{K}'_n \cdot F$, and $\mathscr{L}_n(J') = \mathscr{L}'_n(J)$. Put

$$\mathscr{L}^0_n = \mathscr{L}_n(J') = \mathscr{L}'_n(J).$$

(C) Let E, E', $\lambda : E \to E'$ be as in (B). *Then there is a unique automorphism σ of $\mathscr{L}^0_\infty$ over $\mathscr{F}_p$ satisfying the following properties* (i), (ii), (iii);

(i) $\sigma|_{\mathscr{F}}$ is the Frobenius automorphism over $\mathscr{F}_p$;

(ii) $\sigma(J) = J'$, $\sigma(J') = J$.

(iii) if $\tilde{\sigma}$ is an extension of σ to Ω, $\tilde{\sigma}$ acts as $\pm\lambda$ on E_∞.

It is clear that such σ is at most unique. It is also easy to check that the condition (iii) does not depend on the choice of E and λ. To show the existence, observe first that there exists an automorphism $\tilde{\sigma}$ of Ω satisfying (i), (ii) (symmetricity of the modular equation). Then $E^{\tilde{\sigma}} = E'$; hence $\tilde{\sigma}$ maps E_∞ onto E'_∞. Consider the automorphism $u \rightsquigarrow \tilde{\sigma}^{-1}(\lambda u)$ of E_∞. Its determinant is unity by (i) and by $\deg \lambda = p$. Therefore, by Prop. 1.1.2, we can find an automorphism τ of Ω over $\mathscr{F}(J)$ acting on E_∞ as $\pm\tilde{\sigma}^{-1}\circ\lambda$. Moreover, since $\mathscr{L}_\infty$ and J' are linearly disjoint over $\mathscr{F}(J)$, we can assume that τ is trivial on $\mathscr{F}(J, J')$. Replacing $\tilde{\sigma}$ by $\tilde{\sigma}\tau$, we obtain $\tilde{\sigma} = \pm\lambda$ on E_∞, as desired.

Let σ be as above. Then σ *maps* $\mathscr{K}_n$ *onto* $\mathscr{K}'_n$, $\mathscr{K}'_n$ *onto* $\mathscr{K}_n$, *and* σ^2 *is the identity on* $\mathscr{K}^0_n$. In fact, the first two statements are trivial consequences of (ii). Let $\lambda^{\tilde{\sigma}} : E' \to E$ be the transform of λ by $\tilde{\sigma}$. Then $\lambda^{\tilde{\sigma}}\circ\lambda$ is an endomorphism of E of degree p^2, which must be $\pm p$ (since J is a variable). Therefore, by (iii), σ^2 acts as $\pm p$ on E_∞, i.e., the identity on $\mathscr{K}_n$. Thus, σ induces an involution of $\mathscr{K}^0_n$.

(D) Proposition 5.2.4. *Let σ be the involution of $\mathscr{K}^0_n$ defined above. Then we have the following diagram of discrete valuations of the corresponding fields on the right*:

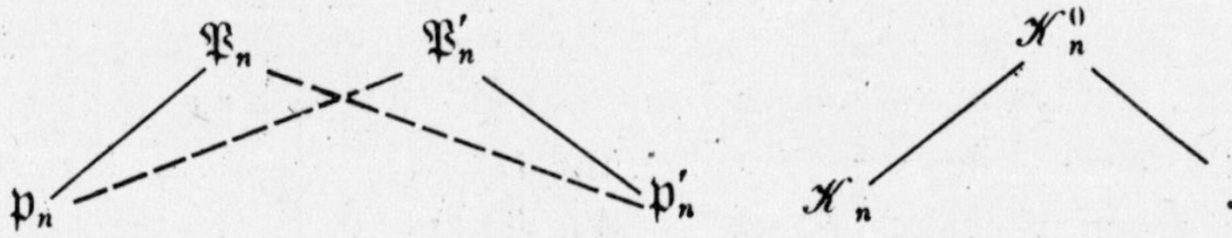

(5.2.5)

where (i) $\mathfrak{p}_n$ *is the discrete valuation of* $\mathscr{K}_n$ *corresponding to the place* pl_n *(see* (A) *), and* $\mathfrak{p}'_n$ *is its conjugate by* σ; (ii) $\mathfrak{P}_n$, $\mathfrak{P}'_n$ *are mutually*

σ-conjugate distinct valuations of $\mathscr{K}_n^0$ each lying on $\mathfrak{p}_n$ and $\mathfrak{p}_n'$; moreover, $\mathfrak{P}_n$, $\mathfrak{P}_n'$ are the only extensions of $\mathfrak{p}_n$ to $\mathscr{K}_n^0$, and also of $\mathfrak{p}_n'$ to $\mathscr{K}_n^0$; (iii) *if x, x' are mutually σ-conjugate elements of $\mathscr{K}_n$, $\mathscr{K}_n'$ that are integral with respect to $\mathfrak{p}_n$, $\mathfrak{p}_n'$, then*

$$x' \equiv x^p \ (\mathrm{mod}\ \mathfrak{P}_n), \qquad x \equiv x'^p \ (\mathrm{mod}\ \mathfrak{P}_n'). \tag{5.2.6}$$

PROOF. Let $\mathfrak{P}$ be any extension of $\mathfrak{p}_n$ to Ω. By the Kronecker congruence relation for the absolute modular equation $\Phi_p(X, J) = 0$, we have $J' \equiv J^{p\pm 1}$ (mod $\mathfrak{P}$) for one choice of the sign. In particular, J' (mod $\mathfrak{P}$) is transcendental over F_p, so that $\mathfrak{P}$ lies on $\mathfrak{p}_n'$. Replacing $\mathfrak{P}$ by $\mathfrak{P}^{\bar\sigma}$ if necessary, we may assume $J' \equiv J^p$ (mod $\mathfrak{P}$). Consider the restriction of σ to $\mathscr{L}_n$, which maps $\mathscr{L}_n$ isomorphically onto $\mathscr{L}_n'$. The restriction of $\mathfrak{P}$ to $\mathscr{L}_n$ is the unique extension of $\mathfrak{p}_n$ to $\mathscr{L}_n$, and the restriction of $\mathfrak{P}$ to $\mathscr{L}_n'$ is the unique extension of $\mathfrak{p}_n'$ to $\mathscr{L}_n'$. Hence $\sigma|_{\mathscr{L}_n}$ leaves $\mathfrak{P}$ invariant. Let E have a good reduction at (5.2.1). Then since $J' \equiv J^p$ (mod $\mathfrak{P}$), λ (mod $\mathfrak{P}$) must be the p-th power map up to the sign. Therefore, by the property (iii) of σ, $\sigma_{\mathscr{L}_n}$ induces the p-th power map of the residue field of $\mathscr{L}_n$ mod $\mathfrak{P}$. (It induces the p-th power map on the base field, i.e., $\mathscr{F}_p(J)$ mod $\mathfrak{P}$, and on the generators of $\mathscr{L}_n$ mod $\mathfrak{P}$ over the base field.) This proves the first congruence of (5.2.6) for $\mathfrak{P} = \mathfrak{P}_n$. It also shows that the residue field of $\mathscr{K}_n'$ mod $\mathfrak{P}_n$ is the p-th power of that of $\mathscr{K}_n$ mod $\mathfrak{P}_n$. Hence the residue extension degree of $\mathfrak{P}_n/\mathfrak{p}_n'$ must be a multiple of p. But since the sum of the products of ramification indices and the residue extension degrees for all extensions of $\mathfrak{p}_n'$ to $\mathscr{K}_n^0$ must be equal to $[\mathscr{K}_n^0 : \mathscr{K}_n'] = p+1$, it follows immediately that the residue extension degree of $\mathfrak{P}_n/\mathfrak{p}_n'$ must be precisely p, that the ramification index is one, and that there is precisely one more extension of $\mathfrak{p}_n'$ to $\mathscr{K}_n^0$. The rest follows immediately by using the transform $\mathfrak{P}_n' = \mathfrak{P}_n^\sigma$.

COROLLARY 1. *The ramification index equals* 1 *for all of the four extensions $\mathfrak{P}_n/\mathfrak{p}_n$, $\mathfrak{P}_n'/\mathfrak{p}_n'$; $\mathfrak{P}_n'/\mathfrak{p}_n$, $\mathfrak{P}_n/\mathfrak{p}_n'$; the residue extension degree equals* 1 *for the first two and equals p for the latter two.*

COROLLARY 2. *Let I, I' be the valuation rings of $\mathfrak{p}_n$, $\mathfrak{p}_n'$, respectively, and let I^0 be the intersection of the valuation rings of $\mathfrak{P}_n$ and of $\mathfrak{P}_n'$.*

Then (i) I^0 *is the integral closure of* I, *and also of* I', *in* $\mathscr{K}_n^0$. (ii) I^0 *is generated by* I *and* I' ;

$$I^0 = I.\,I'. \tag{5.2.7}$$

PROOF. (i) is obvious. To check (ii), it suffices to show that $I^0 = I\,[J']$. Since $\mathscr{K}_n^0 = \mathscr{K}_n\,[J']$, each $z \in \mathscr{K}_n^0$ can be expressed as $z = \sum_{i=0}^{p} a_i J'^i$ $(a_i \in \mathscr{K}_n)$. It suffices to show that "$a_i \in I$ $(0 \leqslant i \leqslant p)$ *and* $z \in pI^0$ *imply* $a_i \in pI$ $(0 \leqslant i \leqslant p)$". Suppose $a_i \in I(0 \leqslant i \leqslant p)$ and $z \in pI^0$. Then $z \equiv 0 \pmod{\mathfrak{P}_n}$ gives $\sum_i \bar{a}_i j^{pi} = 0$, and $z \equiv 0 \pmod{\mathfrak{P}'_n}$ gives $\sum_i \bar{a}_i^p j^i = 0$, where $\bar{a}_i$ is an element of K_n reduced from a_i. Since $1, j, \ldots, j^{p-1}$ are linearly independent over K_n^p, we obtain $\bar{a}_i = 0$ $(1 \leqslant i \leqslant p-1)$ and $\bar{a}_0 + \bar{a}_p j^{p^2} = \bar{a}_0 + \bar{a}_p j = 0$. But j being transcendental over F_p, we have $j^{p^2} \neq j$; hence also $\bar{a}_0 = \bar{a}_p = 0$.

(E) PROPOSITION 5.2.8. *Extend the constant field* $\mathscr{F}_p$ *of* (5.2.3) *to* $\mathbf{C}$ *with respect to the identity embedding* $\mathscr{F}_p \hookrightarrow \mathbf{C}$. *Let* $J(z)$ *be the absolute modular function* (*so normalized as* $J(\sqrt{-1}) = 1728$). *Then the fields* (5.2.3) *over* $\mathbf{C}$ *are isomorphically mapped over* $(J, J') \rightsquigarrow (J(z), J(pz))$ *onto the modular function fields corresponding to the following system of fuchsian groups*;

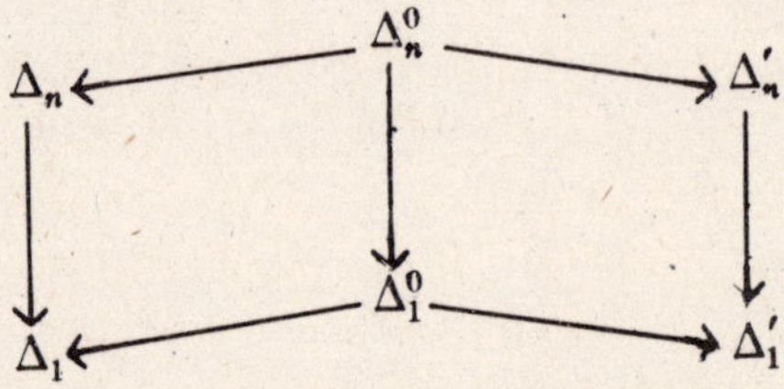

This is obvious.

5.3. *Preliminaries for the construction of schemes.*

(A) In general, let $\mathscr{K}/k$ be an algebraic function field of one variable with exact constant field k. *We assume that k is perfect.* Let $g = g(\mathscr{K}/k)$ be the genus. If D is any divisor of $\mathscr{K}/k$, $L(D)$ will denote the k-module of all multiples of D^{-1} in $\mathscr{K}$. Then the Riemann-Roch theorem gives rank $L(D) = \deg D - g + 1$ for $\deg D > 2g - 2$. By a theorem of Mumford [13] Theorem 6, we have

$$L(DD') = L(D)\, L(D') \quad \text{for } \deg D > 2g - 1,\ \deg D' > 2g. \tag{5.3.1}$$

(We shall only need a far weaker result, that (5.3.1) holds for deg D and deg D' greater than some absolute constant depending only on the genus.) For each finite non-empty set T of prime divisors of $\mathscr{K}/k$ put $\mathfrak{O}^{(T)} = \bigcap_{P \notin T} \mathfrak{O}_P$ where P runs over all prime divisors of $\mathscr{K}/k$ not contained in T and $\mathfrak{O}_P$ is the valuation ring of P. Then $\mathfrak{O}^{(T)} = \bigcup_{\nu \geq 1} L(D^\nu)$ for $D = \prod_{P \in T} P$.

(B) Let v be a non-trivial discrete valuation of k, and V be some extension of v to $\mathscr{K}$. Let i and I denote the valuation rings of v and V respectively, and for each subset $*$ of $\mathscr{K}$ denote by $\bar{*}$ the residue class of $(I \cap *)$ mod V. *Assume that $\bar{k}$ is also perfect.* After Lamprecht [11]†, V is called *regular* if the following three conditions are satisfied; (i) the value groups of $\mathscr{K}^\times$ and of $k^\times$ coincide; (ii) $\bar{\mathscr{K}}/\bar{k}$ is also an algebraic function field of one variable with exact constant field $\bar{k}$; (iii) $g(\mathscr{K}/k) = g(\bar{\mathscr{K}}/\bar{k})$. If V satisfies (i) and $\mathfrak{M}$ is any k-module in $\mathscr{K}$ of a finite rank d, then $\bar{\mathfrak{M}}$ is also a finite $\bar{k}$-module of the same rank d (and $I \cap \mathfrak{M}$ is a finite free i-module of rank d); see [11]. Suppose now that V is regular. Then by [11], there is a unique degree preserving homomorphism $D \rightsquigarrow \bar{D}$ of the group of divisors of $\mathscr{K}/k$ into that of $\bar{\mathscr{K}}/\bar{k}$, with the properties that $D > 1$ implies $\bar{D} > 1$ and that $D = (x)$ with $x \in \mathscr{K}^\times$, $\bar{x} \neq 0, \infty$ implies $\bar{D} = (\bar{x})$. Since $\overline{L(D)} \subset L(\bar{D})$ and rank $\overline{L(D)}$ = rank $L(D)$ are valid for any D, we obtain

$$\overline{L(D)} = L(\bar{D}) \text{ for } \deg D > 2g - 2. \tag{5.3.2}$$

By using (5.3.1) for $\bar{D}$, $\bar{D}'$, and (5.3.2) for DD' (and elementary divisor theory for finite i-modules), we obtain immediately

$$I \cap L(DD') = (I \cap L(D))\,(I \cap L(D')) \tag{5.3.3}$$

† [11] weakened the assumptions in the Deuring theory [2].

for deg $D > 2g-1$, deg $D' > 2g$. Now, for each finite non-empty set T of prime divisors of $\mathcal{K}/k$, put $I^{(T)} = I \cap \mathfrak{O}^{(T)}$. Then $I^{(T \cap T')} = I^{(T)} \cap I^{(T')}$ holds trivially, and

$$I^{(T \cup T')} = I^{(T)} \cdot I^{(T')} \tag{5.3.4}$$

holds due to (5.3.3). For each T, let $\bar{T}$ denote the collection of all prime factors of $\bar{P}$, where P runs over all elements of T. Then (5.3.2) gives

$$\overline{I^{(T)}} = \mathfrak{O}^{(\bar{T})}. \tag{5.3.5}$$

By gluing together the normal affine schemes Spec $I^{(T)}$ for all T in the natural way, we obtain a prescheme $\mathfrak{X}$ which is in fact a scheme due to (5.3.4). By the identity injection $i \hookrightarrow I^{(T)}$, $\mathfrak{X}$ may be regarded as an i-scheme. It is proper and smooth over i. That it is proper can be checked easily by the valuative criterion. (Use the following fact due to Lamprecht [11]. If V is regular, there exists an element $t \in \mathcal{K}$ with the properties that $\mathcal{K}/k(t)$ is a finite separable extension, that $\bar{t}$ is transcendental over $\bar{k}$ and that V is the unique extension of $V|_{k(t)}$ to $\mathcal{K}$.) The generic and the special fibers $\mathfrak{X} \otimes_i k$ and $\mathfrak{X} \otimes_i \bar{k}$ are proper smooth curves over k and $\bar{k}$ with function fields $\mathcal{K}$ and $\bar{\mathcal{K}}$ respectively.

(C) Now let $\mathcal{K}^0$ be a finite separable extension of $\mathcal{K}$ without constant field extensions. Let I^0 be the integral closure of I in $\mathcal{K}^0$. If $V_1, \ldots, V_s$ are all the distinct extensions of V to $\mathcal{K}^0$, I^0 is the intersection of the valuation rings of $V_1, \ldots, V_s$. For each finite non-empty set T of prime divisors of $\mathcal{K}$, put $I^{0(T)} = I^0 \cap \mathfrak{O}^{(T^0)}$, where T^0 is the set of all extensions of prime divisors of $\mathcal{K}/k$ belonging to T. Then $I^{0(T)}$ is the integral closure of $I^{(T)}$ in $\mathcal{K}^0$. Since $I^{(T)}$ is noetherian and $\mathcal{K}^0/\mathcal{K}$ is a finite separable extension, $I^{0(T)}$ is a finite $I^{(T)}$-module. Let $\mathfrak{X}^0$ be the prescheme obtained by gluing together the normal affine schemes Spec $I^{0(T)}$ (for all finite non-empty sets T of prime divisors of $\mathcal{K}/k$) in the natural way. Then the canonical morphism $\Psi : \mathfrak{X}^0 \to \mathfrak{X}$ is *finite*, so that $\mathfrak{X}^0$ is also a proper i-scheme called the integral closure of $\mathfrak{X}$ in $\mathcal{K}^0$.

(D) Now let e_i and f_i be the ramification index and the residue extension degree of V_i/V, respectively. Then since $\mathcal{K}^0/\mathcal{K}$ is a finite separable extension we have

$$\sum_{i=1}^{s} e_i f_i = [\mathscr{K}^0 : \mathscr{K}]. \tag{5.3.6}$$

Let K_i be the residue field of $\mathscr{K}^0$ with respect to V_i, and k_i be the exact constant field of K_i (which is a finite extension of $\bar{k}$). Put $r_i = [k_i : \bar{k}]$, and $g = g(\mathscr{K}^0/k), g_i = g(K_i/k_i)$. Then by Nering [14]†, (or by the "invariance of Euler-Poincaré characteristic"; Grothendieck [4], Ch. III, Theorem 7.9.4), it holds that

$$g - 1 = \sum_{i=1}^{s} r_i e_i (g_i - 1) + \rho \tag{5.3.7}$$

where 2ρ is the degree of the conductor (defined in [14]). (Nering's assumptions are satisfied in view of the existence of such an element t of $\mathscr{K}$ as noted in (B), and of the equality (5.3.6) which implies that the radical degree is one.)

5.4. *The system* $\left\{\mathfrak{X}_n \xleftarrow{\Psi} \mathfrak{X}_n^0 \xrightarrow{\Psi'} \mathfrak{X}_n'\right\}$. Let n be any positive integer with $p \nmid n$, and $\mathscr{K}_n, \mathscr{K}_n', \mathscr{K}_n^0$ be as in § 5.2. They are algebraic function fields of one variable with exact constant field $k = \mathscr{F}_{p^2}$. Let $\mathfrak{p}_n, \mathfrak{p}_n'; \mathfrak{P}_n, \mathfrak{P}_n'$ be the discrete valuations defined in Prop. 5.2.4. The valuation rings of $\mathfrak{p}_n, \mathfrak{p}_n'$ are denoted by I, I', and the intersection of the valuation rings of $\mathfrak{P}_n$ and of $\mathfrak{P}_n'$ is denoted by I^0. Put $i = I \cap k = \mathfrak{z}_{p^2}$. Then, since $\mathfrak{p}_n$ is regular by § 5.2 (A), we can construct a standard i-scheme $\mathfrak{X}_n$ as in § 5.3 (B). Its conjugate $\mathfrak{X}_n'$ is defined in the same way. The integral closures of $\mathfrak{X}_n$ and of $\mathfrak{X}_n'$ in $\mathscr{K}_n^0$ coincide, as they are normal, proper over i, and have the same set of local rings for points of codimension one. Call this integral closure $\mathfrak{X}_n^0$. Let $\Psi : \mathfrak{X}_n^0 \to \mathfrak{X}_n$, $\Psi' : \mathfrak{X}_n^0 \to \mathfrak{X}_n'$ be the canonical morphisms. We shall show that the system $\left\{\mathfrak{X}_n \xleftarrow{\Psi} \mathfrak{X}_n^0 \xrightarrow{\Psi'} \mathfrak{X}_n'\right\}$ thus constructed satisfies the required properties. The schemes are proper over $\mathfrak{z}_{p^2}$ by § 5.3 (B), (C). The isomorphism (5.1.5 C) follows immediately from Prop. 5.2.8. As for the isomorphism (5.1.5 p), first, we can identify $\mathfrak{X}_n \otimes F_{p^2} = X_n, \mathfrak{X}_n' \otimes F_{p^2} = X_n'$. Then, a direct geometric translation of Prop. 5.2.4 and its Cor. 2 gives that the canonical covering $\Pi + \Pi' \to \Pi \cup \Pi'$ is decomposed as:

$$\Pi + \Pi' \longrightarrow \mathfrak{X}_n^0 \otimes F_{p^2} \longrightarrow \Pi \cup \Pi'. \tag{5.4.1}$$

† cf. Popp [20], which contains corrections of mistakes in [14].

To complete the proof of our congruence relation, *it remains to check that the double points of $\mathfrak{X}_n^0 \otimes F_{p^2}$ are precisely at the supersingular points.* This can be proved in at least two different ways. One is obtained by looking carefully at the canonical one-to-one correspondence between the *non*-supersingular F_{p^2}-rational points of X_n and the double points of the image of R_n^0 in $R_n \times R_n'$; and by refining Cor. 2 of Prop. 5.2.4. We shall apply an indirect but shorter proof, using the following

LEMMA 5.4.2. *Let g_n, g_n^0 be the genus of R_n, R_n^0 respectively, and h_n be the number of supersingular prime divisors of K_n. Then*

$$g_n^0 - 1 = 2\,(g_n - 1) + h_n.$$

PROOF. For $n = 1$, so that $g_1 = 0$, the equality $g_1^0 = h_1 - 1$ is well known, and can be checked immediately by comparing the formula for h_1 (quoted in (1.3.2)) with the (well-known) genus formula for "$\Gamma_0(p)$". For $n > 1$, the Hurwitz formula gives

$$g_n^0 - 1 = (p+1)\,(g_n - 1) + \tfrac{1}{2}(p-1)\,\mu_n, \tag{5.4.3}$$

μ_n being the number of cusps of R_n. In fact, on each cusp of R_n, there are precisely *two* cusps of R_n^0, one unramified and the other ramified with index p. (To check this, one needs only note that the largest normal subgroup of Δ_n contained in Δ_n^0 is the principal congruence subgroup of level np, denoted by Δ_{np}, and that a parabolic element of Δ_n is contained in Δ_{np} if and only if it is a p-th power in Δ_n.) Since there are no other ramifications in the covering $\varphi: R_n^0 \to R_n$ for $n>1$, we obtain (5.4.3) by Hurwitz.

Now, (5.4.3) and $(1.3.2)_n$ give

$$\begin{aligned} g_n^0 - 2g_n + 1 = (p-1)\,(g_n - 1 + \tfrac{1}{2}\mu_n) &= \frac{1}{12}\,(p-1)\,(\Delta_1 : \Delta_n) \\ &= \frac{1}{12}\,(p-1)[K_n : K_1] = h_n. \end{aligned}$$

This proves Lemma 5.4.2.

LEMMA 5.4.4. *The number of double points of $\mathfrak{X}_n^0 \underset{\mathfrak{z}_{p^2}}{\otimes} F_{p^2}$ equals h_n.*

PROOF. We shall apply § 5.3 (D) for $\mathscr{K} = \mathscr{K}_n$, $\mathscr{K}^0 = \mathscr{K}^0_n$, $V = \mathfrak{p}_n$, and $\{V_1, V_2\} = \{\mathfrak{P}_n, \mathfrak{P}'_n\}$ (so that $s = 2$). Then $g = g^0_n$ and $g_1 = g_2 = g_n$, $r_i = e_i = 1$, and ρ is the number of double points of $\mathscr{X}^0_n \otimes F_{p^2}$. Therefore, (5.3.7) and Lemma 5.4.2 give $\rho = h_n$.

Now we shall complete the proof of our congruence relation. Denote by S_n, S^*_n and S^{**}_n the sets of supersingular points of X_n, the points of X_n corresponding to the double points of $\mathscr{X}^0_n \otimes F_{p^2}$, and the F_{p^2}-rational points of X_n, respectively. Then

$$S_n \subset S^{**}_n, \qquad S^*_n \subset S^{**}_n, \qquad |S_n| = |S^*_n|,$$

the last equality by Lemma 5.4.4. But by Prop. 1.3.1′, we have

$$S_n = S^{**}_n \qquad (n \in n_0\mathbf{Z}),$$

so that $S_n = S^*_n$ holds for $n \in n_0\mathbf{Z}$. But there is a canonical morphism $\mathscr{X}^0_{n'} \to \mathscr{X}^0_n$ for $n \mid n'$, so that $S_n \subset S^*_n$ holds for all n. Therefore, $S_n = S^*_n$ holds for all n.

6. Completing the proof. The proof of our Main theorem can now be completed by a direct use of the following theorem of Grothendieck.

THEOREM (Grothendieck [4], 18.3.4). *Let A be a noetherian ring, I an ideal of A such that A is separable and complete with respect to the I preadic topology. Put $\bar{A} = A/I$, $S = \operatorname{Spec} A$, $\bar{S} = \operatorname{Spec} \bar{A}$. Let $\mathscr{X}$ be an S-scheme which is proper over S, and put $\bar{\mathscr{X}} = \mathscr{X} \times_S \bar{S}$. Then the functor $\mathscr{Y} \rightsquigarrow \mathscr{Y} \times_{\mathscr{X}} \bar{\mathscr{X}}$ of the category of $\mathscr{X}$-schemes which are finite and étale over $\mathscr{X}$ into that of $\bar{\mathscr{X}}$-schemes which are finite and étale over $\bar{\mathscr{X}}$, is a categorical equivalence.*

6.1. Our congruence relation for the principal congruence subgroup of level n gives a system of $\mathfrak{z}_{p^2}$-schemes $\left\{\mathscr{X}_n \xleftarrow{\Psi} \mathscr{X}^0_n \xrightarrow{\Psi'} \mathscr{X}'_n\right\}$, whose products with $\mathbf{C}$ and F_{p^2} are isomorphic to

$$\left\{R_n \xleftarrow{\varphi} R^0_n \xrightarrow{\varphi'} R'_n\right\} \text{ and } \left\{X_n \xleftarrow{\psi} X^0_n \xrightarrow{\psi'} X'_n\right\}$$

respectively (§ 5.1). We shall replace $\mathfrak{z}_{p^2}$ by its *completion* $\mathbf{Z}_{p^2}$, and consider $\mathscr{X}_n$, $\mathscr{X}'_n$, $\mathscr{X}^0_n \ldots$ as already tensored with $\mathbf{Z}_{p^2}$. (Note that $\mathbf{Z}_{p^2}$ is the ring of Witt vectors over F_{p^2}.) Then $\mathscr{X}_n$, $\mathscr{X}'_n$, $\mathscr{X}^0_n$ are proper over $\mathbf{Z}_{p^2}$. (In particular, they are of finite type over $\mathbf{Z}_{p^2}$, and hence they are noetherian.)

Let $f : Y \to X_n$ be an unramified covering of X_n satisfying the assumptions of $[\mathrm{MT}\ 5]_n$ (§ 4.3). Since f is étale and $\mathscr{X}_n$ is proper over $\mathbf{Z}_{p^2}$, the above theorem of Grothendieck shows that $f : Y \to X_n$ can be lifted to an étale covering $F : \mathscr{Y} \to \mathscr{X}_n$. Since $\mathscr{X}_n$, $\mathscr{X}^0_n$ are noetherian, F gives rise to an étale covering $F^{01} : \mathscr{Y} \underset{\mathscr{X}_n}{\times} \mathscr{X}^0_n \longrightarrow \mathscr{X}^0_n$. In the same way, we can start from $f' : Y' \to X'_n$, and we obtain the diagram

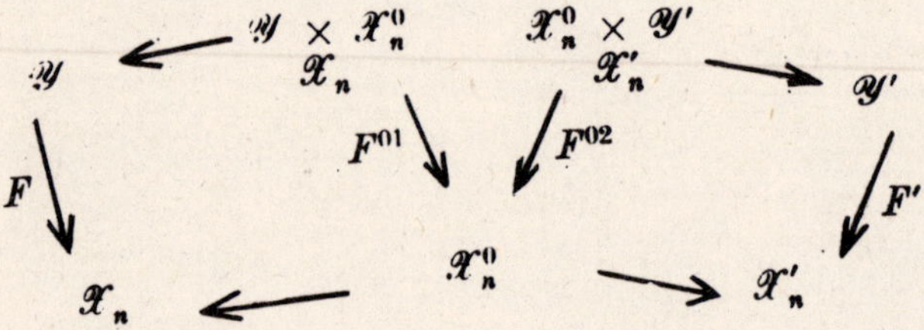

(6.1.1)

of $\mathbf{Z}_{p^2}$-schemes and $\mathbf{Z}_{p^2}$-morphisms (with F, F'; F^{01}, F^{02} étale) with which the special fiber is

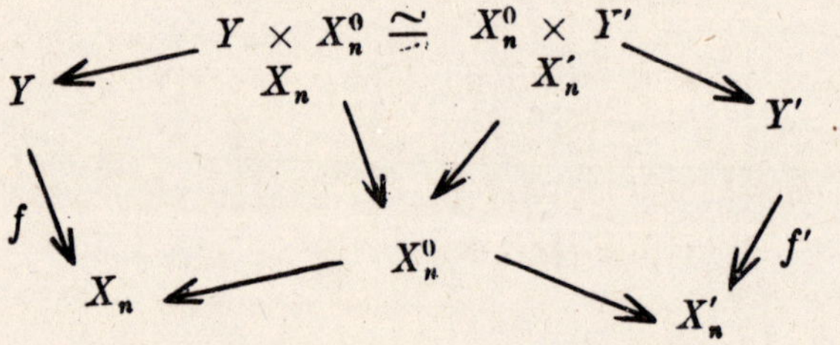

(6.1.2)

But since $\mathscr{X}^0_n$ is proper over $\mathbf{Z}_{p^2}$, the isomorphism (4.3.2) and the uniqueness of liftings in the above Grothendieck theorem give

$$\mathscr{Y} \underset{\mathscr{X}_n}{\times} \mathscr{X}_n^0 \simeq \mathscr{X}_n^0 \underset{\mathscr{X}'_n}{\times} \mathscr{Y}' \text{ over } \mathscr{X}_n^0. \tag{6.1.3}$$

Therefore by (5.1.5 **C**) and Lemma 3.2, $F: \mathscr{Y} \to \mathscr{X}_n$ must be a constant field extension. But then, f must also be a constant field extension. This proves $[\text{MT } 5]_n$.

6.2. In the first version of my proof, whose sketch was circulated (summer 1972), I proved (6.1.3) from (4.3.2) (under a slightly different formulation) by using some p-adic analysis, not knowing much of the Grothendieck theorem (although I used it to lift f). The p-adic analysis that I used was based on the general theory of Krasner, and (a higher level version for) Deligne's p-adic rigidity of the modular transforms of degree p. Professor Deligne kindly pointed out to me that this part can be directly proved by the Grothendieck theorem. I am very much grateful to him for his kind remark, which simplified the proof considerably.

REFERENCES

1. P. Deligne: Formes modulaires et représentations l-adiques, *Sém. Bourbaki* 21e *année n*^{0} 355 (1969).
2. M. Deuring: Reduktion algebraischer Funktionenkörper nach Primdivisoren des Konstantenkörpers, *Math. Z.* 47, 643-654 (1942).
3. M. Eichler: Quaternäre quadratischen Formen und die Riemannsche Vermutung für die Kongruenzzetafunktion, *Arch. Math.* 5, 355-366 (1954).
4. A. Grothendieck: Eléments de géométrie algébriques IV (4e Partie)(with Dieudonné), *IHES*, No. 32 (1967).
5. J. Igusa: Fiber systems of Jacobian varieties III, *Amer. J. Math.* 81, 453-476 (1959).

6. J. IGUSA: Kroneckerian model of fields of elliptic modular functions, *ibid*, 561-577.

7. Y. IHARA: Algebraic curves mod p and arithmetic groups, *Proc. symposia in pure math.* 9, 265-271 (1966).

8. Y. IHARA: On congruence monodromy problems, vol. 1,2, *Lecture Notes.* U. Tokyo; (1968, 69).

9. Y. IHARA: Non-abelian classfields over function fields in special cases, *Actes Congrès intern. math.*, (1970). Tome 1, 381-389.

10. A. G. KUROSH: *The theory of Groups*, New York (1956).

11. E. LAMPRECHT: Restabbildungen von Divisoren, *Arch. Math.*, 8, 30-38 (1957).

12. J. MENNICKE: On Ihara's modular group, *Invent. Math.*, 4, 202-228 (1967).

13. D. MUMFORD: Varieties defined by quadratic forms, *C.I.M.E.*, (1969) III "*Questions on Algebraic Varieties*".

14. E. D. NERING: Reduction of an algebraic function field modulo a prime in the constant field, *Ann. Math.* 67-3, 590-606 (1958).

15. J-P. SERRE: *Groupes discrètes* (in collaboration with H. Bass), Collège de France, 1969 (Ch. II, SL_2).

16. J-P. SERRE: La problème des groupes des congruences pour SL_2, *Ann. Math.* 92, 489-527 (1970).

17. G. SHIMURA: Correspondances modulaires et les fonctions ζ de courbes algébriques, *J. Math. Soc. Japan* 10, 1-28 (1958).

18. G. SHIMURA: Introduction to the arithmetic theory of automorphic functions, *Publ. Math. Soc. Japan* (1971).

19. A. WEIL: *Variétés abéliennes et courbes algébriques*, Paris (1948).

20. H. Popp: Über das Verhalten des Geschlechts eines Funktionenkörpers einer Variablen bei Konstantenreduktion, *Math. Z.* 106, 17-35 (1968).

STRONG RIGIDITY OF DISCRETE SUBGROUPS AND QUASI-CONFORMAL MAPPINGS OVER A DIVISION ALGEBRA

By G. D. MOSTOW

LET G BE a semi-simple analytic linear group having no compact normal subgroup other than (1) and let Γ be a lattice in G, that is, a discrete subgroup such that G/Γ has finite Haar measure. The pair (G, Γ) is called *strongly rigid* if it is uniquely determined by Γ; that is, given two such pairs (G, Γ) and (G', Γ'), and given an isomorphism $\theta: \Gamma \to \Gamma'$, then θ extends to an analytic isomorphism of G to G'. In [3d], I announced the strong rigidity of (G, Γ) if G/Γ is compact and G has no $\mathbf{R}$-rank 1 factors. In this paper, I will indicate the proof of strong rigidity in the case of $\mathbf{R}$-rank 1 groups other than $PL(2, \mathbf{R})$—rigidity being false for this latter group, as is seen from the example of two analytically inequivalent compact Riemann surfaces S and S' of genus greater than one: $(PL\,(2, \mathbf{R}), \Gamma)$ is not rigid if Γ is the fundamental group of the surface S.

My method of proving strong rigidity evolved from an effort to understand the failure of rigidity in $PL\,(2, \mathbf{R})$ from a geometric viewpoint. If X denotes the simply connected covering space of S and S', then we may regard X as the interior of the unit disc in the complex plane. As differentiable transformation groups on X, Γ and Γ' are equivalent. Why then are they inequivalent in the complex analytic sense? or to rephrase the question, why are they not conjugate in $PL\,(2, \mathbf{R})$? The natural conjecture is: because they are not differentiably equivalent on the *boundary* of X.

This fact is quite general. In [3a], I proved:

Let (G, Γ) and (G', Γ') be two pairs as above. Let X be the symmetric Riemannian space associated to G; that is $X = G/K$ where K is a maximal compact subgroup of G. Let X_0 denote the unique compact G-orbit in a maximal Satake compactification of G; or equivalently, the Furstenberg maximal boundary of X (cf. [1], [5], [3c]). Let X'

and X_0' be the corresponding spaces associated with G'. Let $\theta: \Gamma \to \Gamma'$ be an isomorphism and let $\varphi: X \to X'$ be a Γ-space morphism. Then if φ extends to a diffeomorphism of X_0 to X_0', θ extends to an analytic isomorphism of G to G'.

The question then arose: given two pairs (G, Γ) and (G', Γ'), can one always find a Γ-space morphism $\varphi: X \to X'$ which extends to a homeomorphism of X_0 to X_0' ? Under the added hypothesis that G/Γ and G'/Γ' are compact, the answer turned out to be affirmative.

Actually, the hypothesis that G/Γ and G'/Γ' are compact was necessary for establishing only one lemma, albeit a central lemma, which I shall now describe.

It is known that we lose no generality in assuming that Γ is torsion-free. Therefore we assume that Γ operates freely on X. As is well known, the space X is homeomorphic to euclidean space. Thus both X and X' are universal principal Γ-bundles, and therefore there is a Γ-space morphism $\varphi : X \to X'$; that is, a continuous (but not necessarily injective) map such that

$$\varphi(\gamma x) = \theta(\gamma)\, \varphi(x)$$

for all $\gamma \in \Gamma$, $x \in X$. Under the hypothesis that G/Γ and G'/Γ' are compact, one can prove that φ can be chosen to be a *pseudo-isometry*.

I call a map $\varphi : X \to X'$ between two metric spaces a (k, b) pseudo-isometry if

(i) $d_{X'}(\varphi(x), \varphi(y)) \leqslant k\, d_X(x, y)$, all x, y in X

(ii) $k^{-1} d_X(x, y) \leqslant d_{X'}(\varphi(x), \varphi(y))$ for all x, y in X

such that $d_X(x, y) \geqslant b$.

A pseudo-isometry is a map which is a (k, b) pseudo-isometry for some (k, b).

In [3d], we have

THEOREM. *Let G and G' be linear analytic groups, let Γ and Γ' be lattices in G and G' respectively. Let $\theta: \Gamma \to \Gamma'$ be an isomorphism, and let $\varphi: X \to X'$ and $\varphi': X' \to X$ be Γ-space pseudo-isometries. Then φ extends to a homeomorphism $\varphi: X_0 \to X_0'$.*

The proofs of the announcements in [3d] are in [3e]. The question arises, what about the smoothness of φ_0 ? If G has no $\mathbf{R}$-rank 1 factors, then not only is φ_0 smooth, but it in fact provides the desired extension of the isomorphism θ of Γ to Γ'.

THEOREM. *The map φ_0 induces an isomorphism of the Tits geometry of G onto the Tits geometry of G'. Assume G and G' have no compact normal subgroups other than* (1). *If G has no factors of* $\mathbf{R}$-*rank* 1, *then*

$$\varphi_0 \circ G_{X_0} \circ \varphi_0^{-1} = G'_{X'_0}$$

and thus θ is the restriction to Γ of the continuous (*and hence analytic*) *isomorphism*

$$g_{X_0} \to \varphi_0 \circ g_{X_0} \circ \varphi_0^{-1},$$

where g_{X_0} denotes the action of g on X_0.

Two clarifying comments are in order. First, the kernel of the homomorphism $G \to G_{X_0}$ is the maximum normal compact subgroup of G; by hypothesis, it consists only of the identity element. Secondly, the isomorphism between the Tits geometries induced by φ_0 is continuous, and by the generalized Fundamental Theorem of Projective Geometry for Tits geometries, induces an isomorphism of G to G' provided there are no $\mathbf{R}$-rank 1 factors. The reason for this exception is that a continuous incidence-preserving bijective map between two projective n-planes need not be a projective transformation if $n=1$.

We now take up the case of $\mathbf{R}$-rank 1 groups G. In this case the symmetric space X associated to G is hyperbolic space $H^n{}_{\mathbf{K}}$ where $\mathbf{K}$ is the field of real numbers $\mathbf{R}$ or complex numbers $\mathbf{C}$, the quaternions $\mathbf{H}$, or the Cayley numbers $\mathbf{O}$. That is, $\mathbf{K}$ is a composition division algebra over $\mathbf{R}$ with a positive definite norm.

In case $\mathbf{K} = \mathbf{R}$, $H^n{}_{\mathbf{R}}$ is the n-dimensional simply connected space of constant negative curvature. As a model for $H^n{}_{\mathbf{R}}$, we can take the interior of the unit ball in $\mathbf{R}^n$ with the G-invariant metric

$$ds = \frac{|dx|}{1 - |x|^2}$$

where $|dx|^2 = dx_1^2 + \dots + dx_n^2$. This metric ds differs from the euclidean metric by merely a point-function multiplicative factor. Therefore any map $\varphi: H^n{}_{\mathbf{R}} \to H^n{}_{\mathbf{R}}$ which is conformal with respect to ds is conformal with respect to the euclidean metric $|dx|$. More generally, any homeomorphism $\varphi: H^n{}_{\mathbf{R}} \to H^n{}_{\mathbf{R}}$ which is *quasi-conformal* with respect to the G-invariant metric is quasi-conformal with respect to the euclidean metric, where quasi-conformal is defined as follows:

A homeomorphism $\varphi: X \to X'$ between two metric spaces is called *k-quasi-conformal* if: (1) $\limsup_{t\to 0} \dfrac{L(p, t)}{l(p, t)} < \infty$ at all $p \in X$ where $L(p, t)$ (resp. $l(p, t)$) is the radius of the circumscribed (resp. inscribed) ball with center $\varphi(p)$ of the image $\varphi(B(p, t))$, where $B(p, t)$ is the ball with center p and radius t;

(2) $\limsup_{t\to 0} \dfrac{L(p, t)}{l(p, t)} \leqslant k$ for almost all p.

A map φ is called quasi-conformal if it is k-quasi-conformal for some k.

In [3b], I proved: If φ is a quasi-conformal Γ-space morphism of $H^n_{\mathbf{R}}$ to $H^n_{\mathbf{R}}$ then the boundary map $\varphi_0: X_0 \to X_0$ is a möbius map. Thus

$$\varphi_0 \circ G_{X_0} \circ \varphi_0^{-1} = G_{X_0}$$

in this case too, and $\theta: \Gamma \to \Gamma'$ extends to an analytic automorphism of G.

The proof makes essential use of the theory of quasi-conformal mappings for euclidean space.

Now in case $\mathbf{K} = \mathbf{C}$, $\mathbf{H}$, or $\mathbf{O}$, we can again represent $H^n{}_{\mathbf{K}}$ as the interior of the unit ball in $\mathbf{K}^n$, but, as is well known, the G-invariant metric is no longer related to the euclidean metric by a multiplicative point-function factor. If $\mathbf{K} = \mathbf{R}$, $\mathbf{C}$ or $\mathbf{H}$, the metric is

$$ds^2 = (1 - |x|^2)^{-1} (|dx|^2 + (1 - |x|^2)^{-1} |(dx, x)|^2).$$

One can calculate the hyperbolic distance between two points x and y of $H^n_{\mathbf{K}}$:

$$d(x, y) = \cosh^{-1} \frac{|1 - (x, y)|}{(1 - |x|^2)^{1/2} (1 - |y|^2)^{1/2}}.$$

If $\mathbf{K} = \mathbf{O}$, then $n \leqslant 2$ and we have

$$d(x, y) = \cosh^{-1} \frac{(|1 - (x, y)|^2 + R(x, y))^{1/2}}{(1 - |x|^2)^{1/2} \quad (1 - |y|^2)^{1/2}}$$

where $x = (x_1, x_2)$, $y = (y_1, y_2)$, and

$$2^{-1} R(x, y) = \mathrm{Re}\,(x_1 \overline{x_2})(y_2 \overline{y}_1) - \mathrm{Re}\,(\overline{x}_2\, y_2)(\overline{y}_1\, x_1)$$

{Note that for any p, q, r in $\mathbf{O}$, $\mathrm{Re}\, p(q\, r) = \mathrm{Re}\, q(r\, p)$}.

It is of some interest therefore that one can successfully develop an analogue of quasi-conformal mappings for $\mathbf{K}$ as well as for $\mathbf{R}$.

Let d_0 denote the positive real valued function on $\mathbf{K}^n$ defined by the formula

$$\begin{aligned} 4^{-1} d_0\, (x, y)^4 &= |\, 1 - (x, y)^2 \,| + R(x, y) - (1 - |x|^2)(1 - |y|^2) \\ &= |x - y|^2 - (|x|^2 |y|^2 - |(x, y)|^2) + R(x, y). \end{aligned}$$

For any x, y on the unit sphere,

$$d_0(x, y) \geqslant |x - y|.$$

We can regard $\mathbf{K}^n$ as a subset of projective n-space over $\mathbf{K}$, and the $\mathbf{K}$-lines in $\mathbf{K}^n$ are subsets of dimension $k = \dim_{\mathbf{R}} \mathbf{K} = 1, 2, 4$, or 8. The intersection of a $\mathbf{K}$-line with $H^n_{\mathbf{K}}$ is a geodesic subspace. The intersection of $\mathbf{K}$-lines with X_0 are called $\mathbf{K}$-*spheres*. By a real 2-plane in $H^n_{\mathbf{K}}$ we mean a 2-dimensional hyperbolic subspace which lies in no $\mathbf{K}$-line. The boundary of a real two-plane is called an $\mathbf{R}$-*circle* (even though it is not always a euclidean circle). If, however, the $\mathbf{K}$-line (resp. real two-plane) passes through the origin, then the corresponding $\mathbf{K}$-sphere (resp. $\mathbf{R}$-circle) is a great $\mathbf{K}$-sphere (resp. $\mathbf{R}$-circle).

We have

$d_0(x, y) = |x - y|$, if x, y are on the same great $\mathbf{R}$-circle,

$d_0(x, y) = (2\,|x - y|)^{1/2}$, if x, y are on the same great $\mathbf{K}$-sphere.

If we attempted to form a metric from d_0, it would yield the usual geodesic arc length along great $\mathbf{R}$-circles, but would yield infinite arc length along great $\mathbf{K}$-spheres. Thus we call d_0 a *semi-metric.* We set for any $p \in X_0$

$$\mathbf{K}(p, t) = \{\, q \in X_0;\ d_0\,(p, q) < t \,\}$$

and we call $\mathbf{K}(p, t)$ the d_0-ball of center p and radius t.

Given **R**-rank 1 groups G and G' containing isomorphic lattices Γ and Γ', let $\varphi : X \to X'$ be a Γ-space pseudo-isometry as above, and let $\varphi_0 : X_0 \to X'_0$ be the induced boundary homeomorphism.

We set for any $p \in X_0$,

$$L(p, t) = \inf \{s; \varphi_0(\mathbf{K}(p, t)) \subset \mathbf{K}'(\varphi_0(p), s)\}$$

$$l(p, t) = \sup \{s; \varphi_0(\mathbf{K}(p, t)) \supset \mathbf{K}'(\varphi_0(p), s)\}$$

Then one proves

THEOREM. *There is a positive constant K such that for all $p \in X_0$*

$$\limsup_{t \to 0} \frac{L(p, t)}{l(p, t)} < K.$$

We can describe this result by saying that the *boundary map* is *K-quasi-conformal with respect to the semi-metric d_0*.

This theorem has striking consequences. It allows us to prove in the first instance

THEOREM. *The boundary map φ_0 is absolutely continuous on almost all* **R**-*circles, if $H^n_{\mathbf{K}} \neq H^2_{\mathbf{R}}$.*

This result in turn allows us to form non-zero directional derivatives of φ_0 on sets of positive measure along **R**-circles.

Upon utilizing the fact that Γ operates ergodically at the boundary, we can succeed ultimately in proving that the map φ_0 is induced by an isomorphism of G. Thus once again we find

$$\varphi_0 \circ G_{X_0} \circ \varphi_0^{-1} = G'_{X_0}.$$

Putting together our results on **R**-rank greater than one and on **R**-rank 1, our method yields the

THEOREM. *Let G be a semi-simple analytic linear group having no compact normal subgroups other than* 1, *and let Γ be a discrete cocompact subgroup. Assume that there is no continuous homomorphism $\pi: G \to PL(2, \mathbf{R})$ with $\pi(\Gamma)$ discrete. Then the pair (G, Γ) is rigid* (cf. [3e]).

Recent results of Margulis and Raghunathan establish the above theorem if we replace "cocompact" by "non-compact lattice", and add the hypothesis that G has no **R**-rank 1 factors.

As pointed out above, we use the cocompactness of Γ only to construct the pseudo-isometry φ. However, if sufficient information is available about a fundamental domain for Γ, one can construct a pseudo-isometry in the non-compact case as well. This was recently verified by Gopal Prasad for the case of lattices in groups with a $\mathbf{R}$-rank 1 factor.

Thus, putting together the results of Margulis, Raghunathan, Prasad, and [3e] we can weaken the hypothesis of "cocompact" in the above theorem to merely "lattice".

REFERENCES

1. H. Furstenberg: A Poisson formula for semi-simple Lie groups, *Ann. of Math.*, 77 (1963), 335-386.

2. F. I. Mautner: Geodesic flows on symmetric Riemann spaces, *Ann. of Math.*, 65 (1957), 416-431.

3. G. D. Mostow: (a) On the conjugacy of subgroups of semi-simple groups, *Proc. of Symposia in Pure Math.*, 9 (1966), 413-419.

   (b) Quasi-conformal mappings in n-space and the rigidity of hyperbolic space forms, *Publ. I. H. E. S.*, 34, 53-104.

   (c) *Lectures on discrete subgroups*, Tata Institute of Fundamental Research, (1970).

   (d) The rigidity of locally symmetric spaces, *Proc. International Congress of Math.* , (1970), vol. 2, 187-197.

   (e) Strong rigidity of locally symmetric spaces (to appear).

4. G. Prasad : Strong rigidity of $\mathbf{Q}$-rank 1 lattices, *Inv. Math.*, 21 (1973), 255-286.

5. I. Satake: On representations and compactifications of symmetric Riemannian spaces, *Ann. of Math.*, 71 (1960), 77-110.

A NEW APPROACH TO COMPACTIFYING LOCALLY SYMMETRIC VARIETIES

By DAVID MUMFORD

SUPPOSE D IS a bounded symmetric domain and $\Gamma \subset \mathrm{Aut}\,(D)$ is a discrete group of arithmetic type. Then Borel and Baily [2] have shown that D/Γ can be canonically embedded as a Zariski-open subset in a projective variety $\overline{D/\Gamma}$. However, Igusa [6] and others have found that the singularities of $\overline{D/\Gamma}$ are extraordinarily complicated and this presents a significant obstacle to using algebraic geometry on $\overline{D/\Gamma}$ in order to derive information on automorphic forms on D, etc. Igusa [7] for $D = \mathfrak{M}_2$ and $\mathfrak{M}_3$ ($\mathfrak{M}_n$ = Siegel's $n \times n$ upper half-space) and Γ commensurable with $Sp(2n, \mathbf{Z})$, and Hirzebruch [4] for $D = \mathfrak{M}_1 \times \mathfrak{M}_1$ and Γ commensurable with $SL(2, R)$ (R = integers in a real quadratic field) have given explicit resolutions of $\overline{D/\Gamma}$. Independently, Satake [9] and I working in collaboration with Y. Tai, M. Rapaport and A. Ash have attacked the general case, using closely related methods. The purpose of this paper is to give a very short outline of my approach. It builds in an essential way on the construction of "torus embeddings", a theory which has been published in the Springer Lecture Notes [8] (by G. Kempf, F. Knudsen, B. Saint-Donat and myself; some of which has been worked out independently by M. Demazure [3] and M. Hochster [5]). We intend to publish full details of the present research as soon as possible in a sequel "Toroidal Embeddings II" to the Notes [8]. At the present time, however, we cannot claim to have written down complete proofs of our "Main Theorem" and although I definitely believe it is true and not difficult, it is more accurate to describe the ideas below only as a suggested approach to the problem of constructing a non-singular compactification of D/Γ.

1. Let us look first at the familiar case:

$$D = \mathfrak{M}_1$$
$$\Gamma = SL(2, \mathbf{Z}).$$

We know that $D/\Gamma \cong \mathbf{C}$ via the elliptic modular function j and adding one point $j = \infty$, we get the unique non-singular compactification:

$$\begin{array}{ccc} D/\Gamma & \subset & \overline{D/\Gamma} \\ \| & & \| \\ \mathbf{C} & \subset & \mathbf{CP}^1. \end{array}$$

However, let me describe a way of glueing in the point at ∞ that will suggest generalizations:

STEP a. Factor the map $\mathfrak{M}_1 \xrightarrow{j} \mathbf{C}$ as follows

$$\begin{array}{ccccc} \mathfrak{M}_1 & \xrightarrow{\alpha} & \Delta_1^* & \xrightarrow{\beta} & \mathbf{C} \\ \| & & \| & & \\ \{\omega \mid \operatorname{Im}\omega > 0\} & & \{\zeta \mid 0 < |\zeta| < 1\} & & \end{array}$$

where $\zeta = e^{2\pi i\omega}$. If $\Gamma_0 = \left\{ \begin{pmatrix} 1 & b \\ 0 & 1 \end{pmatrix} \middle| b \in \mathbf{Z} \right\} \subset SL(2, \mathbf{Z})$, then

$$\Delta_1^* \cong \mathfrak{M}_1/\Gamma_0.$$

STEP b. *Partially* compactify Δ_1^* by adding the origin

$$\Delta_1^* \subset \Delta_1 = \{\zeta \,\big|\, |\zeta| < 1\}.$$

STEP c. Note that if

$$\mathfrak{M}_1(c) = \{\omega \,\big|\, \operatorname{Im}\omega > c\},$$

then if c is large enough, $SL(2, \mathbf{Z})$-equivalence in $\mathfrak{M}_1$ reduces, in $\mathfrak{M}_1(c)$, to Γ_0-equivalence:

$$\left.\begin{array}{l} \omega_1, \omega_2 \in \mathfrak{M}_1(c) \\ \omega_1 = \gamma(\omega_2),\ \gamma \in SL(2, \mathbf{Z}) \end{array}\right\} \Rightarrow \gamma \in \Gamma_0. \qquad (*)$$

Now $\mathfrak{M}_1(c)$ maps to Δ_b^*, where

$$\Delta_b^* = \{\zeta \,\big|\, 0 < |\zeta| < b\}$$
$$b = e^{-2\pi c},$$

and (*) says:

$$\operatorname{res} \beta : \Delta_b^* \longrightarrow \mathbf{C} \text{ is injective.}$$

STEP d. This gives us the situation:

$$\Delta_b^* \overset{\text{res } \beta}{\hookrightarrow} \mathbf{C}$$

$$\Delta_b^* \hookrightarrow (\text{Interior of closure of } \Delta_b^* \text{ in } \Delta_1) = \Delta_b$$

It is easy to see that $\mathbf{CP}^1$ is nothing but the union of $\mathbf{C}$ and Δ_b, glued on Δ_b^*.

2. Let us look next at how this procedure can be generalized to the $n \times n$ Siegel case:

$$D = \mathfrak{M}_n = \left\{ \Omega \,\middle|\, \begin{array}{l} \Omega \quad n \times n \text{ complex symmetric matrix,} \\ \text{Im } \Omega \text{ positive definite} \end{array} \right\},$$

$$\Gamma = Sp(2n, \mathbf{Z}).$$

Actually, it is usually more convenient to replace $Sp(2n, \mathbf{Z})$ by a subgroup of finite index, or else to allow V-manifold-type singularities on D/Γ and $\overline{D/\Gamma}$, because of the elements of finite order in Γ which need not act freely. We will ignore this technicality. Of course, these V-manifold singularities can also be resolved: but that involves a totally different set of problems.

STEP a′: Factor $\mathfrak{M}_n \longrightarrow \mathfrak{M}_n/\Gamma$ as follows:

$$\mathfrak{M}_n \longrightarrow \mathfrak{J}_n^0 \longrightarrow \mathfrak{M}_n/\Gamma$$

$$\mathfrak{J}_n^0 = \left\{ Z \,\middle|\, \begin{array}{l} Z \;\; n \times n \text{ complex symmetric matrix, } Z_{ij} \neq 0 \\ \text{and} -\log |Z_{ij}| \text{ positive definite} \end{array} \right\}$$

where $Z_{ij} = e^{2\pi i(\Omega_{ij})}$. If $\Gamma_0 = \left\{ \begin{pmatrix} I_n & B \\ 0 & I_n \end{pmatrix} \,\middle|\, \begin{array}{l} B \in M_n(\mathbf{Z}) \\ \text{and symmetric} \end{array} \right\} \subset \Gamma,$

then

$$\mathfrak{J}_n^0 \simeq \mathfrak{M}_n/\Gamma_0.$$

STEP b′: Note that $\mathfrak{Z}_n^0$ is an open set in the algebraic torus group:

$$\mathfrak{Z}_n^0 \subset \mathfrak{Z}_n = \left\{ Z \,\middle|\, \begin{matrix} Z\ n \times n \text{ symmetric} \\ Z_{ij} \neq 0 \end{matrix} \right\}$$

(the group law being componentwise multiplication). The generalization of Δ in the first case is now given by the theory of equivariant torus embeddings which we must now summarize

Torus embedding theory :

T = an algebraic torus of dimension n, *i.e.*, $\simeq (\mathbf{C}^*)^n$,

$N = \pi_1(T)$ a free abelian group of rank n, $N_{\mathbf{R}} = N \otimes R$,

$N_{\mathbf{C}} = N \otimes \mathbf{C}$ so that

$T \simeq N_{\mathbf{C}}/N$ (via $\exp : N_{\mathbf{C}} \longrightarrow T$),

$M = \text{Hom}(N, \mathbf{Z}) \simeq$ [the group of characters $\mathscr{X} : T \longrightarrow \mathbf{C}^*$].

If $\alpha \in M$, write $X^\alpha : T \longrightarrow \mathbf{C}^*$ for the associated character write $\langle x, a \rangle$ for the pairing of $M_{\mathbf{R}}$ and $N_{\mathbf{R}}$.

$\forall\, \sigma$ = closed convex rational polyhedral cone in $N_{\mathbf{R}}$
(*i.e.*, $\sigma = \{x \in N_{\mathbf{R}} \mid \langle x, \alpha_i \rangle \geqslant 0,\ 1 \leqslant i \leqslant N\}$ for some finite set of points $\alpha_i \in M$), which are "proper":
$\sigma \not\supset$ pos. dim. subspace of $N_{\mathbf{R}}$

$$X_\sigma \underset{\text{def}}{=} \text{Spec}\, \{\mathbf{C}[\ldots, X^\alpha, \ldots]_{\alpha \in M \cap \check{\sigma}}\},\ \check{\sigma} = \text{dual of } \sigma \text{ in } M_{\mathbf{R}}.$$

Then X_σ is a normal affine variety, T is an open subset of X_σ and the action of T on itself extends to an action of T on X_σ; moreover all such embeddings $T \subset X$ arise like this. X_σ is non-singular if σ is a simplicial cone generated by a subset of a basis of N.

$\forall\, \{\sigma_\alpha\}$ = collection of such σ's such that every face of a σ_α is some σ_β, and every intersection $\sigma_\alpha \cap \sigma_\beta$ is a common face,

$X_{\{\sigma_\alpha\}} \underset{\text{def}}{=} \bigcup X_{\sigma_\alpha}$, where X_{σ_α} and X_{σ_β} are glued along $X_{\sigma_\alpha \cap \sigma_\beta}$

(which is, in fact, an open subset of each).

Then $X_{\{\sigma_\alpha\}}$ is an irreducible separated normal scheme, locally of finite type over $\mathbf{C}$, T is an open subset of $X_{\{\sigma_\alpha\}}$ and the action of T extends; again all such embeddings arise like this (Sumihiro [10]).

Let $N_{n,\mathbf{R}}$ = vector space of real $n \times n$ symmetric matrices,
N_n = lattice of integral ones,
C_n = cone of positive semi-definite matrices.

Then for all collections $\{\sigma_\alpha\}$, $\sigma_\alpha \subset N_{n,\mathbf{R}}$ being closed convex rational polyhedral proper cones, fitting together as above, we get

$$\mathfrak{I}_n^0 \subset \mathfrak{I}_n \subset X_{\{\sigma_\alpha\}}.$$

It can be shown that for all α:

$$(*)\quad \begin{cases} \exists\, x \in X_{\sigma_\alpha} - \bigcup\limits_{\beta \text{ face of } \alpha} X_{\sigma_\beta} \text{ and a neighborhood} \\ U \text{ of } x \text{ in } X_{\sigma_\alpha} \text{ such that } U \subset \overline{\mathfrak{I}_n^0} \\ \text{if and only if } \sigma_\alpha \subset C_n. \end{cases}$$

For this reason, we assume that $\sigma_\alpha \subset C_n$, all α, and define

$$\mathfrak{I}^0_{n,\{\sigma_\alpha\}} = \text{Interior of closure of } \mathfrak{I}_n^0 \text{ in } X_{\{\sigma_\alpha\}}.$$

Then

$$\mathfrak{I}_n^0 \subset \mathfrak{I}^0_{n,\{\sigma_\alpha\}}$$

is the partial compactification of $\mathfrak{I}_n^0$ which we shall use.

STEP c′. There seem to be several choices for an n-dimensional analog of $\mathfrak{M}_1(c)$ but we take :

$$\mathfrak{M}_n(c) = \left\{ \Omega \in \mathfrak{M}_n \;\middle|\; \begin{array}{l} \forall\, k \in \mathbf{Z}^n,\ k \neq (0), \\ {}^t k \cdot (\operatorname{Im} \Omega) \cdot k > c. \end{array} \right\}$$

By Siegel-Minkowski reduction theory, it can be shown that if c is large enough, $Sp(2n,\mathbf{Z})$-equivalence in $\mathfrak{M}_n$ reduces, in $\mathfrak{M}_n(c)$, not to Γ_0-equivalence but to Γ_1-equivalence; where :

$$\Gamma_1 = \left\{ \begin{pmatrix} A & B \\ 0 & {}^tA^{-1} \end{pmatrix} \;\middle|\; \begin{array}{l} A \in GL(n,\mathbf{Z}),\ B \in M_n(\mathbf{Z}) \text{ and} \\ A^{-1}B \text{ symmetric} \end{array} \right\}$$

$$= \left\{ \begin{pmatrix} A & B \\ 0 & D \end{pmatrix} \;\middle|\; \begin{array}{c} A,\ D \in GL(n,\mathbf{Z}) \\ B \in M_n(\mathbf{Z}) \end{array} \right\} \cap Sp(2n,\mathbf{Z}).$$

Let $\mathfrak{I}_n^c$ = image of $\mathfrak{M}_n(c)$ in $\mathfrak{I}_n$.

Then $\Gamma_1/\Gamma_0 \cong GL(n, \mathbf{Z})$ acts on the torus $\mathfrak{I}_n$, preserving the open subsets $\mathfrak{I}_n^0$ and $\mathfrak{I}_n^C$, and we get the situation :

$$\begin{array}{ccccc} \mathfrak{I}_n^C & \subset & \mathfrak{I}_n^0 & \subset & \mathfrak{I}_n \\ \downarrow & & \downarrow & & \\ \mathfrak{I}_n^C/(\Gamma_1/\Gamma_0) & \subset & \mathfrak{I}_n^0/(\Gamma_1/\Gamma_0) & & \\ & \searrow & \downarrow & & \\ & & \mathfrak{M}_n/\Gamma & & \end{array}$$

where the dotted arrow is injective.

STEP d′ : It is now clear how to finish up : we must assume that the collection $\{\sigma_\alpha\}$ satisfies the condition $\forall\ \gamma \in \Gamma_1/\Gamma_0,\ \forall_\alpha,\ \exists\beta$ such that $\gamma\sigma_\alpha = \sigma_\beta$ (under the natural action of $\Gamma_1 \mid \Gamma_0$ on $\mathfrak{I}_n$, hence on $N_{n,\mathbf{R}}$). Then the action of Γ_1/Γ_0 on $\mathfrak{I}_n$ extends to $X_{\{\sigma_\alpha\}}$. Define

$$\mathfrak{I}^C_{n\ \{\sigma_\alpha\}} = \text{Interior of closure of } \mathfrak{I}_n^C \text{ in } X_{\{\sigma_\alpha\}}Z$$

and consider

$$\mathfrak{I}_n^C/(\Gamma_1/\Gamma_0) \begin{array}{l} \hookrightarrow \mathfrak{M}_n/\Gamma \\ \hookrightarrow \mathfrak{I}^C_{n,\{\sigma_\alpha\}}/(\Gamma_1/\Gamma_0) \end{array}$$

and glue !

Some comments : First of all, there are many things to check in the above procedure, but we will not try to justify them here. Secondly, this glueing alone will never give us something compact ; but what I do claim is that if you take just enough σ_α's, in the sense :

(a) $\bigcup \sigma_\alpha = \left\{ Z \in N_{n,\mathbf{R}} \;\middle|\; \begin{array}{l} Z \text{ positive semi-definite with null-} \\ \text{space defined over } \mathbf{Q} \end{array} \right\}$

(b) Modulo Γ_1/Γ_0, the set of σ_α's is finite, then the resulting partial compactification of $\mathfrak{M}_n/\Gamma$ does cover the entire 0-dimensional boundary component; and moreover it "analytically prolongs" to a compactification $\overline{\mathfrak{M}_n/\Gamma}$ of $\mathfrak{M}_n/\Gamma$ in the following way :

(a) let

$$\mathfrak{I}^o_{n,\{\sigma_\alpha\}} = \text{Interior of closure of } \mathfrak{I}^o_n \text{ in } X_{\{\sigma_\alpha\}},$$

(b) require the existence of a map π:

$$\begin{array}{ccc} \mathfrak{I}^o_n & \subset & \mathfrak{I}^o_{n,\{\sigma_\alpha\}} \\ \big\downarrow & & \vdots\ \pi \\ \mathfrak{M}_n/\Gamma & \subset & \overline{\mathfrak{M}_n/\Gamma} \end{array}$$

where π is surjective and open and $\mathfrak{M}_n/\Gamma$ is open and dense in $\overline{\mathfrak{M}_n/\Gamma}$.

Thirdly, the resulting space is not necessarily non-singular. However, if the σ_α's are chosen satisfying :

(c) $\forall\alpha$, σ_α is the set of positive linear combinations of matrices $A_1,\ldots, A_k \in N_n$, which are part of a basis of the free abelian group N_n,

and if $Sp(2n, \mathbf{Z})$ is replaced by a subgroup Γ of finite index without elements of finite order, *then* we get a non-singular compactification of $\mathfrak{M}_n/\Gamma$. Even without these two conditions, the singularities are quite mild, e.g., rational (and if the σ_α's are merely simplicial cones, the singularities are V-manifolds). Fourthly, an objection may be raised that there is still a huge amount of freedom in the choice of the σ_α's, leading to a whole family of non-singular compactifications rather than one best possible one. This in fact discouraged me for several years and made me think the theory was not useful (the simplest case I know where this non-uniqueness seems really basic is $\mathfrak{M}_4$). But I believe now that this non-uniqueness is a fact of life of higher-dimensional birational geometry and that for many applications, this class of compactifications is just as usable as one canonical one would be.

3. Finally, and with many more gaps, let me sketch how I believe this procedure extends to the general case :

$$D = \text{any bounded symmetric domain,}$$
$$\Gamma = \text{an arithmetic subgroup of } \mathrm{Aut}(D).$$

Assume for simplicity that Γ has no elements of finite order.

STEP a″. For every rational boundary component F, we get groups:

$$\mathrm{Aut}(D)^0 \supset N(F) \supset U(F)$$

where

$$N(F) = \left\{ g \in \mathrm{Aut}(D^0) \,\middle|\, gF = F \right\}$$
$$U(F) = \text{center of unipotent radical of } N(F) :$$
this is just a real vector space under addition.

Let $\Gamma_0 = \Gamma \cap U(F)$: a lattice in $U(F)$ and $\Gamma_1 = \Gamma \cap N(F)$.

We factor $D \longrightarrow D/\Gamma$ via :

$$D \longrightarrow D/\Gamma_0 \longrightarrow D/\Gamma_1 \longrightarrow D/\Gamma.$$

STEP b″: To describe D relative to F suitably, embed D in $\check{D}$, its compact dual, so that the complexification $G_{\mathbf{C}}$ of $\mathrm{Aut}(D)^0$ acts on $\check{D}$. Moving D around only by $U(F)_{\mathbf{C}}$, we get an intermediate open set :

$$D \subset U(F)_{\mathbf{C}} \cdot D \subset \check{D}.$$

This gives us a description of D as a *Siegel Domain of 3rd kind* as follows : I claim

$$U(F)_{\mathbf{C}} \cdot D \simeq U(F)_{\mathbf{C}} \times \mathscr{E}(F)$$

for some complex vector bundle $p \,.\, \mathscr{E}(F) \to F$ over F itself (the isomorphism being complex analytic and taking the action of $U(F)_{\mathbf{C}}$ on the left to translations in the 1st factor on the right), and that this isomorphism restricts to:

$$D \simeq \left\{ (u,x) \,\middle|\, \mathrm{Im}\, u \in C(F) + h(x) \right\}$$

for some open convex cone $C(F) \subset U(F)$ and real analytic map $h : \mathscr{E}(F) \to U(F)$. Let $T(F) = U(F)_{\mathbf{C}}/\Gamma_0$: this is an algebraic torus group over $\mathbf{C}$. We get

$$D/\Gamma_0 \subset (U(F)_{\mathbf{C}} \cdot D)/\Gamma_0 \simeq T(F) \times \mathscr{E}(F).$$

We now choose a collection $\{\sigma_\alpha\}$ of rational polyhedral cones in $\overline{C(F)}$ and note that these define, by our general theory, an embedding

$$T(F) \subset X_{\{\sigma_\alpha\}}.$$

Define

$$(D/\Gamma_0)_{\{\sigma_\alpha\}} = \text{Interior of closure of } D/\Gamma_0 \text{ in } X_{\{\sigma_\alpha\}} \times \mathscr{E}(F).$$

STEP c″. If $\gamma \in C(F)$ and $K \subset F$ is a compact set, let

$$D(\gamma, K) = \Gamma_1 \cdot \left\{ (u,x) \;\middle|\; \operatorname{Im} u \in C(F) + h(x) + \gamma,\ p(x) \in K \right\}.$$

Then, I believe, *for all K, if c is large enough*, the composition

$$D(c,K)/\Gamma_1 \hookrightarrow D/\Gamma_1 \longrightarrow D/\Gamma$$

is injective.

STEP d″. Assume now that the collection $\{\sigma_\alpha\}$ satisfies the conditions:

(a) $\forall\, \gamma \in \Gamma_1/\Gamma_0$, and $\forall \sigma_\alpha$, $\gamma\sigma_\alpha =$ some σ_β; and modulo this action, there are only finitely many σ_α's.

(b) $C(F) \subset \bigcup_\alpha \sigma_\alpha \subset \overline{C(F)}$.

It requires proof at this point that such $\{\sigma_\alpha\}$ exist — this seems quite likely. Define

$$(D(c, K)/\Gamma_0)_{\{\sigma_\alpha\}} = \text{Interior of closure of } D(c,K)/\Gamma_0 \text{ in } X_{\{\sigma_\alpha\}} \times \mathscr{E}(F)$$

and consider :

$$\begin{array}{ccc} & & D/\Gamma \\ & \nearrow & \\ D(c, K)/\Gamma_1 & & \\ & \searrow & \\ & & (D(c, K)/\Gamma_0)_{\{\sigma_\alpha\}}/(\Gamma_1/\Gamma_0) \end{array}$$

(both arrows are inclusions $\hookrightarrow$)

and glue! The whole set-up is summarized in the figure overleaf.

Summary of spaces

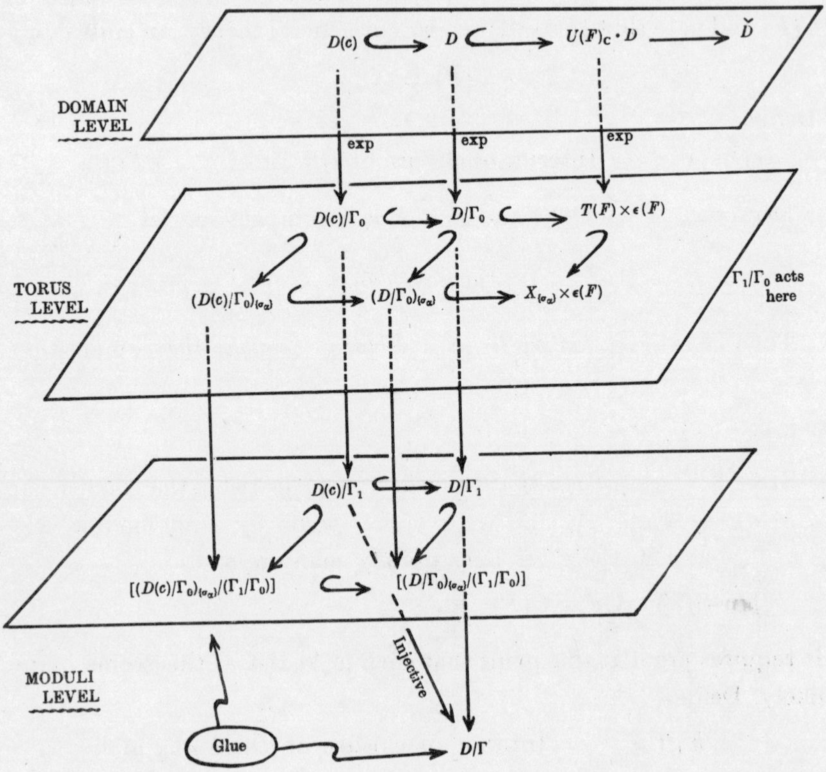

STEP e″. Finally, if we let F range over the finite set of Γ-inequivalent rational boundary components, we must check that if suitably compatible collections $\{\sigma_\alpha\}$ are chosen for each F, then these partial compactifications are compatible in the sense that they are all part of one big compact Hausdorff space D/Γ containing $\overline{D/\Gamma}$ as an open dense set (this D/Γ being uniquely determined by these requirements) and such that there are even unramified maps π in the diagram

$$\begin{array}{ccc} D/\Gamma_0 & \subset & (D/\Gamma_0)_{\{\sigma_\alpha\}} \\ \downarrow & & \downarrow \pi \\ D/\Gamma & \subset & \overline{D/\Gamma} \end{array}$$

The compatibility of the $\{\sigma_\alpha\}$'s can be expressed as follows :

Say $F_1, F_2 \subset \bar{D}$ are 2 rational boundary components and $F_1 \subset \bar{F}_2$. Then

$$U(F_1) \supset U(F_2) \quad \text{and} \quad C(F_2) \cong \text{face of } \overline{C(F_1)}.$$

Then we require that the set of cones $\sigma_\alpha{}^{(2)} \subset \overline{C(F_2)}$ be exactly the set $\sigma_\alpha{}^{(1)} \cap \overline{C(F_2)}$.

4. In order to express more clearly what our compactification depends on, and to relate it to the theory of toroidal embeddings ([8], Ch. II), it is convenient to introduce the following interesting abstract cone :

$$\Sigma = \Big(\coprod_{\substack{\text{rat. boundary}\\ \text{Comp. } F}} C(F)\Big)/\Gamma = \coprod_{\substack{\Gamma\text{-equiv. classes}\\ \text{of rat. } F}} \big(C(F)/\Gamma \cap N(F)\big)$$

where $\gamma \in \Gamma$ acts on $\coprod_F C(F)$ by the natural maps $C(F) \xrightarrow{\approx} C(\gamma F)$ for all F. To express the structure that Σ has, we use the definition.

Definition. *A 'conical polyhedral complex'*† *is a topological space X, plus a finite stratification $\{S_\alpha\}$ of X, (i.e., a partition of X into disjoint locally closed pieces S_α such that each $\bar{S}_\alpha$ is a union of various*

†This definition is a slight modification of that used in [8] to allow 2 faces of the same polyhedra to be identified. Thus

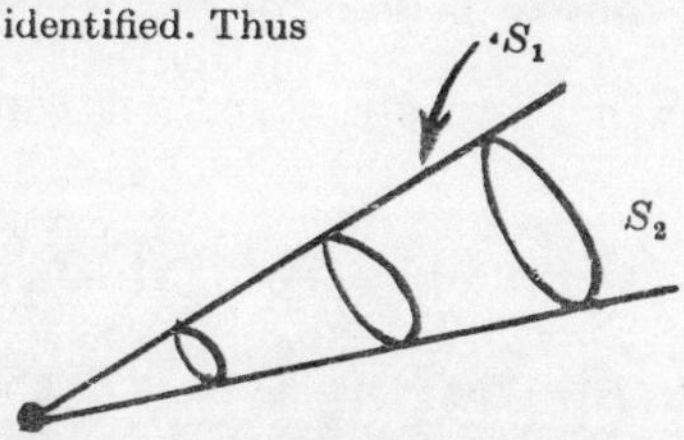

is allowed, as well as the previously allowed :

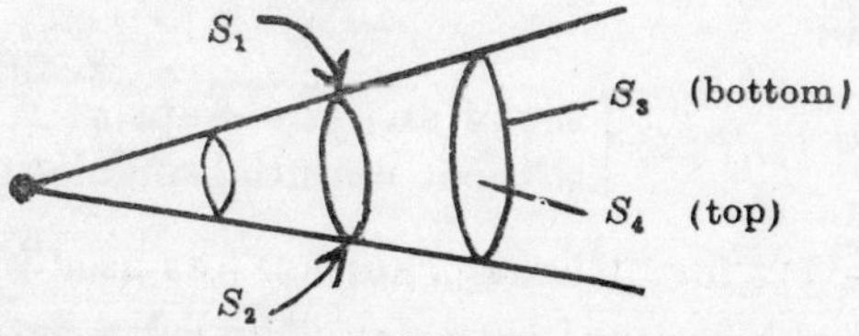

S_β's), plus for each α a finite-dimensional vector-space V_α of real-valued continuous functions on S_α such that :

(a) *if $n_\alpha = \dim V_\alpha$ and $f_1, \ldots, f_{n_\alpha}$ is a basis of V_α, then $(f_i) : S_\alpha \to \mathbf{R}^{n_\alpha}$ is a homeomorphism of S_α with an open convex polyhedral cone $C_\alpha \subset \mathbf{R}^{n_\alpha}$,*

(b) *$(f_i)^{-1}$ extends to a continuous surjective map*

$$(f_i)^{-1} : \bar{C}_\alpha \to \bar{S}_\alpha$$

mapping the open faces $C_\alpha^{(\beta)}$ of $\bar{C}_\alpha$ (= closed faces less their own faces) homeomorphically to the strata S_β in $\bar{S}_\alpha$ and inducing isomorphisms

$$\operatorname{res}_{C_\alpha^{(\beta)}} (\text{lin. fcns. } on\ \mathbf{R}^{n_\alpha}) \xrightarrow{\approx} V_\beta.$$

Now for each compatible set of decompositions $\{\sigma_{\alpha,F}\}$, Σ becomes a conical polyhedral complex : just take the S_α's to be the images of the sets ($\sigma_{\alpha,F}$—faces of $\sigma_{\alpha,F}$). In particular, this makes Σ into a topological space with piecewise-linear structure; these structures are easily seen to be independent of the choice of $\{\sigma_{\alpha,F}\}$'s. Note that conversely, the structure $\{S_\alpha, V_\alpha\}$ of conical polyhedral complex on Σ determines the $\{\sigma_{\alpha,F}\}$'s: they are just the closures of the connected components of the inverse images in the various $C(F)$'s of the strata S_α. We shall call the structures $\{S_\alpha, V_\alpha\}$ on Σ that arise from choices of $\{\sigma_{\alpha,F}\}$'s, *admissible conical polyhedral subdivisions of Σ*.

Moreover Σ has even more structure : it contains the abstract "lattice"

$$\Sigma_{\mathbf{Z}} = (\coprod_F C(F) \cap \Gamma)/\Gamma$$

(here regard $C(F) \subset U(F) \subset \operatorname{Aut}(D)^0$, so that $C(F) \cap \Gamma$ makes sense), which plays the role of the set of *orders of approach to ∞ in D/Γ*. In fact, let

$$\text{R. S.}(D/\Gamma) = \left\{ \begin{array}{l} \text{set of analytic maps } \varphi\colon \Delta^* \to D/\Gamma \\ \text{without essential singularity at } 0 \in \Delta \end{array} \right\}.$$

(R. S. is short for "Riemann surface" as used by Zariski in higher-dimensional birational geometry). We get a natural surjective map:

$$\text{ord}: \ \text{R.S.}\ (D/\Gamma) \to \Sigma_{\mathbf{Z}}$$

by the procedure: lift φ to $\tilde{\varphi}: H \to D, \quad H = \{z \mid \text{Im}\, z > 0\}$, such that

$$\tilde{\varphi}(z) \bmod \Gamma = \varphi(e^{2\pi i z})$$

hence for some $\gamma_0 \in \Gamma$:

$$\tilde{\varphi}(z+1) = \gamma_0 \tilde{\varphi}(z), \qquad \forall z \in H.$$

Then γ_0 can be shown to lie in $C(F)$ for some F, hence it determines an element of $\Sigma_{\mathbf{Z}}$.

We can now state the main result we hope to prove :

Main Theorem (?). *Let D be a bounded symmetric domain. $\Gamma \subset \text{Aut}\, D^0$ an arithmetic group without elements of finite order, and Σ the piecewise-linear topological space defined by D and Γ as above. Then there is a map*

$$\left[\begin{array}{c} \textit{Admissible conical} \\ \textit{polyhedral subdivisions} \\ \{S_\alpha, V_\alpha\} \textit{ of } \Sigma \end{array}\right] \longmapsto \left[\begin{array}{l} \textit{toroidal embeddings} \\ D/\Gamma \subset \overline{D/\Gamma}, \textit{ where} \\ \overline{D/\Gamma} \textit{ is a compact} \\ \textit{algebraic space} \end{array}\right]$$

such that if $\Sigma(\overline{D/\Gamma})$ is the conical polyhedral complex associated by the theory in [8] *to this toroidal embedding*†, *there is a unique isomorphism φ making the diagram*

$$\begin{array}{ccc} & & \Sigma \\ & \overset{\text{ord}}{\nearrow} & \\ \text{R.S.}\ (D/\Gamma) & & \wr\!\wr \ \downarrow \varphi \\ & \underset{\text{ord}}{\searrow} & \\ & & \Sigma(\overline{D/\Gamma}) \end{array}$$

†In general, $D/\Gamma \subset \overline{D/\Gamma}$ may be a toroidal embedding *with* self-intersection, however, it is without twisting in the sense that for all strata T, the branches of $\overline{D/\Gamma} - D/\Gamma$ through T are not permuted by going around loops in T. This makes it possible to associate a complex of the type defined above to this embedding by a generalization of the procedure in ([8], Ch. II).

commute. The map is a functor in the sense that if (subd)$_1$ *is finer than* (subd)$_2$, *then* $(\overline{D/\Gamma})^{(1)}$ *dominates* $(\overline{D/\Gamma})^{(2)}$. *An integrality condition on the subdivision* $\{S_\alpha, V_\alpha\}$ *characterizes which* $\overline{D/\Gamma}$*'s are nonsingular* (see p. 11 above).

I also expect that certain convexity properties of the subdivision imply $\overline{D/\Gamma}$ projective.

REFERENCES

1. M. Artin: Algebraic Spaces, *Yale Math. Monographs.*
2. A. Borel and W. Baily: Compactification of arithmetic quotients of bounded symmetric domains, *Annals of Math.*, 84 (1966), 442-528.
3. M. Demazure: Sous-groups algébriques des rang maximum du groupe de Cremona, *Annales de ENS*, 3 (1970).
4. F. Hirzebruch: Hilbert modular surfaces, *L'Enseigement Math.* (1973).
5. M. Hochster: Rings of invariants of tori, Cohen-Macauley rings generated by monomials and polytopes, *Annals of Math.* 96, (1972).
6. J.-I. Igusa: On the desingularization of Satake compactifications in Algebraic Groups, *Proc. of AMS Symposia in Pure Math.*, vol. 9, (1966).
7. J.-I. Igusa: A desingularization problem in the theory of Siegel modular functions, *Math. Annalen*, 168 (1967), 228-260.
8. G. Kempf, F. Knudsen, D. Mumford and B. Saint-Donat: Toroidal Embeddings I, *Springer Lecture Notes*, 339, (1973).
9. I. Satake: On the arithmetic of tube domains, *Bull. Amer. Math. Soc.*, 79(1973), 1076.
10. H. Sumihiro: Equivariant Completion, *Kyoto Math. J.*, 14. (1974).

DISCRETE GROUPS AND Q-STRUCTURES ON SEMI-SIMPLE LIE GROUPS

By M. S. RAGHUNATHAN

INTRODUCTION. The main aim of this paper is to establish the following result.

MAIN THEOREM. *Let* $\mathbf{G}$ *be a connected linear semisimple algebraic group defined over* $\mathbf{R}$ *of* $\mathbf{R}$*-rank* $\geqslant 2$ *and with trivial centre. Let* G *be the connected component of the identity in* $\mathbf{G}_{\mathbf{R}}$, *the group of* $\mathbf{R}$*-rational points of* $\mathbf{G}$. *Assume that* G *has no non-trivial compact connected normal subgroup. Let* $\Gamma \subset G$ *be a discrete subgroup such that* G/Γ *is non-compact and has finite Haar measure. Assume further that* $\Gamma \cap G'$ *is finite for every connected proper normal subgroup* G' *of* G. *Then there exists a* $\mathbf{Q}$*-algebraic group* $\mathbf{G}^*$ *and an isomorphism (defined over* $\mathbf{R}$) $f\colon \mathbf{G}^* \to \mathbf{G}$ *such that* $f^{-1}(\Gamma) \subset G^*_{\mathbf{Q}}$ *and for every proper* $\mathbf{Q}$*-parabolic-subgroup* $\mathbf{P}$ *of* $\mathbf{G}^*$, $\mathbf{P} \cap f^{-1}(\Gamma)$ *is an arithmetic subgroup of* $\mathbf{P}$.

If in addition, there exists a unipotent $\theta \in \Gamma$ *which is contained in two distinct maximal unipotent subgroups of* Γ, *then* $f^{-1}(\Gamma)$ *is an arithmetic subgroup of* $\mathbf{G}^*$; *in this case,* $\mathbf{Q}$*-rank* $\mathbf{G}^* \geqslant 2$.

The main theorem with the exception of the last assertion was announced by Margolis [1] in 1969. (In a recent letter Margolis informs the author that he has proved the arithmeticity of $f^{-1}(\Gamma)$ in all cases.) As no proofs were available for a long time the present author pursued the problem on his own despite the announcement. Most of the work in this paper was done while the author was visiting at Osaka University during March-June 1972. The author would like to take this opportunity to express his warmest thanks to the J.S.P.S. and Professor Murakami and his colleagues for their kind hospitality. Later, in January 1973, the author spoke on the work at the International Colloquium on "Discrete subgroups of Lie groups" held in Bombay.†

Margolis' proof is apparently based on the following result:

†The proofs given here are somewhat different from those outlined in the talk.

THEOREM. *Let $X \in M(n, \mathbf{R})$ be a nilpotent matrix. Then the map $f: \mathbf{R} \to SL(n, \mathbf{R})/SL(n, \mathbf{Z})$ defined by $f(t) = \pi(\exp tX)$, where $\pi: SL(n, \mathbf{R}) \to SL(n, \mathbf{R})/SL(n, \mathbf{Z})$ is the natural map, is not proper.*

(A proof of the theorem by Margolis [2] has appeared. The result was conjectured first by Pjatetski-Shapiro [1] and some years later, independently by H. Garland and the author (see Raghunathan [4])).

The author tried without success to prove this result and eventually obtained the main theorem without the use of the lemma. Some parts of this paper especially in §2 can perhaps be shortened with the aid of this theorem; but it does not help shorten the paper on the whole significantly. A theorem proved by the author (Raghunathan [1], Theorem 11.16) can in some ways be regarded as the starting point. Repeated use is also made of the basic result on the existence of unipotent elements in discrete groups due to Kazdan-Margolis [1].

We now indicate how the main result is proved explaining in the process also how the paper is organised.

The main result above is formulated for lattices. However, we will work with what we call "L-subgroups". The class of L-subgroups is *a priori* a wider class than that of lattices. (However, at the end of the paper one finds that L-subgroups are indeed lattices). In § 1 we generalise Borel's density theorem (Borel [1]) to L-subgroups. A justification for including this material here can perhaps be found in the following: that arithmetic subgroups are L-subgroups is essentially the Godement compactness-criterion which is only the first step towards proving that arithmetic groups are indeed lattices.† The impatient reader can skip this section beyond §1.8 altogether and go on to §2. He need only read "lattice" for "L-subgroup" everywhere.

The material in §2 contains the most difficult steps. We prove here the following (notation as in the statement of the main theorem):

There exists an $\mathbf{R}$-parabolic subgroup $\mathbf{N}$ of $\mathbf{G}$ and a normal $\mathbf{R}$-subgroup $\mathbf{N}_0$ of codimension 1 in $\mathbf{N}$ containing the unipotent

†Curiously enough, we never use the fact, that arithmetic groups are lattices, in this paper.

radical **F** of **N** with the following properties: let **L** be the Zariski-closure of $\mathbf{N} \cap \Gamma$ in **G** and **E** the centre of **F**; then **L**/**E** and **F** carry natural **Q**-structures such that (a) $\mathbf{F} \cap \Gamma$ (resp. image of $\mathbf{L} \cap \Gamma$ in **L**/**E**) is arithmetic and (b) the natural action of **L**/**E** on **F** is defined over **Q**; moreover if $L = \mathbf{L} \cap G$ and $N_0 = \mathbf{N}_0 \cap G$, then $L \subset N_0$ and N_0/L is compact.

The proof is involved and rather difficult to outline. It makes use of the techniques—not merely the final results—of Kazdan and Margolis [1]. In this section, we also introduce the notion of rank for an L-subgroup. L-subgroups of rank 1 and rank > 1 are treated separately in the subsequent sections.

§ 3 begins with the following characterisation of L-subgroups of rank 1.

$\Gamma \subset G$ is (an L-subgroup) of rank 1 if and only if every unipotent ($\neq$ identity) in Γ is contained in a unique maximal unipotent subgroup of Γ.

The case of rank 1 L-subgroups is then handled as follows. One starts with a parabolic subgroup **N** as described above and shows that a conjugate of **N** by any element of Γ not in **N** is opposed to **N**; as a consequence one obtains a semi-direct product decomposition $\mathbf{N} = \mathbf{M}.\mathbf{F}$ with **F** the unipotent radical such that $(\mathbf{M} \cap \Gamma)(\mathbf{F} \cap \Gamma)$ has finite index in $\mathbf{N} \cap \Gamma$. Also the assumption that **R**-rank $\mathbf{G} > 1$ guarantees that $\mathbf{M} \cap \Gamma$ is sufficiently big (to guarantee that its centraliser in **F** is in the centre of $\mathscr{F}$). If $\mathbf{M}_1$ is the Zariski-closure of $\mathbf{M} \cap \Gamma$, then $\mathbf{D} = \mathbf{M}_1.\mathbf{F}$ has a natural **Q**-structure in which $\mathbf{D} \cap \Gamma$ is arithmetic. Now let $\theta \in \Gamma - N$; then $\mathbf{M}_1^\theta = \theta\, \mathbf{N}\, \theta^{-1} \cap \mathbf{D}$ is shown to be a maximal reductive **Q**-subgroup of **D** so that one can find $x, y \in \mathbf{F}_\mathbf{Q}$ such that $x\, \mathbf{M}_1^\theta\, x^{-1} = \mathbf{M}_1$, $y\, \mathbf{M}_1 y^{-1} = \theta^{-1}(\mathbf{M}_1^\theta)\, \theta$, leading to: $x\, \theta^{-1} y$ is in the normaliser of $\mathbf{M}_1$. From this we immediately obtain, what one expects should be the Bruhat-decomposition of elements of Γ. Using this decomposition, it is then shown that trace (ad (Ad $\theta(X)$) ad Y) $\in \mathbf{Q}$, for every pair $X, Y \in \mathfrak{F}$ (= Lie algebra of **F**) such that $\exp X$, $\exp Y \in \mathbf{F} \cap \Gamma$.

§§ 4-5 are independent of §3 (and can be read without reading §3).

In the higher rank case we start again with a parabolic group $\mathbf{N}$ as above. The problem of finding $\mathbf{M} \subset \mathbf{N}$ such that $\mathbf{M} \cap \Gamma . \mathbf{F} \cap \Gamma$ has finite index in $\mathbf{N} \cap \Gamma$, now presents certain difficulties (because no conjugate of $\mathbf{N}$ need be opposed to $\mathbf{N}$). However in §4, we prove the main theorem for L-subgroups of rank $\geqslant 2$ under the additional hypothesis that we can indeed find such an $\mathbf{M}$ in $\mathbf{N}$. Under this additional assumption, one constructs a complete set of what one would expect to be the conjugacy classes maximal $\mathbf{Q}$-parabolics—if there was a $\mathbf{Q}$-structure on $\mathbf{G}$ with Γ arithmetic. Once this is done, there are "enough" unipotents to play around with to obtain the main theorem (including the arithmeticity of Γ).

The problem of finding $\mathbf{M}$ as described in the paragraph above is taken up in §5. It is interpreted as a cohomology-vanishing result and the requisite vanishing theorems are proved when $\mathbf{N}$ is not conjugate to any opposing parabolic subgroup. Where $\mathbf{N}$ is conjugate to an opposing group, the problem is handled as in the rank 1 case.

The appendix contains the proofs of two results used in the main body of the paper.

The following notational conventions are used. As usual $\mathbf{Q}$(resp. $\mathbf{R}$, resp. $\mathbf{C}$) denotes the field of rational (resp. real, resp. complex) numbers and $\mathbf{Z}$, the ring of integers. By an algebraic group we mean a complex algebraic subgroup of $GL(n, \mathbf{C})$ in general. However (especially in §1 and §2) the term algebraic group sometimes refers to the group of $\mathbf{R}$-rational points $\mathbf{G}_{\mathbf{R}}$ of an algebraic group $\mathbf{G}$ defined over $\mathbf{R}$ or even a subgroup of $\mathbf{G}_{\mathbf{R}}$ containing the identity component of $\mathbf{G}_{\mathbf{R}}$. Correspondingly the meaning of terms like "Zariski-dense" and "Zariski-closure" will depend on the context. To avoid (or atleast reduce the possibilities of) confusion, we adopt the following convention: bold face capital roman letters are used to denote complex algebraic groups while real Lie groups are usually denoted by ordinary capital *italic* letters. Further, if $\mathbf{G}$ is an algebraic group defined over $\mathbf{R}$ and $H \subset G_{\mathbf{R}}$ is any "algebraic" subgroup, $\mathbf{H}$, the corresponding bold face letter is used to denote its Zariski-closure in $\mathbf{G}$. Greek capital letters are usually used to denote discrete groups: as far as possible we use corresponding roman (bold face) capital letters to denote their "real" (complex) Zariski-closures (in algebraic groups that contain them). Lie algebras over **R** (resp. **C**) are denoted by gothic lower case (resp. capital gothic) letters. The Lie algebra of a Lie group is denoted by the gothic equivalent of the roman letter used to denote the group. For a Lie group H, H^0 denotes its identity component. (On the whole the notation used is inevitably somewhat unwieldy and, possibly a little confusing; the author seeks the reader's indulgence for this).

Standard results on algebraic groups are used without giving any references. They can in any event be found in Borel [3] or Borel-Tits [2]. Proofs of most of the results on discrete groups used in this paper can be found in Raghunathan [1]. Some results of which we make frequent use are listed below for convenient reference.

I.1. LEMMA. *Let G be a Lie group and $\Gamma \subset G$ a discrete subgroup. Let $\Phi \subset \Gamma$ be any finite set and Z be the centraliser of Φ in G. Then the map $Z/Z \cap \Gamma \to G/\Gamma$ is proper.*

(PROOF. It suffices to show that $Z\Gamma$ is closed in G. Let $z_n \in Z$, $\gamma_n \in \Gamma$ be sequences such that $z_n \gamma_n$ converges to a limit $x \in G$. For $\theta \in \Phi$, $\gamma_n^{-1} z_n^{-1} \theta z_n \gamma_n = \gamma_n^{-1} \theta \gamma_n$ is a convergent sequence contained in Γ. Since Φ is finite, there exists m such that $\gamma_n^{-1}\theta\gamma_n = \gamma_m^{-1}\theta\gamma_m$ for all $\theta \in \Phi$ and $n \geqslant m$ i.e. $\gamma_n\gamma_m^{-1} \in Z$ consequently $x = x\gamma_m^{-1}\gamma_m \in Z\Gamma$).

I.2. THEOREM (Malcev [1]). *Let N be a nilpotent connected simply connected Lie group and $\mathfrak{n}$ its Lie algebra and* exp: $\mathfrak{n} \to N$, *the exponential map. Let $\Gamma \subset N$ be any discrete subgroup. Let $\mathbf{N}$ be a unipotent algebraic subgroup such that $N \subset \mathbf{N}_{\mathbf{R}}$. Then we have the following results.*

(i) *Γ is finitely generated.*

(ii) *N/Γ is compact if and only if Γ is Zariski-dense in $\mathbf{N}$.*

(iii) *If N/Γ is compact, the $\mathbf{Z}$-linear span of* $\exp^{-1}(\Gamma)$ *is a lattice in the (vector space) $\mathfrak{n}$. The $\mathbf{Q}$-Linear span of* $\exp^{-1}(\Gamma)$ *is a $\mathbf{Q}$-Lie subalgebra of $\mathfrak{n}$. Consequently $\mathbf{N}$ acquires a $\mathbf{Q}$-structure. For a subgroup $U \subset N$, $U/U \cap \Gamma$ is compact if and only if the Zariski-closure*

$\mathbf{U}$ *of* U *is defined over* $\mathbf{Q}$. *In particular if* E *denotes the centre of* N, $E/E \cap \Gamma$ *is compact.*

Proofs are also available in Raghunathan ([1], Chapter II).

I.3. THEOREM (Auslander [1], Wang [1]). *Let* G *be a connected Lie group and* Γ *a lattice in* G. *Let* R *(resp.* N*) be the radical (resp. maximal connected nilpotent normal subgroup of* G*). Let* S *be a maximal connected semisimple subgroup of* G *and* S' *the kernel of the action of* S *on* R. *If* S' *has no nontrivial compact connected normal subgroup, then* $R/R \cap \Gamma$ *and* $N/N \cap \Gamma$ *are compact.*

A proof is given in Chapter VIII of Raghunathan ([1], Corollary 8.28).

I.4. THEOREM (Garland-Raghunathan [1]). *Let* G *be a connected linear semisimple Lie group and* $U \subset G$ *a connected unipotent subgroup. Let* Z *be the centraliser of* U *in* G *and* B *a maximal reductive subgroup of* Z. *Then the kernel of the action of* B *on* U *is normal in* G. *If* U *is not contained in any proper normal subgroup of* G, *this kernel is discrete (hence central).*

A proof can also be found in Chapter XI of Raghunathan([1], Proposition 11.19).

I.5. THEOREM (Borel-Tits [2]). *Let* G *be a connected linear semisimple Lie group and* U *be a unipotent subgroup. Let* P *be the normaliser of* U. *Let* $\mathfrak{p}$ *(resp.* $\mathfrak{u}$*) be the Lie algebra of* P *(resp.* U*). If* U *is the unipotent radical of* P, *we can find* $X \in \mathfrak{g}$ *with the following properties:*

(i) ad X *is semisimple and all its eigen-values are real*;

(ii) *for* $\lambda \in \mathbf{R}$, *let* $\mathfrak{g}(\lambda) = \{v \in \mathfrak{g} \mid [X, v] = \lambda v\}$; *then*

$$\mathfrak{u} = \sum_{\lambda > 0} \mathfrak{g}(\lambda), \ \mathfrak{p} = \sum_{\lambda \geq 0} \mathfrak{g}(\lambda).$$

For a proof see Borel-Tits [1] (The theorem proved in this reference is for algebraic groups, but the present formulation is an immediate consequence; see also Mostow [1] and Raghunathan ([1], Chap. XII).

I.6. COROLLARY. *Let* G *be a connected linear semisimple Lie group and* $H \subset G$ *an algebraic subgroup with a non-trivial unipotent radical* V.

Let B be a maximal reductive subgroup. Then there exists $X \in \mathfrak{g}$ *(= Lie algebra of G) such that*

(i) ad X *is semisimple and has all eigen-values real,*

(ii) *for* $\lambda \in \mathbf{R}$, *let* $\mathfrak{g}(\lambda) = \{v \in \mathfrak{g} \mid [X, v] = \lambda . v\}$; *then the Lie algebra of V is contained in* $\sum_{\lambda > 0} \mathfrak{g}(\lambda)$,

(iii) $\{\exp tX\}_{-\infty < t < \infty}$ *centralises B.*

I.7. Theorem (Mahler's criterion). *Let*

$$\pi: SL(n,\mathbf{R}) \to SL(n,\mathbf{R})/SL(n,\mathbf{Z})$$

denote the natural map. Let $\{x_n\}_{1 \leqslant n < \infty}$ *be any sequence in* $SL(n,\mathbf{R})$. *Then the following two conditions are equivalent:*

(i) $\pi(x_n)$ *has no convergent subsequence;*

(ii) *there exists* $v_n \in \mathbf{Z}^n - \{0\}$ *such that* $x_n(v_n)$, *tends to* 0.

I.8. Theorem (Godement-criterion). *Let* $\mathbf{G} \subset GL(n, \mathbf{C})$ *be a semi-simple algebraic* $\mathbf{Q}$*-subgroup and* $G = \mathbf{G}^0_{\mathbf{R}}$. *Let* $\Gamma \subset G$ *be a discrete subgroup such that* $\Gamma \cap \mathbf{G}_{\mathbf{Z}}$ *has finite index in* Γ *and* $\mathbf{G}_{\mathbf{Z}}$ *where* $\mathbf{G}_{\mathbf{Z}} = G \cap SL(n,\mathbf{Z})$. *Let* $\pi: G \to G/\Gamma$ *be the natural map. Then the following two conditions on a sequence* $\{x_n\}_{1 \leqslant n < \infty}$ *in G are equivalent:*

(i) $\pi(x_n)$ *has no convergent subsequence,*

(ii) *There exist unipotents* $\theta_n \in \Gamma - \{e\}$ *such that* $x_n \theta_n x_n^{-1}$ *tends to e.*

For a proof, see for instance, Raghunathan ([1], Chapter X).

I.9. Theorem (Zassenhaus [1], Kazdan-Margolis [1]). *Let G be a Lie group. Then there exists a neighbourhood* Ω *of* e *in G with the following property: if* $\Gamma \subset G$ *is any discrete subgroup,* $\Gamma \cap \Omega$ *is contained in a connected nilpotent subgroup of G.*

A neighbourhood Ω of e in G as in the above theorem will be called a *Zassenhaus neighbourhood* of e in G.

A proof is given in Raghunathan ([1], Chapter VIII).

1. *L*-subgroups : a density theorem and consequences.

1.1. Let G be a connected linear semisimple Lie group without compact factors. Let $\mathfrak{g}$ be the Lie algebra of G. We fix once for all a norm, $\| \ \|$, on $\mathfrak{g}$. For $a > 0$, let $V(a) = \{X \in \mathfrak{g} \ \|X\| < a\}$. Fix also a constant $a_0 > 0$ such that the exponential map $\exp: \mathfrak{g} \to G$ maps $V = V(a_0)$ diffeomorphically onto an open subset U of G. For $a < a_0$, let $U(a) = \exp V(a)$. For $g \in U$, we set $|g| = \|X\|$ where $X \in V$ is the unique element such that $\exp X = g$. We need two definitions for the formulation of the first result.

1.2. Definition. *For constants $C > c > 1$ and $a > 0$ with $0 < a < b = aC < a_0$ and a discrete subgroup Φ of G, an element $g \in G$ is (a, c, C)-adapted to Φ if the following conditions hold:*

(i) $\mathrm{Ad}\, g(V(a)) \cup \mathrm{Ad}\, g^{-1}(V(a)) \subset V(b)$,

(*hence* $g\,U(a)\,g^{-1} \cup g^{-1}\,U(a)\,g \subset U(b)$) *where* $b = aC$;

(ii) *for* $x \in \Phi \cap U(b)$, $gxg^{-1} \in U$ *and* $c|x| < |gxg^{-1}| < C|x|$.

1.3. Definition. *A discrete subgroup $\Gamma \subset G$ is an L-subgroup if it has the following property:*

Let $p: G \to G/\Gamma$ be the natural map and $E \subset G$ any subset; then $p(E)$ is relatively compact if and only if there exists a neighbourhood W of e in G such that $g\Gamma g^{-1} \cap W = \{e\}$ for all $g \in E$.

(*Equivalently, for a sequence $g_n \in G$, $p(g_n)$ has no convergent subsequence if and only if there exists a sequence $\theta_n \in \Gamma - \{e\}$ such that $g_n \theta_n g_n^{-1}$ converges to e*).

1.4. Remark. A lattice in G is an L-subgroup. An arithmetic subgroup of G is an L-subgroup (this last fact can be proved directly without appealing to the fact that arithmetic subgroups are lattices).

The next lemma is essentially due to Kazdan and Margolis [1]. We give a proof, as the technique of the proof itself is needed later. Kazdan and Margolis [1] formulate and prove the lemma for lattices; the present formulation is to be found in Raghunathan ([1], Ch. XI).

1.5. LEMMA. *Fix constants C, c, a with $C > c > 1$ and $0 < aC = b < a_0$. Let $\Gamma \subset G$ be an L-subgroup and E a compact subset of G such that*

$$E\Gamma \supset \{g \in G | g\Gamma g^{-1} \cap U(a) = (e)\}.$$

Let S' be the finite set $\{\gamma \in \Gamma | E\gamma \cap EU(b) \neq \varnothing\}$ and S the set of unipotents in S'. Let $d' > 0$, $a > d'$, be a constant such that no element of $S' - S$ has a conjugate in $U(d')$. Let m be the least positive integer such that $c^{m+1} d' \geqslant a$ and let $d = C^{-(m)+1}.a$. Suppose now that we are given elements g, $\{g_r | \leqslant r < \infty\}$ with the following properties: let $h_r = g_r h_{r-1}$ with $h_1 = g_1$ and $\Gamma_r = h_r\, g\, \Gamma g^{-1} h_r{}^{-1}$; then g_{r+1} is (C,c,a)-adapted to Γ_r and $g_{r+1} = (e)$ if $\Gamma_r \cap U(a) = (e)$.

Then, there exists a minimal integer $k \geqslant 0$ such that $\Gamma_{k+1} \cap U(a) = (e)$. Further we can find $\theta \in \Gamma$ such that $h_{k+1}\, g\theta \in E$. Moreover if $g\Gamma g^{-1} \cap U(d) \neq (e)$, then for any $x \in \Gamma$ with $h_{k-1} g x g^{-1} h_{k-1}^{-1} \in U(a)$, $gxg^{-1} \in U(d')$ and $\theta^{-1} x \theta \in S$. In particular, x is unipotent.

PROOF. Let $a' = \inf\{|gxg^{-1}| \,|\, x \in \Gamma - (e),\ gxg^{-1} \in U(a)\}$. Let p be an integer such that $c^p a' > a$. We then claim that $\Gamma_p \cap U(a) = (e)$. For this we observe first that if for some integer $r > 0$, we have $h_r g x g^{-1} h_r^{-1} \in U(a)$ then $|gxg^{-1}| \leqslant c^{-r} |h_r g x g^{-1} h_r^{-1}|$. To see this we argue by induction on r. In fact if

$$h_r\, g x g^{-1}\, h_r^{-1} \in U(a),$$

we have

$$h_{r-1}\, g x g^{-1} h_{r-1}^{-1} = g_r^{-1}(h_r\, g x g^{-1}\, h_r{}^{-1})\, g_r \in U(b).$$

Since g_r is (C,c,a)-adapted to Γ_{r-1} we have

$$c\,|h_{r-1}\, g x g^{-1}\, h_{r-1}^{-1}| < |h_r\, g x g^{-1} h_r^{-1}|. \tag{*}$$

Thus $h_{r-1} g x g^{-1} h_{r-1}^{-1} \in U(a)$ so that by induction hypothesis,

$$|h_{r-1}\, g x g^{-1} h_{r-1}^{-1}| \geqslant c^{r-1} |g x g^{-1}|. \tag{**}$$

The claim now follows from (*) and (**). Consider now $\Gamma_p \cap U(a)$. If $h_p\, g x g^{-1}\, h_p^{-1} \in U(a)$, we conclude that $|g x g^{-1}| < c^{-p} a < a'$ (by virtue of our choice of p); this implies that $g x g^{-1} = (e)$ in view of the definition of a'.

This proves the first assertion.

Let k be the minimal integer such that $\Gamma_{k+1} \cap U(a) = (e)$. This means that

$$(h_{k+1}\, g\, \Gamma\, g^{-1}\, h_{k+1}^{-1}) \cap U(a) = (e)$$

and by the definition of E, we can find $\theta \in \Gamma$ such that $q = h_{k+1}\, g\, \theta \in E$. This proves the second assertion.

Suppose now $g\Gamma g^{-1} \cap U(d) \neq (e)$. Let r be the smallest integer such that for $x \in \Gamma$ with $g x g^{-1} \in U(d)$, $h_{r+1}\, g x g^{-1}\, h_{r+1}^{-1} \notin U(a)$. Evidently $r \leqslant k$. Moreover since $h_r\, g\, x\, g^{-1}\, h_r \in U(a)$, we conclude that $h_p\, g x g^{-1}\, h_p^{-1} \in U(a)$ for all $p \leqslant r$ and hence that for $0 \leqslant p \leqslant r$, we have

$$|h_p\, g x g^{-1}\, h_p^{-1}| \leqslant C^p\, |x|.$$

It follows that $C^{r+1}\, |x| > a$. Since $|x| < d$, we conclude that $r \geqslant m$. Evidently $k \geqslant r$ so that $k \geqslant m$. Now if $x \in \Gamma$ is such that $h_k\, g x g^{-1}\, h_k^{-1} \in U(a)$, we see that $|g x g^{-1}| < c^{-k}.\ a \leqslant c^{-m}.\ a \leqslant d'$. This proves that $|g x g^{-1}| < d'$ i.e., $g x g^{-1} \in U(d')$. Finally we note that we have

$$h_{k+1}\, g\, x\, g^{-1}\, h_{k+1}^{-1} = q\theta^{-1}\, x\, \theta q^{-1} \in U(b)\ (= U\ (aC))$$

where $q \in E$. Thus

$$E\ \theta^{-1} x\, \theta \cap U(b)\ E \neq \varnothing$$

i.e., $\theta^{-1} x\, \theta \in S'$. Since $\theta^{-1} x\, \theta$ has a conjugate $g\, x\, g^{-1}$ in $U(d')$, $\theta^{-1} x\, \theta \in S$.

1.6. Remark. According to Kazdan and Margolis [1] we can find C,c,a with $1 < c < C$, $0 < aC = b < a_0$ such that for *any* discrete subgroup Φ of G, we can find $g \in G$ (a,c,C)-adapted to Φ.

As a consequence we observe that we have

1.7. Theorem. *Let* $\Gamma \subset G$ *be an* L*-subgroup and* $\pi\colon G \to G/\Gamma$ *the natural map. Let* $g_n \in G$ *be any sequence. Then* $\pi(g_n)$ *has no convergent subsequence if and only if we can find unipotents* $\theta_n \in \Gamma - \{e\}$ *such that* $g_n \theta_n g_n^{-1}$ *converges to e.*

(Theorem 1.7 again was established by Kazdan and Margolis when Γ is a lattice.)

This theorem guarantees the existence of nontrivial unipotent elements in an L-subgroup $\Gamma \subset G$ such that G/Γ is not compact.

We will now obtain a generalisation to L-subgroups of a theorem of Borel for lattices.

1.8. THEOREM. *An L-subgroup $\Gamma \subset G$ is Zariski-dense in G.*

PROOF. Let H be the Zariski-closure of G. Then H has finitely many connected components. Replacing Γ by a subgroup of finite index, we can assume that H is connected: note that a subgroup of finite index in Γ is again an L-subgroup. Assume that $H \neq G$. We assert then that H is contained in a proper reductive subgroup of G. Now according to Mostow [1], a maximal connected subgroup M of G is either reductive or its Lie algebra $\mathfrak{m}$ is of the following form: there exists $X \in \mathfrak{g}$ such that ad X is semisimple and has all eigen-values real and $\mathfrak{m}$ is the span of the eigenspaces of Ad X corresponding to the non-negative eigen-values of ad X. Thus to prove the last assertion we may without loss of generality assume that there exists $X \in \mathfrak{g}$ with the following properties;

(i) ad X is semisimple;

(ii) the eigen-values of ad X are real;

(iii) for $\lambda \in \mathbf{R}$, let $\mathfrak{g}^\lambda = \{z \in \mathfrak{g} \mid [X, Z] = \lambda Z\}$ and $\mathfrak{m} = \sum_{\lambda \geqslant 0} \mathfrak{g}^\lambda$; then $\mathfrak{h} \subset \mathfrak{m}$ where $\mathfrak{h}$ is the Lie algebra of H.

Now let $\mathfrak{n} = \sum_{\lambda > 0} \mathfrak{g}^\lambda$ and $\mathfrak{m}_0 = \mathfrak{g}^0$. Let M, M_0 and N be the Lie subgroups corresponding to $\mathfrak{m}, \mathfrak{m}_0$ and $\mathfrak{n}$ respectively. Then M is the semi-direct product of M_0 and N. It is a closed subgroup of G containing H. Further N is unipotent and the exponential maps $\mathfrak{n}$ homeomorphically onto N. Let $g = \exp X$. Clearly then if $\| \ \|$ denotes a norm on $\mathfrak{g}$ with respect to which ad X is symmetric, then

$$\left.\begin{aligned} &\|\mathrm{Ad}\, g^r (Z)\| \geqslant \|Z\| \\ \text{and} \quad & \\ &g^r x g^{-r} = x \quad \text{for } x \in M_0. \end{aligned}\right\} \qquad (*)$$

Consider now the sequence $\{g^r\}_{1 \leqslant r < \infty}$ in G. From (*) one sees easily that $g^r \Gamma g^{-r} \cap W = \{e\}$ for a suitable neighbourhood W of e in G: note that $\Gamma \subset H$ so that every element of Γ is of the form $x.y$ with $x \in M_0$ and $y \in N$. Since Γ is an L-subgroup we can find $\{\alpha_n \in \Gamma(\subset H)\}_{1 \leq n < \infty}$ such that $g^n \alpha_n \in E$, where E is a compact subset of G. Passing to a subsequence, we see that we obtain a sequence λ_n of integers such that $g^{\lambda_n} \theta_n$ converges to a limit x $(\in H)$ where $\theta_n = \alpha_{\lambda_n}$. Now let $\Phi \subset \Gamma$ be a finite set and consider the sequences $g^{\lambda_n} \theta_n \varphi \theta_n^{-1} g^{\lambda_n}$, $\varphi \in \Phi$; these sequences converge to limits. On the other hand for $\varphi \in \Phi$, $\theta_n \varphi \theta_n^{-1} \in H \subset M$, so that we have $\theta_n \varphi \theta_n^{-1} = x_n \exp Y_n$ where $x_n \in M_0$, $Y_n \in \mathfrak{n}$. It follows that

$$g^{\lambda_n} \theta_n \varphi \theta_n^{-1} g^{-\lambda_n} = x_n \cdot \exp \operatorname{Ad} g^{\lambda_n}(Y_n).$$

Thus we conclude that x_n and $\operatorname{Ad} g^{\lambda_n}(Y_n)$ must be convergent sequences. In view of (*), this implies that Y_n must converge as well. But then $\theta_n \varphi \theta_n^{-1}$ must converge as well. Since $\theta_n \varphi \theta_n^{-1} \in \Gamma$, we conclude that for all large n and $\varphi \in \Phi$, we have

$$\theta_n \varphi \theta_n^{-1} = \theta_{n+1} \varphi \theta_{n+1}^{-1} = \varphi^*, \text{ say.}$$

Now assume that Φ is so chosen that the centraliser of Φ in G is the same as that of H(Such a finite set Φ can be found: in fact if $\rho : G \to \operatorname{Aut}_{\mathbf{R}} F$ is a faithful linear representation, we need only choose Φ such that $\rho(\Phi)$ generates the associative subalgebra of $\operatorname{End}_{\mathbf{R}}(F)$ generated by $\rho(H)$). Clearly then an element $h \in G$ commutes with H if and only if h commutes with $\Phi^* = \{\varphi^* \mid \varphi \in \Phi\}$. Consider now the sequence $g^{\lambda_n} \varphi^* g^{-\lambda_n}$. Let $\varphi^* = x^* \exp Y^* \in \Phi^*$ where $x^* \in M_0$ and $Y^* \in \mathfrak{n}$. Then

$$g^{\lambda_n} \varphi^* g^{-\lambda_n} = x^* . \exp\ (\operatorname{Ad} g^{\lambda_n}\ (Y^*)).$$

From (*) we conclude then that this sequence converges if and only if $Y^* = 0$. Since $g^{\lambda_n} \varphi^* g^{-\lambda_n} = g^{\lambda_n} \theta_n \varphi \theta_n^{-1} g^{-\lambda_n}$ for all large n, this sequence does converge so that $Y^* = 0$ i.e. $\varphi^* = x^* \in M_0$. Thus g commutes with Φ^* i.e. H commutes with g. From this one concludes that $H \subset M_0$. We have thus shown that if $H \neq G$, then H is contained in a proper reductive subgroup of G.

Now assume that $H \neq G$ and let M be a proper reductive subgroup of G containing H. M being *reductive*, we can find a representation $\rho : G \to \operatorname{Aut}_{\mathbf{R}} F$ of G and a vector $f_0 \in F$ such that

$$M = \{x \in G \mid \rho(x) f_0 = f_0\}.$$

Now if G/M is compact,

$$E = \{f \in F \mid \rho(G) f \text{ is relatively compact in } F\}$$

is a G-stable subspace which is *non-zero*. The representation of G on E is then equivalent to a unitary representation. Since G has no compact factors, G acts trivially on E, a contradiction since $f_0 \in E$. Thus G/M is not compact. Since $M \supset \Gamma$, G/Γ is not compact. Thus Γ contains a unipotent. Now let $S(\neq e)$ be the minimal connected normal semisimple subgroup of M containing all the unipotents in Γ. We can then write $M = S\,C$ where C is the centraliser of S in M and $S \cap C$ is finite. Let $\mathfrak{s}$ be the Lie subalgebra of $\mathfrak{g}$ corresponding to S. We have then

$$\mathfrak{g} = \mathfrak{s} \oplus \mathfrak{k} = \mathfrak{s} \oplus \mathfrak{c} \oplus \mathfrak{k}'$$

where $\mathfrak{k}$ and $\mathfrak{k}'$ are S-stable and $\mathfrak{c}$ is the Lie subalgebra corresponding to C and

$$\mathfrak{k} = \mathfrak{c} \oplus \mathfrak{k}'.$$

Now if $\mathfrak{k}'$ is a trivial $\mathfrak{s}$-module, $\mathfrak{s}$ is an ideal in $\mathfrak{g}$ so that $G = S.S'$ where S' is the centraliser of S in G. Since G/M is non-compact, we can find $g_n \in S'$ such that $p(g_n)$ (where $p : G \longrightarrow G/\Gamma$ is the natural map) has no convergent subsequence; then there exist unipotents $\theta_n \in \Gamma - (e)$ such that $g_n \theta_n g_n^{-1}$ converges to e, a contradiction since $\theta_n \in \Gamma \subset S$ and S and S' commute. Thus we conclude $\mathfrak{k}'$ is a non-trivial $\mathfrak{s}$-module. Now from the representation theory of real semisimple Lie algebras one concludes easily the following:

Let $\Phi \subset \Gamma$ be a maximal unipotent subgroup. Then there exists $Y \in \mathfrak{k}'$, $Y \neq 0$, such that (i) ad Y is nilpotent and (ii) Ad $\varphi(Y) = Y$ for all $\varphi \in \Phi$. Consider now again a representation $\rho : G \longmapsto \text{Aut } F$ such that there exists a vector $f_0 \in F$ such that

$$M = \{x \in G \mid \rho(x) f_0 = f_0\}.$$

Now the orbit map $t \longmapsto \rho(\exp tY) f_0$ gives a homeomorphism of R onto a closed subset of F (see Rosenlicht [2]). We conclude from this that $\{(\exp tY)\, M\}_{-\infty < t < \infty}$ is a closed subset of G. In particular, the sequence $\{\exp nY\}_{1 \leqslant n < \infty}$ has no convergent subsequent modulo M hence, *a fortiori*, modulo Γ. Let $g_n = \exp nY$. We assume, as we may without loss of generality, that a finite system Σ of generators for Φ is contained in a Zassenhaus neighbourhood Ω of e in G. Now in view of Theorem 1.7, we can find unipotents $\theta_n \in \Gamma - \{e\}$ such that $g_n \theta_n g_n^{-1}$ converges to e. For large n, thus $g_n \theta_n g_n^{-1} \in \Omega$. Since $g_n \Sigma g_n^{-1} = \Sigma \subset \Omega$, θ_n and Σ generate a unipotent subgroup of Γ. In view of the maximality of Φ, $\theta_n \in \Phi$. But then $g_n \theta_n g_n^{-1} = \theta_n$, a contradiction. This completes the proof of Theorem 1.8.

As a consequence, we obtain the following corollaries. We omit their proofs which are analogous to those in the case of lattices. For proofs for lattices see for instance Raghunathan [1, Chapter V]; the proofs given there carry over with minor verbal changes.

1.9. Corollary. *Let $\Gamma \subset G$ be an L-subgroup. Then the centraliser of Γ is the centre of G. The normaliser of Γ in G is discrete.*

1.10. Corollary. *Let $\Gamma \subset G$ be an L-subgroup and H a connected normal subgroup of G. Let H' be the connected centraliser of H in G. Let f (resp. f') be the natural map of G on G/H (resp. G/H'). The following conditions on Γ are equivalent:*

(1) *$f(\Gamma)$ is a discrete subgroup of G/H.*

(2) *$f(\Gamma')$ is a discrete subgroup of G/H'.*

(3) *$\Gamma \cap H$ is an L-subgroup of H.*

(4) *$\Gamma \cap H'$ is an L-subgroup of H'.*

(5) *Γ contains $(\Gamma \cap H).(\Gamma \cap H')$ as a subgroup of finite index.*

(6) *$\Gamma \cap H$ is Zariski-dense in H.*

(7) *$\Gamma \cap H'$ is Zariski-dense in H'.*

1.11. Definition. *An L-subgroup $\Gamma \subset G$ is* reducible *if there exist proper normal subgroups H, H' in G as in Corollary* 1.10 *such that Γ satisfies one of the equivalent conditions* (1) — (7) *of that corollary. If Γ is not reducible we will say that Γ is* irreducible.

We have then

1.12. Corollary. *The following conditions on an L-subgroup* Γ *of* G *are equivalent:*

(1) Γ *is irreducible,*

(2) *if* H *is any proper connected normal subgroup of* G, $H \cap \Gamma$ *is not an L-subgroup of* H,

(3) *if* H *is any proper connected normal subgroup of* G, $H \cap \Gamma$ *is central in* H,

(4) *if* H *is any proper connected normal subgroup of* G, $H\Gamma$ *is not closed,*

(5) *under the same hypothesis as in* (4), $H\Gamma$ *is dense in* G.

With the Definition 1.11 we can formulate a decomposition theorem.

1.13. Corollary. *Let* $\Gamma \subset G$ *be an L-subgroup; then we can find connected normal subgroups* $\{H_i \mid 1 \leqslant i \leqslant r\}$ *of* G *such that*

(i) $H_1.H_2 \ldots H_r = G$,

(ii) $H_i \cap H_1.H_2 \ldots H_{i-1}.\ H_{i+1} \ldots H_r$ *is finite,*

(iii) $H_i \cap \Gamma$ *is an irreducible L-subgroup of* H_i,

(iv) $\prod_{i=1}^{r} (H_i \cap \Gamma)$ *has finite index in* Γ.

1.14. Corollary. *Let* $\Gamma \subset G$ *be a non-uniform L-subgroup. Then* Γ *is irreducible if and only if no unipotent element of* Γ *belongs to a proper (connected) normal subgroup of* G.

2. *P*-subgroups. From now on, we fix once for all a *non-uniform* irreducible L-subgroup Γ of G. We state below two results which are proved for lattices in Raghunathan ([1], Chapter XI). The proofs given there carry over with minor verbal changes to the case of L-subgroups.

2.1. Theorem. *Let* Δ *be a subgroup of* Γ *and* Φ *a unipotent subgroup of* Γ *maximal among all unipotent subgroups of* Γ *normalised by* Δ. *Assume that* $\Phi \neq (e)$. *Let* D*(resp.* F*) be the Zariski-closure of* Δ *(resp.* Φ*). Let* $\mathfrak{f}$ *be the Lie algebra of* F *and* $\sigma : D \longmapsto GL(\mathfrak{f})$ *the* *adjoint representation of D on* $\mathfrak{f}$. *Let M be a maximal reductive subgroup of D. Then* $M \cap$ (*kernel* σ) *is finite.*

The second result we need is a consequence of the above theorem (cf. 1.1 for notation not explained above).

2.2. THEOREM. *There exists a constant* $\beta > 0$, $\beta < a_0$ *such that an element* $\gamma \in \Gamma$ *has a conjugate in* $U(\beta)$ (*by an element of G*) *if and only if* γ *is unipotent.*

We will now sharpen Theorem 2.1 further.

2.3. THEOREM. *The hypothesis and notation are those of Theorem* 2.1. *Then Centraliser* $F =$ *Centre of G. Centre of F. Consequently* $M \cap \ker \sigma \subset$ *Centre of G.*

We begin with

2.4. CLAIM. *Let* $Z =$ *Centraliser of F. Then* $Z/Z \cap \Gamma$ *is compact.*

We observe first that the map

$$Z/Z \cap \Gamma \to G/\Gamma$$

is proper (in fact $Z =$ Centraliser of Φ and Φ is finitely generated; (Theorem I.2); hence Lemma I.1 applies). Suppose now $z_n \in Z$ is any sequence such that $p(z_n)$, $p\colon Z \to Z/Z \cap \Gamma$ being the natural map, has no convergent subsequence; then we can find $\theta_n \in \Gamma - e, \theta_n$ unipotent, such that $z_n\, \theta_n z_n{}^{-1}$ converges to e. Now we assume, as we may, that a finite set of generators Σ of Φ is contained in a Zassenhaus neighbourhood Ω of e in G. Then for $\theta \in \Sigma$, $z_n\, \theta\, z_n^{-1} = \theta$ so that for all large n, $z_n \theta_n z_n^{-1}$ and $\Sigma = z_n\, \Sigma\, z_n^{-1}$ generate a nilpotent (hence unipotent) subgroup. Forming successive commutators one obtains a sequence $\varphi_n \in Z \cap \Gamma - (e)$ of unipotents such that $z_n\, \varphi_n\, z_n^{-1}$ converges to e. It is evidently sufficient to prove the following now.

2.5. ASSERTION. *Any unipotent element in* $Z \cap \Gamma$ *belongs to* Φ.

To prove this, we argue as follows. Let S be the Zariski-closure of $Z \cap \Gamma$. Since Δ normalises Φ, Δ normalises Z as well as $Z \cap \Gamma$. Let $\Delta' = \Delta.\ Z \cap \Gamma$ and D' the Zariski-closure of Δ'. Now Φ is evidently maximal among unipotent subgroups normalised by Δ'. It follows that any maximal reductive subgroup of D' and hence, *a fortiori*, any maximal reductive subgroup of S acts on F with finite kernel. Since S acts trivially on F we conclude that any maximal reductive subgroup of S is finite. It follows that the set of all unipotent elements in $Z \cap \Gamma$ generate a *unipotent* subgroup Ψ. Evidently Ψ normalises Φ so that $\Psi \cdot \Phi$ is a unipotent subgroup of Γ. The group Ψ is normalised by Δ so that $\Psi \cdot \Phi$ is normalised by Δ. In view of the maximality of Φ, $\Psi\Phi = \Phi$ i.e. $\Psi \subset \Phi$. This proves the assertion and hence the claim.

2.6. To establish Theorem 2.3 we argue as follows: $M \cap \ker \sigma \subset Z$; let B be a maximal reductive subgroup of Z containing this finite group. Then the kernel of the action of B on the unipotent radical N of Z is a normal subgroup of G (cf. Theorem I), hence a semisimple subgroup without compact factors. It follows from Theorem I that (since $Z/Z \cap \Gamma$ is compact) $N/N \cap \Gamma$ is compact. Now $N \cap \Gamma \subset \Psi \subset \Phi$ and since $B \subset Z$, B acts trivially on Φ. Further $N \cap \Gamma$ is the Centre of Φ hence non-trivial. According to Corollary 1.14, $N \cap \Gamma$ is not contained in any proper connected normal subgroup of G. It follows now (again from Theorem I) that B is normal in G and hence discrete and central. This proves Theorem 2.3.

2.7. Suppose now that Φ is a unipotent subgroup of Γ and F is the Zariski-closure of Φ. Assume that the centraliser of F is (centre of G. centre of F) (this holds in particular when Φ ($\neq (e)$) is maximal among unipotent groups normalised by a subgroup Δ of Γ: Theorem 2.3). Let N^* denote the normaliser of F and N the identity component of N^*. Let σ denote the adjoint action of N^* on F, as well as on the Lie algebra $\mathfrak{f}$ of $F: \sigma: N^* \to GL(\mathfrak{f})$. Let

$$N_0^* = \{x \in N^* \mid \det \sigma(x) = \pm 1\}$$

$$N_0 = \{x \in N \mid \det \sigma(x) = 1\}$$

(note that $N_0 = N_0^* \cap N$). Finally, let $\mathscr{L} \subset \mathfrak{f}$ denote the $\mathbf{Z}$-span of $\exp^{-1}\Phi$; $\mathscr{L}$ is a lattice in $\mathfrak{f}$ (see Theorem I.2). We have then

2.8. PROPOSITION. *A sequence $x_n \in N_0$ is relatively compact modulo Γ if and only if $\sigma(x_n)$ (in $GL(\mathfrak{f})$) is relatively compact modulo $GL(\mathscr{L})$.*

The proofs below, of this proposition as well as the next one, were obtained in collaboration with H. Garland.

2.9. Let $\{x_n\}_{1\leqslant n<\infty}$ be a sequence in N_0 such that $\sigma(x_n)$ is relatively compact modulo $GL(\mathscr{L})$. Assume that x_n is *not* relatively compact modulo Γ. Replacing $\{x_n\}_{1\leqslant n<\infty}$ by a subsequence if necessary, we can assume that there exist $\theta_n \in \Gamma - e$, θ_n unipotent such that $x_n\,\theta_n\,x_n^{-1}$ converges to e. On the other hand, we can find $u_n \in GL(\mathscr{L})$ such that $\sigma(x_n)u_n$ is relatively compact. It follows that we can find a sequence of bases $\{e_n(i)|1\leqslant i\leqslant p\}_{1\leqslant n<\infty}$ of $\mathscr{L}$ and a compact subset E of $\mathfrak{f}$ such that $\sigma(x_n)\,e_n(i) \in E$ for $1\leqslant i\leqslant p$ and $1\leqslant n<\infty$. Now we can find an integer $r>0$ such that $\exp\, r.x\in\Phi$ for all $x\in\mathscr{L}$. Let $\varphi_n(i)=\exp\, r.\,e_n(i)$. Evidently then, we can find a compact subset $E'\subset F$ such that $x_n\,\varphi_n(i)\,x_n^{-1}\in E'$ for $1\leqslant i\leqslant p$ and $1\leqslant n<\infty$. Let $g\in G$ be an element such that $gE'g^{-1}\subset\Omega$, a Zassenhaus neighbourhood of e in G. Then for large n, $gx_n\varphi_n(i)x_n^{-1}g^{-1}$ as well as $gx_n\theta_nx_n^{-1}g^{-1}$ belong to Ω. Thus $x_n\varphi_n(i)x_n^{-1}$ and θ_n generate together a nilpotent, hence unipotent group. Forming successive commutators of θ_n with the $\varphi_n(i)$ we obtain a unipotent sequence $\Psi_n(\neq e)$ of elements in Γ centralising the $\{\varphi_n(i)\,|\,1\leqslant i\leqslant \mathrm{p}\}$ such that $x_n\Psi_nx_n^{-1}$ tends to e as n tends to ∞. Now the group generated by $\varphi_n(i)$, $1\leqslant i\leqslant p$, it is easily seen, is Zariski-dense in F. Thus $\Psi_n\in$ centraliser of F. In view of Theorem 2.3, $\Psi_n\in F$. Let $X_n\in\mathscr{L}$ be the unique (non-zero) element such that $\exp X_n=\Psi_n$. Then $\sigma(x_n)X_n$ tends to zero as n tends to ∞. On the other hand $\sigma(x_n).\,X_n=\sigma(x_n).u_n.\,u_n^{-1}X_n$ where $\sigma(x_n).\,u_n$ is relatively compact in $GL(\mathfrak{f})$ and $u_n\in GL(\mathscr{L})$ so that $u_n^{-1}X_n\in\mathscr{L}$. Since $\mathscr{L}$ is a lattice in $\mathfrak{f}$, $\sigma(x_n)X_n$ cannot converge to e as n tends to ∞, a contradiction. Thus if $\sigma(x_n)$ is relatively compact modulo $GL(\mathscr{L})$, x_n is relatively compact modulo Γ. Conversely if x_n is relatively compact modulo Γ, $\sigma(x_n)$ is relatively compact modulo $GL(\mathscr{L})$. In fact if $\sigma(x_n)$ is not relatively compact, passing to a subsequence if necessary, we can find $e_n\in\mathscr{L}$, $e_n\neq 0$, such that $\sigma(x_n)e_n$ converges to 0. This follows from Mahler's compactness criterion (note that $\sigma(N_0)\subset SL(\mathfrak{f})$). Fixing an integer r such that $\exp\, r\mathscr{L}\subset\Gamma$ and setting $\varphi_n=\exp\, r\,e_n$, we find that $x_n\,\varphi_nx_n^{-1}$ converges to e; a contradiction, since Γ is an L-subgroup. This proves Proposition 2.8.

2.10. Proposition. *The natural maps*

$$N_0/N_0\cap\Gamma\to G/\Gamma$$

and

$$N_0/N_0 \cap \Gamma \to GL(\mathfrak{f})/GL(\mathscr{L})$$

are both proper.

2.11. In view of Proposition 2.8, it suffices to show that

$$N_0/N_0 \cap \Gamma \to G/\Gamma$$

is proper. Let $x_n \in N_0$, $\gamma_n \in \Gamma$ be sequences such that $x_n\gamma_n$ converges to a limit. According to Proposition 2.8, we can find a sequence of bases $\{e_n(i)|1 \leqslant i \leqslant p\}$ of $\mathscr{L}$ such that $\sigma(x_n)e_n(i) \in E (1 \leqslant i \leqslant p)$ where E is a fixed compact subset of $\mathfrak{f}$. Let $r > 0$ be an integer such that $\exp r\mathscr{L} \subset \Phi \subset \Gamma$ and set $\varphi_n(i) = \exp r\ e_n(i)$. Then there is a compact subset $E' \subset F$ such that $x_n\ \varphi_n\ (i)\ x_n^{-1} \in E'$. Now we have

$$x_n\varphi_n(i)x_n^{-1} = x_n\gamma_n^{-1}\gamma_n\varphi_n(i)\ \gamma_n^{-1}\gamma_n x_n^{-1} \in E'.$$

It follows that $\gamma_n\varphi_n(i)\gamma_n^{-1} \in E''$ where E'' is a relatively compact subset of G. Since $\varphi_n(i) \in \Gamma$, we conclude that we can find a finite number $\{\theta_m(i)|1 \leqslant i \leqslant p,\ 1 \leqslant m \leqslant q\}$ such that for $1 \leqslant n < \infty$, we have $\gamma_n\ \varphi_n\ (i)\ \gamma_n^{-1} = \theta_{m(n)}\ (i)$ for some $m(n)$ with $1 \leqslant m(n) \leqslant q$. For each m with $1 \leqslant m \leqslant q$, choose an integer $\nu(m)$ such that $m(\nu(m)) = m$. Then we have

$$\gamma_n\varphi_n(i)\ \gamma_n^{-1} = \gamma_\nu\varphi_\nu(i)\gamma_\nu^{-1}$$

where $\nu = \nu(m(n))$. Since the $\varphi_n(i)$, $1 \leqslant i \leqslant p$, generate a Zariski-dense subgroup of F, we see that $\xi_n = \gamma_n \cdot \gamma_\nu^{-1}$ normalises F i.e. $\xi_n \in N^*$. Evidently $x_n \cdot \xi_n$ is relatively compact. The group $N^* \cap \Gamma$ leaves Φ stable. Thus $\sigma(N^* \cap \Gamma)(\mathscr{L}) = \mathscr{L}$. It follows that $\det \sigma(x) = \pm 1$ for all $x \in N^* \cap \Gamma$ i.e. $N^* \cap \Gamma \subset N_0^*$ so that $\xi_n \in N_0^*$. Finally, since N_0 has finite index in N_0^*, $N_0 \cap \Gamma$ has finite index in $N_0^* \cap \Gamma$. It follows that we can find $\xi_n' \in N_0 \cap \Gamma$ such that $x_n\xi_n'$ is relatively compact. This completes the proof of Proposition 2.10.

2.12. From now on G *is assumed to have trivial centre.* Let

$\mathscr{U}^*$ = set of all subgroups Δ of Γ which normalise a nontrivial unipotent subgroup of Γ.

$\mathscr{U} = \{\Delta \in \mathscr{U}^* | \Delta$ is generated by unipotents$\}$.

For $\Delta \in \mathscr{U}^*$, let

$\mathscr{U}(\Delta) = \{\Phi \subset \Gamma | \Phi$ unipotent subgroup normalised by $\Delta\}$.

Note that we have for $\Delta \in \mathscr{U}^*$ and $\Phi \in \mathscr{U}(\Delta)$,

$$\Phi \in \mathscr{U}(\Delta\Phi) \subset \mathscr{U}(\Delta).$$

Also, if in addition $\Delta \in \mathscr{U}$, we have $\Delta\Phi \in \mathscr{U}$.

2.13. Definition. *A subgroup* $\Delta \in \mathscr{U}^*$ *is* full *if every* Φ *in* $\mathscr{U}(\Delta)$ *is a subgroup of* Δ. *Equivalently the maximal unipotent normal subgroup of* Δ *is a* (*and hence the* unique) *maximal element in* $\mathscr{U}(\Delta)$.

2.14. Lemma. *Given* $\Delta' \in \mathscr{U}^*$ (*resp.* $\mathscr{U}$), *there exists a full* Δ *in* $\mathscr{U}^*$ (*resp.* $\mathscr{U}$) *such that* $\Delta' \subset \Delta$.

Proof. In fact, let Φ be a maximal element in $\mathscr{U}(\Delta')$. Then $\Delta'\Phi \in \mathscr{U}^*$ (resp. $\mathscr{U}$ if $\Delta' \in \mathscr{U}$). Evidently Φ is the maximum unipotent normal subgroup of $\Delta = \Delta'\Phi$. Hence the lemma.

2.15. Suppose now that $\Delta \in \mathscr{U}^*$ is any element with Zariski-closure D. Then $D = M.U$ where M is reductive algebraic, U is the unipotent radical of D and $M \cap U = (e)$. Moreover the e-component of M itself decomposes further into an almost direct product $S.C$ where S is semisimple and C is central in M. Also, SC has finite index in M. Further if $\Delta \in \mathscr{U}$, it is easily seen that D and M are connected, $M = S$ and $C = \{e\}$. For $\Delta \in \mathscr{U}$, we define the *content* of Δ as the dimension of $M(=S)$ and denote it $c(\Delta)$. Finally let $p(\Gamma) = \max\{c(\Delta) | \Delta \in \mathscr{U}\}$. (Note that $c(\Delta)$ has *not* been defined for $\Delta \in \mathscr{U}^*$). Occasionally, we refer to $p(\Gamma)$ as the *parabolic content* of Γ.

2.16. Definition. *A subgroup* Δ *of* Γ *is a P-subgroup if*

(i) $\Delta \in \mathscr{U}$,

(ii) $c(\Delta) = p(\Gamma)$,

and (iii) Δ *is full.*

2.17. Notation. There always exist P-subgroups: in fact, we first choose $\Delta' \in \mathscr{U}$ with $c(\Delta') = c(\Gamma)$. Evidently any full subgroup $\Delta \in \mathscr{U}$ containing Δ' is a P-subgroup. In the sequel we choose and *fix once for all a P-subgroup* Δ *of* Γ. We continue with some of the notation already introduced: thus

Φ = maximum normal unipotent subgroup of Δ,

D = Zariski-closure of Δ, $\mathfrak{d}$ its Lie algebra,

$F \;= $ Zariski-closure of Φ, $\mathfrak{f}$ its Lie algebra,
$N^* =$ Normaliser of F, $\mathfrak{n}$ its Lie algebra,
$N \;=$ Identity component of N^*,
σ^* (resp. σ) $=$ the adjoint representation of N^* (resp. N) on $\mathfrak{f}$,
$N_0^* = \{x \in N \mid \det \sigma(x) = \pm 1\}$,
$N_0 = N \cap N_0^* \,(= \{x \in N \mid \det \sigma(x) = 1\})$,
$H \;=$ kernel σ ($=$ centre of F: Theorem 2.3),
$\Theta \;= H \cap \Phi$,
$\Lambda^* = N^* \cap \Gamma \;(= N_0^* \cap \Gamma)$,
$\Lambda \;= N \cap \Gamma \;(= N_0 \cap \Gamma)$,
$L^* =$ Zariski-closure of Λ^*, $\mathfrak{l}$ its Lie algebra,
$L \;= L^* \cap N$,
$\mathscr{L} = \mathbf{Z}$-span of $\exp^{-1} \Phi$ in $\mathfrak{f}$.

Θ is a lattice in H. Also for a subgroup $B \subset N^*$, we denote by B' the subgroup $\sigma^*(B)$ in $GL(\mathfrak{f})$.

It is known that the $\mathbf{Q}$-span $\mathfrak{f}_{\mathbf{Q}}$ of $\mathscr{L}$ in $\mathfrak{f}$ is a $\mathbf{Q}$-Lie algebra and that the inclusion $\mathfrak{f}_{\mathbf{Q}} \hookrightarrow \mathfrak{f}$ induces an isomorphism $\mathfrak{f}_{\mathbf{Q}} \otimes_{\mathbf{Q}} \mathbf{R} \to \mathfrak{f}$. This $\mathbf{Q}$-structure on $\mathfrak{f}$ gives a structure of $\mathbf{Q}$-algebraic group on the group F. The structure of $\mathfrak{f}$ as a vector space defined over $\mathbf{Q}$, gives a natural $\mathbf{Q}$-structure on $GL(\mathfrak{f})$ (isomorphic over $\mathbf{Q}$ to $GL(p, \mathbf{R})$, $p = \dim \mathfrak{f}$). These $\mathbf{Q}$-structures have some obvious properties. Evidently Φ is contained in the $\mathbf{Q}$-rational points of F. Secondly, $\sigma(N_0 \cap \Gamma)$ consists of elements of $GL(\mathfrak{f})$ which leave $\mathscr{L}$ stable; they are, thus, in particular, $\mathbf{Q}$-rational points. It follows that $L^{*\prime}\,(= \sigma^*(L^*))$ and D' $(= \sigma^*(D))$ are $\mathbf{Q}$-subgroups of $GL(\mathfrak{f})$.

Let S' be a maximal reductive $\mathbf{Q}$-subgroup of D' and $R^{*\prime} \supset S'$ a maximal reductive $\mathbf{Q}$-subgroup of $L^{*\prime}$. We denote the Lie algebras of $L^{*\prime}$, D', S' and $R^{*\prime}$ by $\mathfrak{l}'$, $\mathfrak{d}'$, $\mathfrak{s}'$ and $\mathfrak{r}'$ respectively:

$R^* = \sigma^{*-1}(R^{*\prime})$, $\mathfrak{r}$ its Lie algebra,
$R \;= R^* \cap N \;(= R^* \cap N_0)$,
$S \;= \sigma^{*-1}(S') \;(= \sigma^{-1}(S))$, $\mathfrak{s}$ its Lie algebra,
$\Sigma \;= S \cap \Gamma$, $\Sigma_u =$ subgroup generated by unipotents in Σ,
$P \;= R \cap \Gamma$.

The two assertions in parenthesis follow from the fact that $N \cap \Gamma \subset N_0 \cap \Gamma$ and that D and D' are connected (as $\Delta \in \mathscr{U}$). Note that S/H (resp. R^*/H) is isomorphic to S' (resp. $R^{*\prime}$). Finally we remark that $\dim S' = c(\Delta) = p(\Gamma)$, (since H is unipotent). We will now prove

2.18. PROPOSITION. *Λ' is an arithmetic subgroup of L' and Σ' is an arithmetic subgroup of S'. The group Σ' is Zariski-dense in S' and Σ_u is Zariski-dense in S. Finally, if $p(\Gamma) = 0$ then $S' = (e)$ and L'/Λ' is compact.*

PROOF. L' is the e-component of the algebraic $\mathbf{Q}$-group $L^{*\prime}$. Moreover, $\Lambda' \subset L' \subset N_0'$ so that (in view of Proposition 2.10) the map

$$L/(\Lambda' \to GL(\mathfrak{f})/GL(\mathscr{L})$$

is proper. We conclude that Λ' is arithmetic in L'. Consider now the map $\pi : L' \to L'/V$ where V is the unipotent radical of L'. The Zariski-closure of Δ' in $GL(\mathfrak{f})$ is $D' = S'.F'$. Clearly $\pi(\Delta')$ is Zariski-dense in $\pi(S')$. On the other hand, $\pi(\Delta') \subset \pi(\Lambda')$ and $\pi(\Sigma')$ being an arithmetic subgroup of $\pi(D')$, $\pi(\Sigma') \cap \pi(\Lambda')$ has finite index in $\pi(\Lambda')$. Thus we can find an integer r such that for $x \in \pi(\Delta')$, $x^r \in \pi(\Sigma')$. If x is unipotent then so is $x^r = y \in \pi(\Sigma')$. Let $z \in \Sigma'$ be such that $\pi(z) = y$. Then z is unipotent i.e., $z \in \Sigma_u'$. We conclude that $\pi(\Sigma_u')$ is Zariski-dense in $\pi(S')$—note that $\pi(\Delta')$ is generated by unipotents. That Σ_u is Zariski-dense in S follows from the exactness of the sequence

$$(e) \to H \to S \to S' \to (e)$$

since $\Theta \subset \Sigma$ is Zariski-dense in H.

If $p(\Gamma) = 0$, $S' = (e)$. The Zariski-closure D of Δ is unipotent. Hence $\Delta = \Phi$ is a maximal unipotent subgroup of Γ. We conclude from this easily that the arithmetic subgroup $\pi(\Lambda')$ in the reductive group Λ'/V contains no unipotents. Hence $(\Lambda'/V)/\pi(\Lambda')$ is compact. Since $V/V \cap \Lambda'$ and H/Θ are compact, L/Λ is compact. Hence the proposition.

2.19. As was seen in the course of the proof given above, the $\mathbf{Q}$-rational unipotents in the $\mathbf{Q}$-group S' generate a Zariski-dense subgroup of S'. It follows that S' is semisimple and has no connected normal **Q**-subgroup anisotropic over **Q**. Let $r = r(\Delta)$ denote the **Q**-rank of S'. Evidently $S' = (e)$ if and only if $r(\Delta) = 0$ i.e. $p(\Gamma) = 0$ if and only if $r(\Delta) = 0$.

2.20. DEFINITION. *The integer $r(\Delta)$ will be called the* rank *of the P-subgroup Δ. The rank of Γ is the integer*

$$r(\Gamma) = \sup \{r(\Delta + 1) \mid \Delta \text{ a } P\text{-subgroup of } \Gamma\}.$$

(Observe that $r(\Gamma) = 1$ if and only if $p(\Gamma) = 0$.)

We will next establish two propositions which will be needed in the proof of the main result (Theorem 2.25 below). The first of these is

2.21. PROPOSITION. *If Ψ is a unipotent subgroup of Γ normalised by Σ_u, then Ψ is contained in Φ.*

2.22. Since Θ normalises Ψ (note that $\Theta \subset \Sigma_u$) and Σ_u normalises Θ as well as Ψ, Σ_u normalises the *unipotent* group $\Theta\Psi$. Replacing Ψ by $\Theta\Psi$, we can thus assume that $\Psi \supset \Theta$. Clearly $\Phi \cap \Psi \supset \Theta$. Now let

$$\mathcal{M} = \{ \Psi' \in \mathcal{U}(\Sigma_u) \mid \Psi' \supset \Theta,\ \Psi' \not\subset \Phi \}.$$

Evidently, it suffices to show that $\mathcal{M}$ is empty. Assume that $\mathcal{M} \neq \varnothing$. For $\Psi' \in \mathcal{M}$, let $d(\Psi')$ denote the dimension of the Zariski-closure of $\Phi \cap \Psi'$. Choose $\Phi_1 \in \mathcal{M}$ such that

$$d(\Phi_1) \geqslant d(\Psi')$$

for all $\Psi' \in M$. We will then show that $\Phi_1 \subset \Phi$ i.e. $\Phi_1 \notin \mathcal{M}$, a contradiction.

Let F_1 (resp. F^*) denote the Zariski-closure of Φ_1 (resp. $\Phi^* = \Phi \cap \Phi_1$). Let I be the normaliser of F^*. Let Γ_1 be the subgroup of Γ generated by Σ_u, $I \cap F_1 \cap \Gamma$ and $I \cap F \cap \Gamma$ and G_1 its Zariski-closure. Then $\Gamma_1 \in \mathcal{U}$. Let V be the unipotent radical of G_1 and $p : G_1 \to G_1/V$ the natural map. Evidently G_1/V is reductive and $\dim G_1/V = c(\Gamma_1)$ and $c(\Sigma_u) = c(\Delta) \geqslant c(\Gamma_1)$. On the other hand, $S \cap V = H$, as is easily seen so that $c(\Sigma_u) = \dim S/H \leqslant \dim G_1/V$. Thus p maps S onto G_1/V. Now $p(I \cap F_1)$ and $p(I \cap F)$ are evidently unipotent subgroups normalised by $p(S) = G_1/V$ i.e. they are normal in the reductive group G_1/V. Thus $p(I \cap F) = p(I \cap F_1) = \{e\}$. $I \cap F$ and $I \cap F_1$ are contained in V, i.e., they generate a unipotent subgroup.

Consequently the group Φ_2 generated by $I \cap F_1 \cap \Gamma$ and $I \cap F \cap \Gamma$ is unipotent. Now $\Phi_2 \supset \Phi \cap \Phi_1 \supset \Theta$. We assert next that $F_1 \neq F^*$: if $F_1 = F^*$, $\Phi_1 \subset F^* \cap \Gamma \subset F \cap \Gamma = \Phi$ so that $\Phi_1 \notin \mathscr{M}$ (in view of the maximality property of Φ). It follows that $I \cap F_1 \underset{\neq}{\supset} F^*$. Also $I \cap F_1 \not\subset F$ and $(I \cap F_1)/(I \cap F_1 \cap \Gamma)$ is compact as is easily seen. We conclude therefore that $\Phi_2 \not\subset \Phi$. Hence $\Phi_2 \in M$ on the other hand $\Phi_2 \cap \Phi \supset I \cap F \cap \Gamma$ (and $(I \cap F)/(I \cap F \cap \Gamma)$ is compact. Hence

$$d(\Phi_2) = \dim (I \cap F) \geqslant \dim F^* = d(\Phi_1).$$

It follows that $I \cap F = F^*$ (by the maximality of $d(\Phi_1)$). But if $F^* \neq F$, $I \cap F \neq F^*$ (since F is unipotent). We conclude therefore that $F^* = F$.

2.23. Now Φ^* and Φ have the same Zariski-closure F. Thus Φ^* is a subgroup of finite index in Φ. Consider now the normaliser Λ^* of Φ in Γ. This group contains Δ. On the other hand it also contains $I \cap F_1 \cap \Gamma$. Let Δ_1 be the subgroup generated by Δ and $I \cap F_1 \cap \Gamma$ and D_1 its Zariski-closure. Let V_1 be the unipotent radical and $q : D_1 \to D_1/V_1$ be the natural map. Then q map S *onto* D_1/V_1 and since $I \cap F_1$ is Σ_u-stable, $q(I \cap F_1)$ is normal in D/V_1; being unipotent, it must be trivial. It follows that the *unipotent* group $V \cap \Gamma$ contains $I \cap F_1 \cap \Gamma$ as well as Φ and is normalised by Δ. It follows that $\Phi \supset V \cap \Gamma$ i.e. $I \cap F_1 \cap \Gamma \subset \Phi$ leading to $I \cap F_1 \subset F = F^*$, a contradiction since the normaliser of the *proper* subgroup F^* of F_1 in F_1 is strictly bigger than F^*.

2.24. Proposition. *Let $\Phi_1 \in \mathscr{U}(\Sigma_u)$. Then if $\Delta_1 \in \mathscr{U}$ normalises Φ_1, Δ_1 normalises Φ.*

Proof. Let $\Delta_2 =$ subgroup generated by Σ_u and Δ_1. Then $\Delta_2 \in \mathscr{U}$. Let Φ_2 be a maximal element of $\mathscr{U}(\Delta_2)$ containing Φ_1 (note that $\Phi_1 \in \mathscr{U}(\Delta_2)$). Then $\Phi_2 \subset \Phi$ (Proposition 2.19). Let F_2 be the Zariski-closure of Φ_2; clearly $\Phi_2 = F_2 \cap \Gamma$. Let I be the normaliser of F_2 and Δ_3 the subgroup generated $I \cap F \cap \Gamma$ and Δ_2. Let D_3 be the Zariski-closure of Δ_3 and V the unipotent radical of D_3 and $q : D_3 \to D_3/V$ the natural map. Then $q(S) = D_3/V$: in fact we have

$$c(\Delta) = c(\Sigma_u) = \dim q(S) \leqslant \dim D_3/V = c(\Delta_3).$$

Thus $q(I \cap F)$ is a unipotent normal subgroup of $q(S) = q(D_3/V)$. Hence $q(I \cap F) = (e)$ (note that $I \cap F/I \cap F \cap \Gamma$ is compact so that $I \cap F \subset D_3$). We have proved therefore that $I \cap F \cap \Gamma \subset V \cap \Gamma$. On the other hand $\Phi_2 \subset V \cap \Gamma$ and $V \cap \Gamma \in \mathscr{U}(\Delta_2)$ so that in view of the maximality of Φ_2, $\Phi_2 = V \cap \Gamma$. In particular, $I \cap F \cap \Gamma \subset \Phi_2$. It follows that $I \cap F \subset F_2$ i.e. F_2 is its own normaliser in F since F is unipotent, this means that $F = F_2$ i.e. $\Phi = \Phi_2$. Hence Δ_1 normalises Φ.

We will now establish the main result of this section.

2.25. THEOREM. *N_0'/L' is compact.*

2.26. Consider the decomposition of $\mathfrak{n}_0'$ as an S'-module. We have

$$\mathfrak{n}_0' = \mathfrak{l}' \oplus E \oplus E^*$$

where S' operates trivially on E and non-trivially on every irreducible S'-submodule of E^*. We then assert that we have

2.27. CLAIM. $E^* = 0$.

To prove the claim, we first observe that S' is a semisimple $\mathbf{Q}$-group without any $\mathbf{Q}$-component anisotropic over $\mathbf{Q}$: in fact Δ' is Zariski-dense in D' and it is generated by $\mathbf{Q}$-*rational unipotents*; moreover S' is a maximal $\mathbf{Q}$-reductive subgroup of D'. Assume $E^* \neq 0$. Fix a maximal $\mathbf{Q}$-split torus $T \subset S'$ and introduce a lexicographic ordering on the group of ($\mathbf{Q}$-) characters on T and let $0 \neq v \in E^*$ be the highest weight vector for T of an S'-irreducible (over $\mathbf{R}$) subspace of E^*. Let λ be the corresponding character on T. Then it is easily seen that ad v is a nilpotent endomorphism and that the map

$$R \times L' \xrightarrow{u} N'_0$$

defined by $u(t, l) = \exp.\, t\, v.\, l$, maps $\mathbf{R} \times L'$ homeomorphically onto a closed subset of N'_0. The map $N_0/\Lambda \to GL(\mathfrak{f})/GL(\mathscr{L})$ being proper, so is the map

$$N_0'/\Lambda' \to GL(\mathfrak{f})/GL(\mathscr{L}).$$

Since $\Lambda' \subset L'$, it follows that $\exp n v \in N'_0$ has no convergent subsequence modulo $GL(\mathscr{L})$. By Mahler's compactness criterion, we can find $e_n \in \mathscr{L} - (0)$ such that $\exp nv(e_n)$ converges to zero. Now we can decompose $\mathfrak{k}$ into eigen-spaces for T. This decomposition respects the **Q**-structure on $\mathfrak{k}$ so that we find that a suitable subgroup $\mathscr{L}'$ of $\mathscr{L}$ of finite index in $\mathscr{L}$ decomposes into a direct sum,

$$\mathscr{L}' = \sum_{\alpha \in \Psi} \mathscr{L} \cap \mathfrak{k}_\alpha$$

where Ψ is a finite set of characters on T and for $\alpha \in \Psi$

$$\mathfrak{k}_\alpha = \{X \in \mathfrak{k} \mid t.X = \alpha(t).\ X \text{ for } t \in T\}$$

Clearly we may assume that $e_n \in \mathscr{L}'$ so that

$$e_n = \sum_{\alpha \in \Psi} e_n(\alpha).$$

Now passing to a subsequence $e_{\lambda_n} = f_n$ we can assume that

$$f_n = \sum_{\substack{\alpha \in \Psi \\ \alpha \geqslant \beta}} f_n(\alpha),$$

$f_n(\alpha) \in \mathscr{L}' \cap f_\alpha$ and $\mathfrak{k}_n(\beta) \neq 0$. Applying exp $\lambda_n v$ to f_n, we obtain (in view of the fact that $\lambda_n > 0$)

$$\exp \lambda_n v(f_n) = f_n(\beta) + \xi_n$$

with $\xi_n \in \sum_{\alpha > \beta} \mathfrak{k}_\alpha$. Now $f_n(\beta) \in \mathscr{L}'$ so that $f_n(\beta)$ cannot tend to zero. Thus exp $\lambda_n v f_n$ cannot converge to zero, a contradiction. This establishes Claim 2.27.

We will now prove

2.28. Claim. *Let Z' be the centraliser of S' in N_0'. Then $Z'/Z' \cap \Lambda'$ is compact.*

Assume for the moment that the claim is proved. In view of Claim 2.27, we see that $\mathfrak{n}_0' = \mathfrak{l}' \oplus E$ and E is evidently contained in the Lie algebra of Z'. Since $Z'/Z' \cap \Lambda'$ is compact, we see that $Z'/Z' \cap L'$ is compact. Since $\mathfrak{n}_0' = \mathfrak{l}' + \mathfrak{z}'$, $\mathfrak{z}'$ = Lie algebra of Z', the Z' orbit of the coset L' in N_0'/L' is open. On the other hand, this orbit is compact, since $Z'/L' \cap Z'$ is. It follows from the connectedness of N_0'/L', that N_0'/L' is compact.

2.29. We now begin with some preliminary remarks on the function $|g|$ (defined in a neighbourhood $U(a_0)$ of e in G) introduced in 1.1. Recall that we fixed a norm $\| \ \|$ on $\mathfrak{g}$ (the Lie algebra of G); choose a constant $a_0 > 0$ such that

$$\exp : V(a_0) \ (=\{X \in \mathfrak{g} \mid \|X\| < a_0\}) \longrightarrow G$$

is a diffeomorphism onto an open subset $U(a_0)$ of G. For $g \in U(a_0)$, we set $|g| = \|X\|$ where $X \in V(a_0)$ and $\exp X = g$.

Finally recall that for $a \leqslant a_0$, we set

$$U(a) = \{g \in U(a_0) \mid |g| < a\}$$

and

$$V(a) = \{X \in g \mid \|X\| < a\}.$$

We will now establish

2.30. LEMMA. *There exists a constant $a_1, 0 < a_1 < a_0$ and $M > 0$ such that for $x, y \in U(a_1)$, $xyx^{-1}y^{-1} \in U(a_0)$ and*

$$|xyx^{-1}y^{-1}| < M . |x| . |y|.$$

We can assume moreover that $U(a_1)$ is a Zassenhaus-neighbourhood of e.

PROOF. We assume that $G \subset GL(p, \mathbf{R})$ and identify $\mathfrak{g}$ with a Lie subalgebra of $M(p, \mathbf{R})$. For $X = \{X_{ij}\}_{1 \leqslant i,j \leqslant p}$, let $\|X\|' = (\Sigma X_{ij}^2)^{\frac{1}{2}}$. Then we have for $X, Y \in M(p, \mathbf{R})$, $\|XY\|' \leqslant \|X\|' \|Y\|'$ and $\|[X, Y]\|' \leqslant 2\|X\|' \|Y\|'$. Let $W = \{A = I + X \in GL(p) \mid \|X\|' < \frac{1}{2}\}$. Then for A $(= I + X)$ in W, the series $\Sigma_{1 \leqslant u < \infty}(-1)^n X^n/n$ converges to a limit $\log A = Y \in M(p, \mathbf{R})$ such that $A = \exp Y$. We have then

$$\|Y\|' \leqslant \sum_{1 \leqslant n < \infty} \|X\|'^n/n < \|X\|' \sum_{0 \leqslant n < \infty} 1/2^n < 2\|X\|' < 1 \tag{1}$$

and

$$\|X\|' \leqslant \sum_{1 \leqslant n \leqslant \infty} \|Y\|'^n/n! < \|Y\|' \sum_{0 \leqslant n \leqslant \infty} 1/n! < 3 . \|Y\|'. \tag{2}$$

Next we observe that for $A = I + X \in W$, the Neumann series $\Sigma_{1 \leqslant n < \infty}(-1)^n X^n$ converges to A^{-1} so that we have

$$\|A^{-1}\|' \leqslant \|I\| + \sum_{1 \leqslant n < \infty} \|X\|'^n < \sqrt{p} + 1. \tag{3}$$

Now for $A, B \in W$, we have, setting $A = I + X$, $B = I + Y$,

$$\begin{aligned}\|AB\,A^{-1}B^{-1}-I\|' &= \|(AB-BA)\,A^{-1}\,B^{-1}\|' \\ &\leqslant (\sqrt{p}+1)^2.\,\|AB-BA\|' \\ &= (\sqrt{p}+1)^2\,\|XY-YX\|' \\ &\leqslant 2(\sqrt{p}+1)^2.\,\|X\|'\,\|Y\|' \\ &= Q\,\|X\|'\,\|Y\|'.\end{aligned} \tag{4}$$

Next, let $\alpha > 0$ be a constant such that for all $X \in g$,

$$\alpha^{-1}\,\|X\|' < \|X\| < \alpha\,\|X\|'. \tag{5}$$

Suppose now $A_1, A_2 \in W$ and $A_3 = A_1A_2A_1^{-1}A_2^{-1}$. Let $Y_i = \log A_i$ and $X_i = A_i - I$ for $1 \leqslant i \leqslant 3$. We then have

$$\|Y_3\|' \leqslant 2\,\|X_3\|' \leqslant 2Q\,\|X_1\|'.\,\|X_2\|' \leqslant 18\,Q\,\|Y_1\|'\,\|Y_2\|'. \tag{6}$$

(The first inequality results from (1), the second from (4) and the last from (3)). Now let $a_2 > 0$, $a_2 < a_0$ be a constant such that $V(a_2) \subset \{\log A | A \in W\}$ (and hence $U(a_2) \subset W$). Let $a_1 > 0$ be chosen such that $A_1\ A_2\ A_1^{-1}\ A_2^{-1} \in U(a_2)$ for all $A_1, A_2 \in U(a_2)$. Suppose then that A_1, A_2 are in $U(a_1)$ and $A_1A_2A_1^{-1}A_2^{-1} = A_3 (\in U\ (a_2))$, we have in view of (5) and (6) (setting $Y_i = \log A_i$ $(i = 1, 2, 3)$),

$$|A_3| = \|Y_3\| < \alpha\|Y_3\|' < 18Q\,\alpha\|Y_1\|'\,\|Y_2\|' < 18\,Q\,\alpha^3|\ |A_1|\,|A_2|. \tag{7}$$

This proves the lemma.

A second technical fact we need is

2.31. LEMMA. *Fix constants a_1, M as in Lemma 2.30. Assume $M > 1 > a_1$ and that $U(a_1)$ is a Zassenhaus neighbourhood of e (cf. I.9) with the following property: an element $x \in \Gamma$ has a conjugate in $U(a_1)$ if and only if x is unipotent. Assume that $g\,\Phi\,g^{-1} \cap U(a_1) \neq \{e\}$ and define inductively a sequence $\{\varphi_n \mid 1 \leqslant n < \infty\}$ of elements of Λ^* as follows : $\varphi_1 \neq e$ is an element of Λ^* such that $g\varphi_1g^{-1} \in U(a_1)$ and for any ψ $(\neq e)$ in Λ^* with $g\,\psi\,g^{-1} \in U(a_1)$,*

$$|g\,\varphi_1\,g^{-1}| \leqslant |g\psi g^{-1}|;$$

assume $\varphi_1, \ldots, \varphi_r \in \Lambda^*$ choosen and let Φ_r be the subgroup generated by $\varphi_1, \ldots, \varphi_r$; let F_r be the Zariski closure of Φ_r; set $\varphi_{r+1} = \varphi_r$ if $F_r = F$ or if $g\psi g^{-1} \notin U(a_1)$ for all *$\psi \in \Lambda^* - F_r$; if neither alternative holds, choose φ_{r+1} in $\Lambda^* - F_r$ such that*

$$|g\,\varphi_{r+1}\,g^{-1}| \leqslant |g\,\psi\,g^{-1}|$$

for all $\psi \in \Lambda^* - F_r$ *with* $g\,\psi\,g^{-1} \in U(a_1)$.

Suppose $k > 0$ *is an integer such that* $\varphi_i \in \Phi$ *for* $1 \leqslant i \leqslant k$ *and* F_k *is* Σ_u*-stable. Then for* $\gamma \in \Gamma$ *if*

$$g\,\gamma\,g^{-1} \in U\,(inf\,(a_1, M^{-\dim G}\ B^{-1})$$

where $B = |g\,\varphi_k\,g^{-1}|/|g\,\varphi_1\,g^{-1}|$, *then* $\gamma \in \Lambda^*$.

Proof. Since $U(a_1)$ is a Zassenhaus neighbourhood and the groups Φ_i (and hence F_i) are unipotent for $1 \leqslant i \leqslant k$. The group Ψ generated by γ and Φ_k is necessarily unipotent. Define inductively the sequence γ_n, $0 \leqslant n < \infty$ as follows: $\gamma_0 = \gamma$; assume γ_r defined; if γ_r commutes with F_k, let $\gamma_{r+1} = e$; if not choose $i = i(r)\ 1 \leqslant i \leqslant k$, such that γ_r does not commute with φ_i and set $\gamma_{r+1} = [\gamma_r, \varphi_i]$. Since Ψ is unipotent $\gamma_r = e$ for $r \geqslant \dim G$. Let p be the minimal integer such that $\gamma_{p+1} = e$.

Claim. $p = 0$.

Assume that $p > 0$; then $\gamma_r \neq e$ for $1 \leqslant r \leqslant p$. We will establish the following inequalities inductively for $1 \leqslant r \leqslant p$: let $\varphi_{i(r)} = \theta_r$; then

$$A_r\colon\ |g\gamma_r g^{-1}| < M^{r-\dim G}\,B^{-1}\,|\,g\theta_{r-1}\,g^{-1}| \\ < M^{r-\dim G}\,|\,g\varphi_1\,g^{-1}\,| < M^{r-\dim G}.$$

Assume that A_i holds: then we have

$$|g\gamma_{i+1}\,g^{-1}| < M.\ |g\gamma_i\,g^{-1}|\ |g\theta_i\,g^{-1}| \\ < M^{i+1-\dim G}\,B^{-1}\,|g\theta_{i-1}\,g^{-1}|\,|g\theta_i\,g^{-1}|;$$

since $|g\theta_{i-1}g^{-1}| < a_1 < 1$, and $|g\theta_i g^{-1}| \leqslant |g\varphi_k g^{-1}|$ for all i, the inequality A_{i+1} holds. At the start of the induction we have

$$g\gamma_1 g^{-1} = M\,|g\gamma g^{-1}|\,|g\theta_1 g^{-1}| < M.M^{-\dim G}B^{-1}.\ |g\theta_1 g^{-1}|$$

and A_1 therefore holds. Since $\dim G \geqslant p$ and $M \geqslant 1$, we obtain the following inequality:

$$|\,g\gamma_p g^{-1}\,| < |\,g\varphi_1 g^{-1}\,|. \tag{B}$$

Now γ_p centralises $F_k \cap \Gamma$. Since Σ_u normalises $\Gamma \cap F_k$, γ_p normalises Φ (Proposition 2.24), i.e. $\gamma_p \in \Lambda^*$. But then (B) contradicts our choice of φ_1. Thus $p = 0$ and $\gamma_p = \gamma$ is in Λ^*.

2.32. Lemma. *There is a finite set* $\Xi \subset \Sigma_u$ *such that* $E\, \mathfrak{g}$ *is a* Σ_u*-stable subspace if and only if* $\operatorname{Ad} x\, E = E$ *for all* $x \in \Xi$.

We need only choose Ξ to be a finite set such that $\operatorname{Ad} \Xi$ spans the same finite dimensional subspace as $\operatorname{Ad} \Sigma_u$ in End g.

2.33. Lemma. *Let* $R^* = A.\, H$ *be a semi-direct product decomposition of* R^* *(*H *is the unipotent radical of* R^* *and* A *a maximal reductive subgroup). Then there exists* $X \in \mathfrak{g}$ *with the following properties:*

(i) ad X *is semisimple and has all eigen-values real*

(ii) $\left\{ \exp tX \,\middle|\, -\infty < t < \infty \right\}$ *commutes with* A *and*

(iii) $\mathfrak{f}$ *(resp.* $\mathfrak{n}$*) is contained in the linear span of the eigen-spaces corresponding to the positive (resp. non-negative) eigen-values of* ad X.

This lemma follows easily from Corollary I.6.

2.34. Lemma. *Let* Ξ *be as in Lemma* 2.32 *and* $g = \exp X$. *Then there exists a constant* $\lambda > 0$ *such that the following holds: for* $z \in Z$, $(= \sigma^{-1}(Z'))$ $x \in \Phi$ *and* k *an integer if* $g^k z x z^{-1} g^{-k} \in U(\lambda^{-1} a_0)$, $g^k z^{\xi} \xi x \xi^{-1} z^{-1} g^{-k} \in U(a_0)$ *for all* $\xi \in \Xi$ *and*

$$|g^k z^{\xi} \xi x \xi^{-1} z^{-1} g^{-k}| < \lambda |g^k z x z^{-1} g^{-k}|.$$

Proof. Each element $\xi \in \Xi$ can be written in the form $\xi_1 . \eta_1$ where $\eta_1 \in H$ and $\xi_1 \in A$. We observe then that

$$g^k z \xi x \xi^{-1} z^{-1} g^{-1} = g^k z \xi_1 x \xi_1^{-1} z^{-1} g^{-k}.$$

Now $z \xi_1 x \xi_1^{-1} z^{-1} = \xi_1 z x z^{-1} \xi_1^{-1}$ since $x \in \Phi$: in fact $\sigma(\xi_1)\sigma(z) = \sigma(z)\sigma(\xi_1)$ since Z' and $\sigma(A) = S'$ commute. Since g commutes with ξ_1 we conclude that

$$g^k z \xi x \xi^{-1} z^{-1} g^{-1} = \xi_1 g^k z x z^{-1} g^{-k} \xi_1^{-1}$$

for all $\xi \in \Xi$. Since Ξ is a finite set, we can find a λ of the desired kind: in fact we can find λ such that for all $x \in U(\lambda^{-1} a)$,

$$\xi_1 x \xi_1^{-1} \in U(a) \text{ and } |\xi_1 x \xi_1^{-1}| < \lambda |x|.$$

Hence the lemma.

2.35. LEMMA. *There is a constant* $b > 0$, $b < a_0$ *such that for all integer* k, *for all* $z \in Z$ *and* $\varphi \in \Lambda^*$, $\varphi \notin \Phi$, $g^k z \varphi z^{-1} g^{-k} \notin U(b)$.

PROOF. We observe first that $L^{*\prime} = R^{*\prime}.F'$ and F' is the unipotent radical of $L^{*\prime}$. In fact if V' is the unipotent radical of $L^{*\prime}$ and $V = \sigma^{-1}(V')$, $V' \cap \Lambda^{*\prime}$ and (hence) $V \cap \Gamma$ are lattices in V' and V respectively. Clearly $V' \supset F'$ so that $V \supset F$. V is unipotent and $V \cap \Gamma \supset \Phi$; hence $V \cap \Gamma = \Phi$ (note that Δ normalises $V \cap \Gamma$) leading to $V = F$ and $V' = F'$. Next $R^{*\prime}$ decomposes into an almost direct product $R^{*\prime} = S'.T'$ where T' is a maximal normal $\mathbf{Q}$-subgroup of $R^{*\prime}$ anisotropic over $\mathbf{Q}$. Since $\Lambda^{*\prime}$ is arithmetic in $L^{*\prime}$, the group $\Lambda_1' = (S' \cap \Lambda^{*\prime}).(T' \cap \Lambda^{*\prime}).(F' \cap \Lambda^{*\prime})$ is a subgroup of finite index in $\Lambda^{*\prime}$. From this one sees easily that one need prove the assertion of the lemma under the additional assumption that $\sigma^*(\varphi) = \alpha.\beta.\Psi$ where $\alpha \in S' \cap \Lambda^{*\prime}$, $\beta \in T \cap \Lambda^*$ and $\Psi \in F' \cap \Lambda^{*\prime}$. Now $\alpha.\beta \neq e$ so that atleast one of them does not equal e. We consider first the case when $\alpha \neq e$, $\beta = e$. Let α^* (resp. β^*) be the unique element of A such that $\sigma(\alpha^*) = \alpha$ (resp. $\sigma(\beta^*) = \beta$. Then we find that we have

$$\varphi = \alpha^*.\beta^*.\Psi^*$$

where Ψ^* is some element of F with $\sigma(\Psi^*) = \Psi$. Clearly then $g^k z \varphi z^{-1} g^{-k} = \alpha^*.z\beta^* z^{-k}.g^k z \psi^* z^{-1} g^{-k} (= \alpha^*.g^k z \psi^* z^{-1} g^{-k}$ *if* $\beta = e)$.

Now consider the Lie subgroup G' whose Lie algebra is the linear span $\mathfrak{g}'$ of the eigen-spaces of ad X corresponding to nonnegative eigen-values. Let $\mathfrak{f}'$ be the Lie subalgebra spanned by the eigen-spaces corresponding to positive (resp. 0) eigen-values. Let F' be the corresponding Lie subgroup of G. Let A' = Centraliser of $\{\exp tX \mid -\infty < t < \infty\}$ in G. Then G' is a semidirect product $A'.F'$ with F' as the unipotent radical. We see thus that when $\beta = e$, the element $g^k z \varphi z^{-1} g^{-k}$ has α^* as its projection on the factor A'. Since $\sigma|A'$ is a bijection and $\sigma(\alpha^*) \in \Lambda^* \cap S'$, a discrete group, the desired conclusion follows in the case of elements of the form φ with $\sigma(\varphi) = \alpha.\Psi$, $\alpha \neq e$, $\alpha \in S'$, $\Psi' \in \Phi'$. When $\sigma(\varphi) = \alpha.\beta.\Psi$, $\alpha \in S'$, $\beta \in T'$, $\beta \neq e$, $\Psi \in \Phi$, the characteristic polynomial of β is a monic integral polynomial; since T' is anisotropic over $\mathbf{Q}$, this polynomial is different from the characteristic polynomial of the identity element. These considerations, it is easily seen, imply the lemma.

2.36. LEMMA. *Let* $p = \dim G$. *Fix constants* $M, a_1 > 0$ *as in Lemma* 2.30, $\lambda > 1$ *as in Lemma* 2.34 *and* $b > 0$ *as in Lemma* 2.35. *Let* $X \in \mathfrak{g}$ *be chosen as in Lemma* 2.38 *and a subset* $\Xi \subset \Sigma_u$ *as in Lemma* 2.32. *Let* $\mathfrak{g} = \exp X$. *Let* k *be any integer,* $z \in Z$ *any element and* $h = g^k.z$. *Assume that* $h\, \Phi\, h^{-1} \cap U(\lambda^{-p} \operatorname{Inf}(a_1, b)) \neq (e)$. *Then we have*

$$\{\gamma \in \Gamma \mid h\gamma h^{-1} \in U(\operatorname{Inf}\{M^{-p}\, \lambda^{-p}, b, a_1\})\} \subset \Phi.$$

PROOF. Define inductively a sequence of elements $\{\varphi_n \mid 1 \leqslant n < \infty\}$ in Φ as follows, φ_1 is an element of Φ, different from e such that $h\varphi_1 h^{-1} \in U(a_0)$ and for all $\Psi \in \Phi$ with $(h \Psi h^{-1}) \in U(a_0)$,

$$|h \varphi_1 h^{-1}| \leqslant |h \Psi h^{-1}|.$$

Assume $\varphi_1, \ldots, \varphi_r$ defined and let Φ_r be the group generated by $\varphi_1, \ldots, \varphi_r$. Let F_r be its Zariski-closure. If $F_r = F$, set $\varphi_{r+1} = \varphi_r$. If $F_r \neq F$ and $h(\Phi - F_r)\, h^{-1} \cap U(a_0) = (e)$, again set $\varphi_{r+1} = \varphi_r$. Finally if there exists Ψ with $h \Psi h^{-1} \in U(a_0)$, $\Psi \in \Phi - F_r$, choose for φ_{r+1} one such Ψ taking the least possible value for $|h\Psi h^{-1}|$. We observe then that we have for all $\Psi \in \Lambda^*$ with $h\Psi h^{-1} \in U(a_0)$,

$$|h\varphi_1 h^{-1}| < |h \Psi h^{-1}|.$$

This follows from our choice of b (Lemma 2.35). We next remark that there exists an integer r such that F_r is Σ_u-stable. To see this, first observe that if $|h\varphi_r h^{-1}| < \lambda^{-1} a_0$, then we have for $s \leqslant r$,

$$h\, \xi\, \varphi_s\, \xi^{-1} h^{-1} \in U(a_0) \text{ and}$$

$$|h\, \xi\, \varphi_s\, \xi^{-1} h^{-1}| < \lambda\, |h\, \varphi_s h^{-1}| \leqslant \lambda\, |h\varphi_r h^{-1}|.$$

If F_r is not Σ_u-stable, one atleast of the elements $\{\xi\, \varphi_r \xi^{-1} \mid \xi \in \Xi\}$ does not belong to F_r. It follows that if F_r is not Σ_u-stable and $|h\, \varphi_r\, h^{-1}| < \lambda^{-1} a_0$, then $\varphi_{r+1} \neq \varphi_r$ and

$$|h\varphi_{r+1} h^{-1}| < \lambda\, |h\, \varphi_r h^{-1}|. \qquad (*)$$

Now $|h\varphi_1 h^{-1}| < \lambda^{-p}. \inf(a_1, b)$. Also since $p = \dim \mathfrak{g}$, $F_p = F_{p+k}$ for all $k > 0$. Suppose none of the F_i, $1 \leqslant i \leqslant p$ are Σ_u-stable. Applying (*) repeatedly, one obtains $h\varphi_p h^{-1} \in U(a_0)$ and

$$|h\varphi_p h^{-1}| < \lambda^{-1}(\inf(a_1, b)).$$

But then as seen from the argument given above $\varphi_{p+1} \neq \varphi_p$, a contradiction. Thus there exists a minimal integer k with $1 \leqslant k \leqslant p$ such that F_k is Σ_u-stable — in particular, F_j for $j < k$ are *not* Σ_u-stable. Applying (*) again we conclude that we have

$$|h\varphi_k h^{-1}| < \lambda^{(k-1)} |h\varphi_1 h^{-1}| < \lambda^p |h\varphi_1 h^{-1}|.$$

We are now in the situation of Lemma 2.31. In fact for $\varphi \in \Lambda^* - \Phi$, we have $h \varphi h^{-1} \notin U(b)$ so that $\varphi_1, \varphi_2 \dots \varphi_k$ have the properties required in that lemma. Suppose $\gamma \in \Gamma$, $h\gamma h^{-1} \in U(a_0)$ and

$$|h\gamma h^{-1}| < \inf\{M^{-p}\lambda^{-p}, b, a_1\},$$

then we have

$$|h\gamma h^{-1}| \in U(\inf(a_1, M^{-\dim G}.(|h\varphi_1 h^{-1}|/|h\varphi_k h^{-1}|).$$

Applying Lemma 2.31, we conclude that $\gamma \in \Lambda^*$. Since $h\gamma h^{-1} \in U(b)$, this shows that $\gamma \in \Phi$ (Lemma 2.35). This proves Lemma 2.36.

2.37. Proof of Claim 2.28. Lemma 2.36 guarantees the existence of a constant $a' > 0$, $a' < a_0$ such that for $h = g^k z$, $z \in Z$, k an integer, we have

$$\{\gamma \in \Gamma \mid h\gamma h^{-1} \in U(a)\} = \{\varphi \in \Phi \mid h\varphi h^{-1} \in U(a)\}$$

for all $a < a'$ (in fact we can take $a' = \lambda^{-p}(\inf M^{-p}, b, a_1)$). Now there exists an integer $r > 0$ and constant $c > 1$ such that

$$|g^r x g^{-r}| > c|x| \; (g^r x g^{-r} \in U(a_0) \qquad (*)$$

for all $x \in F$ with $|x| < c^{-1}a_0$, $(x \in U(a_0))$. Replacing X by rX, we can assume that (*) holds with $r = 1$:

$$|g x g^{-1}| > c|x| \qquad (**)$$

for all $x \in U(c^{-1}a_0) \cap F$. Let $C > c$ be a constant such that $g^j x g^{-j} \in U(a_0)$ for $x \in U(C^{-1}a_0)$ and $|g^j x g^{-j}| < C|x|$ for $j = \pm 1$. Now take $a < a'$ to be any constant such that $aC < a' < a_0 C^{-1} < a_0$, we then have the following:

2.38. *Assertion. If $h = g^k z$, k integer, $z \in Z$ is such that*

$$h\Phi h^{-1} \cap U(a) \neq (e),$$

then g is (C, c, a)-adapted to $h\Gamma h^{-1}$.

We remark that if $hxh^{-1} \in U(aC)$, $x \in \Gamma$, then $x \in \Phi$ since $aC < a'$. Moreover since $aC^2 < a_0$, we have

$$| g\, h\, x\, h^{-1}\, g^{-1} | > c\, |\, h\, x\, h^{-1}|$$

for all $x \in \Gamma$ with $hxh^{-1} \in U(aC)$. This proves the assertion.

2.39. Now the map $Z'/Z' \cap \Lambda' \to N_0^*/\Lambda^{*'}$ is proper. This follows from the fact that Z' is the centraliser of a finite set for instance $\Xi' = \sigma(\Xi)$. We conclude from this that the map $Z/Z \cap \Gamma \to N_0/N_0 \cap \Gamma$ is proper as also the map $Z/Z \cap \Gamma \to GL(\mathfrak{f})/GL(\mathscr{L})$. If $Z/Z \cap \Gamma$ is not compact, we can thus find $z_n \in Z$ and $\varphi_n \in \Phi$, $\varphi_n \neq (e)$ such that $z_n \varphi_n z_n^{-1}\, U(a)$ converges to e.

2.40. Using the assertion above and Lemma 1.5 we conclude that we can find a sequence λ_n of integers and elements $\gamma_n \in \Gamma$ such that $g^{\lambda_n} z_n \gamma_n \in E$ where E is a compact set such that

$$E\,\Gamma \supset \{\, g \in G \,|\, g \Gamma g^{-1} \cap U(a) = (e)\, \}.$$

Further

$$g^{\lambda_n - 1} z_n\, \Gamma z_n^{-1}\, g^{-\lambda_n + 1} \cap U(a) \neq (e).$$

Now let θ_n be chosen such that $\theta_n \in \Phi - \{e\}$ and

$$g^{\lambda_n - 1} z_n \theta_n\, z_n^{-1}\, g^{-\lambda_n + 1} \in U(a).$$

Then we have for $\xi = e$ or $\xi \in \Xi$,

$$g^{\lambda_n - 1}\, z_n\, \xi\, \theta_n \xi^{-1}\, z_n^{-1} g^{-\lambda_n + 1} \in U(\lambda a)$$

i.e.

$$g^{-1} g^{\lambda_n} z_n \gamma_n \gamma_n^{-1} \xi\, \theta_n\, \xi^{-1} g_n^{-1} \cdot \gamma_n z_n^{-1} g^{-\lambda_n} g \in U(\lambda a).$$

We thus find

$$\gamma_n^{-1} \xi \theta_n\, \xi^{-1} \gamma_n^{-1} \in \{\gamma \in \Gamma \,|\, E\gamma \cap EU(\lambda a) \neq \varnothing\, \}\,(= Y).$$

We thus find that the entire set $\{\xi\theta_n\, \xi^{-1} | \xi \in \Xi \cup \{e\}\} = Y_n$ is conjugated into a fixed finite set Y of unipotents by γ_n. Let Φ_n denote the group generated by $\{\, \xi\theta_n\, \xi^{-1} | \xi \in \Xi \cup \{e\}\}$ and F_n its Zariski-closure. The Lie algebra $\mathfrak{f}_n$ of F_n is generated by the linear span $\mathfrak{f}'_n$ of $\{\mathrm{Ad}\, \xi \log \theta_n \,|\, \xi \in \Xi \cup \{e\}\}$. $\mathfrak{f}'_n$ being Ξ stable is S-stable. Thus F_n is S-stable. Since the set of finite subsets of Y is finite, we can assume, passing to a subsequence, that

$$\gamma_n\, Y_n\, \gamma_n^{-1} = Y_0 \subset Y$$

is a fixed subset. In other words $\gamma_n F_n \gamma_n^{-1} = \gamma_{n+1} F_{n+1}^{-1} \gamma_{n+1}^{-1}$ or $\gamma_{n+1}^{-1} \gamma_n F_n \gamma_n^{-1} \gamma_{n+1} = F_{n+1}$.

2.41. ASSERTION. $\gamma_{n+1}^{-1} \gamma_n \in \Lambda^*$. *Hence* $\gamma_{n+1}^{-1} \gamma_1 \in \Lambda^*$.

For a subgroup $\Phi' \subset \Phi$ define inductively Φ'_k, F'_k and Λ'_k as follows: $\Phi'_0 = \Phi'$, F'_0 is the Zariski-closure of Φ'_0; assume Φ'_k defined and let F'_k be the Zariski-closure of Φ'_k; then Λ'_k is defined as the normaliser of F'_k in Γ and Φ'_{k+1} as the maximum unipotent normal subgroup of Λ'_k. Then it is easily seen that for $k>0, \Phi'_k = \Gamma \cap F'_k$. Moreover we have the following inclusions:

$$\Phi'_0 \subset \Phi'_1 \ldots \subset \Phi'_{k-1} \subset \Phi'_k \ldots$$

$$\Lambda'_0 \subset \Lambda'_1 \ldots \subset \Lambda'_{k-1} \subset \Lambda'_k \ldots$$

Let $\tilde{\Phi}' = \bigcup_{0 \leqslant k \leqslant \infty} \Phi'_k$ and $\tilde{\Lambda}' = \bigcup_{0 \leqslant k \leqslant \infty} \Lambda'_k$ and $\tilde{F}' = \bigcup_{0 \leqslant k \leqslant \infty} F'_k$. Then since Φ' is unipotent, it is noetherian and we conclude that $\tilde{\Phi}' = \Phi'_k$, $\tilde{F}' = F'_k$ and $\tilde{\Lambda}' = \Lambda'_k$ for all large k. In particular, $\tilde{\Phi}'$ is the maximum unipotent normal subgroup of its normaliser $\tilde{\Lambda}'$ in Γ. Also $\tilde{\Phi}' = \tilde{F}' \cap \Gamma$.

Suppose now that Σ_u normalises Φ'. Then Σ_u normalises Φ'_k for all k; it follows that Σ_u normalises $\tilde{\Phi}'$. Hence by Propositions 2.20 and 2.24, $\tilde{\Phi}' \subset \Phi$ and $\tilde{\Lambda}' \subset \Lambda^*$. Hence $\tilde{F}' \subset F$. If $\tilde{F}' \neq F$, we can find $\theta \in F\text{-}\tilde{F}' \cap \Gamma$ normalising $\tilde{F}'$. But then $F \cap \Lambda'$ is unipotent, normal and contains $\tilde{\Phi}'$ *properly*, a contradiction. Thus $\tilde{\Phi}' = \Phi$. Applying these observations to $F_n \cap \Gamma$ and $F_{n+1} \cap \Gamma$ we obtain Assertion 2.41.

2.42. From Assertion 2.41 and the definition of λ_n, we conclude that there exists $x_n (= \gamma_n \gamma_1^{-1}) \in \Lambda^*$ such that $g^{\lambda_n} z_n x_n$ converges to a limit (passing to a subsequence if necessary). But this means that if $v \in \wedge^r \mathfrak{f} \subset \wedge^r \mathfrak{g}$, $r = \dim \mathfrak{f}$, is any non-zero vector,

$$g^{\lambda_n} z_n x_n v = \pm g^{\lambda_n} v$$

converges to a limit (note that $\det \sigma(\gamma) = \pm 1$ for $\gamma \in \Lambda^*$ and $z_n \in N_0$). On the other hand $\mathfrak{f}$ is contained in the span of eigen spaces of X corresponding to positive eigen-values of ad X. Thus this sequence can converge only if λ_n is bounded. On the other hand since $z_n \varphi_n z_n^{-1}$ tends to e and $g^{\lambda_n} \varphi_n g^{-\lambda_n} \notin U(a)$, this cannot happen, a contradiction. This completes the proof of Claim 2.28 and hence Theorem 2.25.

We obtain the following result as a consequence of Theorem 2.25 and Proposition 2.18.

2.43. THEOREM. *F is the unipotent radical of N.*

2.44. According to Theorem 2.25, N_0/L is compact. We claim that every $X \in \mathfrak{n}_0$ such that ad X is nilpotent belongs to $\mathfrak{l}$. To see this we argue as follows. Let ρ be a faithful representation of $GL(\mathfrak{f})$ defined over $\mathbf{Q}$ on a vector space W and $w \in W$, a $\mathbf{Q}$-rational vector such that

$$L^{*\prime} = \{x \in GL(\mathfrak{f}) \mid \rho(x)\, w \in \mathbf{R} w\}$$

($L^{*\prime}$ being the Zariski-closure of $\Lambda^{*\prime}$ is defined over $\mathbf{Q}$). Now for $x \in \Lambda^{*\prime}$, $\rho(x)(w) = \chi(x)\, w$ with $\chi(x) \in \mathbf{Q}$ since ρ is defined over $\mathbf{Q}$ and $\Lambda^{*\prime}$ consists of $\mathbf{Q}$-rational points. Further $\chi(x)$ is an algebraic integer since $\Lambda^{*\prime} \subset GL(\mathscr{L})$. We conclude that $\rho(L^{*\prime})(w) \subset \{\pm w\}$. Now if $X \in \mathfrak{n}_0$ is such that ad X is nilpotent, $\sigma(X) = X'$ is nilpotent. The orbit $\{\rho(\exp tX)(w) \mid -\infty < t < \infty\}$ is then closed. On the other hand this orbit is contained in $\rho(N_0^{*\prime})(w)$. Since $\rho(L^{*\prime})(w) \subset \{\pm w\}$ and $N_0^{*\prime}/L^{*\prime}$ is compact, we conclude that $\{\rho(\exp tX')(w) \mid -\infty < t < \infty\}$ is compact. Since X' is unipotent, one sees easily, that we have $\exp tX.\, w = w$ for all t, $-\infty\ t < \infty$. This means that $\exp tX' \in L^{*\prime}$ for all t, $-\infty < t < \infty$ i.e. $X' \in \mathfrak{l}'$. Hence $X \in \mathfrak{l}$. This proves our claim.

2.45. What we have just proved implies in particular that the unipotent radical of N_0 is contained in L. Let V be the unipotent radical of L. Since Λ normalises Φ, and its Zariski-closure contains L, L normalises F, so that $V \supset F$. Now $\sigma(V) = V'$ is the unipotent radical of the $\mathbf{Q}$-group $L^{*\prime}$ so that $V'/V' \cap \Lambda^{*\prime}$ is compact ($\Lambda^{*\prime}$ is arithmetic in $L^{*\prime}$) (Proposition 2.18). Since the kernel of σ is H and $H/H \cap \Gamma$ is compact, we conclude that $V/V \cap \Gamma$ is compact. Hence $V \cap \Gamma = \Phi$ (maximality of Φ). We conclude from this that $V = F$. We have therefore proved that N_0 has F as its unipotent radical. Now for $x \in N$, $x \notin N_0$ if and only if det $\sigma(x) \neq 1$. In particular, every unipotent element of N is contained in N_0. Further N_0 is normal in N so that F is normal in N. We conclude from this that F is the unipotent radical of N. This proves the theorem.

Appealing now to a result of Borel and Tits (Corollary I.6) we have

2.46. COROLLARY. *There exists an element $X \in \mathfrak{g}$ and an integer m such that*

(i) ad X *is semisimple with real eigen-values,*

(ii) $\mathfrak{n}$ (*resp.* $\mathfrak{f}$) *is the span of the eigen-spaces of* ad X *corresponding to non-negative* (*resp. positive*) *eigen-values of* ad X,

(iii) X *is orthogonal to* $\mathfrak{n}_0$ *w.r.t. the Killing form. Moreover if* $B \subset N$ *is any reductive algebraic subgroup, we can choose* X *such that* B *commutes with* $\{\exp tX \mid -\infty < t < \infty\}$.

2.47. COROLLARY. *L contains a normal subgroup L_1 of N_0 such that N_0/L_1 is compact.*

PROOF. Let ρ be a representation of $GL(\mathfrak{f})$ defined over $\mathbf{Q}$ on a vector space V and $v_0 \in V$ a vector such that

$$L^{*\prime} = \{x \in GL(\mathfrak{f}) \mid \rho(x)\, v_0 \in \mathbf{R}\, v_0\}. \qquad (*)$$

Then, as in 2.44 we conclude that $\rho\,(L^{*\prime})\, v = \pm\, v$ so that $\rho\,(N'_0)\, v_0$ is *compact*. Let

$$E = \{v \in V \mid \rho\,(N'_0)\; v \text{ is relatively compact in } V\}.$$

Then $v_0 \in E$. One checks easily that E is stable under N_0' and the representation ρ' of N_0' on E is equivalent to an orthogonal representation. Now in view of (*) kernel $\rho' \subset L^{*\prime}$. Since ρ' is unitary this kernel contains F. Since N_0/F is semisimple, we see that $\rho'(N_0)$ is semisimple, hence closed and compact. We can clearly take $L_1 =$ kernel ρ'.

3. *L*-subgroups of rank 1. As hitherto, G will denote a connected semisimple group with trivial centre and no compact factors and Γ a *non-uniform irreducible* L-subgroup. We will study the special case when rank $\Gamma = 1$ in this section. We begin with a characterisation of L-subgroups of rank 1.

3.1. THEOREM. *An L-subgroup Γ in G has rank* 1 *if and only if the following holds: every* (*non-trivial*) *unipotent element of* Γ *is contained in a* unique *maximal unipotent subgroup of* Γ.

3.2. Assume first that rank $\Gamma = 1$. Let $x (\neq e)$ be any unipotent element in Γ. Consider the set $\mathscr{P}$ of all pairs (Φ_1, Φ_2), $\Phi_1 \neq \Phi_2$ of maximal unipotent subgroups of Γ containing x. It suffices to show that $P = \varnothing$. For each such pair (Φ_1, Φ_2) let F_i be the Zariski-closure of Φ_i, $i = 1,2$, and $d\,(\Phi_1, \Phi_2) =$ dimension of $F = F_1 \cap F_2$. Let $r = \operatorname{Sup}\{d\,(\Phi_1, \Phi_2) \mid (\Phi_1, \Phi_2) \in \mathscr{P}\}$. If $\mathscr{P} \neq \varnothing$, $r > 0$ (F always contains x). Fix now a pair (Φ_1, Φ_2) with $F_i =$ Zariski-closure of Φ_i, $i = 1,2$, and dim F $(= F_1 \cap F_2) = r$. Since $\Phi_1 \neq \Phi_2$ and Φ_i are maximal F is a proper subgroup $(\neq e)$ of F_1 as well as F_2. Let F'_i be the normaliser of F in F_i. Since F_i are unipotent, $F'_i \neq F$ for $i = 1, 2$. Now F_i/Φ_i is compact for $i = 1, 2$ (cf. Theorem 1. 2, of Introduction) and if $\Phi'_i = F'_i \cap \Gamma$, F'_i/Φ'_i is again compact. Let Φ_3 be the group generated by Φ'_1 and Φ'_2. Let F_3 be the Zariski-closure of Φ_3. Then, evidently $x \in \Phi$ $(= F_1 \cap F_2 \cap \Gamma) \subset \Phi_3$ and $F_3 \cap F_1 \supset F'_1$. Since dim $F'_1 >$ dim $F = r$, we see that F'_3 cannot be unipotent. On the other hand Φ_3 is generated by unipotents and normalises the unipotent group Φ $(\ni x)$ contradicting our assumption that rank $\Gamma = 1$.

3.3. Suppose now that rank $\Gamma > 1$. One sees then directly from Proposition 2.18, that there exist elements which are contained in two distinct maximal unipotent subgroups: in fact, in the notation introduced in 2.17 let $x \in \Phi$ be any element $(\neq e)$; if rank $\Gamma > 1$, then $S' \neq \{e\}$; let V'_1, V'_2 be two *distinct* maximal unipotent subgroups of S' and $V_i = \sigma^{-1}(V'_i)$ $i = 1, 2$; then $\Phi_1 = V_i \cap \Gamma$ are unipotent groups containing x; since moreover, one sees easily that $V_i/V_i \cap \Gamma$ is compact, $\sigma(\Phi_i)$ is Zariski dense in V'_i, $= 1, 2$, so that Φ_1 and Φ_2 cannot be contained in the same unipotent subgroup of G.

3.4. In the sequel we assume that Γ is an L-subgroup of rank 1. From the definition of P-subgroups (Definition 2.16) of Γ, one sees immediately that any maximal unipotent subgroup Φ of Γ is a P-subgroup. Combining Proposition 2.18 and Theorem 2.25 (note that $p(\Gamma) = 0$ in the present situation) we obtain the following result.

3.5. THEOREM. *Let Φ be any maximal unipotent subgroup of Γ and F the Zariski-closure of Φ in G. Let $\mathfrak{f}$ be the Lie algebra of F, N*

the connected normaliser of F in G and σ the adjoint representation of N in $\mathfrak{f}$. Let $N_0 = \{x \in N \mid \det \sigma(x) = 1\}$. Then $N_0/N_0 \cap \Gamma$ is compact.

3.6. Remark. Theorem 3.1 and 3.5 show that Γ is an L-subgroup of G satisfying Properties $R1$ and $R2$ in the sense of Raghunathan ([1], Chapter XIII) in fact Theorem 3.5 shows that $R2$ is itself a consequence of $R1$. According to Theorem 13.18 of that book, this implies that Γ is a *lattice* in G. We will see later on that any L-subgroup (of arbitrary rank) is a lattice.

3.7. We fix once for all a semisimple connected $\mathbf{R}$-algebraic group $\mathbf{G}$ with trivial centre such that the identity component of $\mathbf{G}_{\mathbf{R}}$, the $\mathbf{R}$-rational points of $\mathbf{G}$, is isomorphic to G and treat G itself as a subgroup of $\mathbf{G}$ through this isomorphism. We adopt the following notational convention. If $H \subset G$ is any subgroup $\mathbf{H}$ will denote its Zariski-closure in $\mathbf{G}$. $\mathfrak{G}$ will denote the Lie algebra of $\mathbf{G}$ and for a subalgebra $\mathfrak{h}$ of $\mathfrak{G}$, $\mathfrak{h}$ will denote its $\mathbf{C}$-linear span.

We fix from now as once and for all groups Φ, F, N and N_0 as in Theorem 3.5 above. Let $\theta \in \Gamma$ be any element not contained in N. According to Corollary 2.45 and Theorem 2.42, $\mathbf{N}$ is a parabolic subgroup of $\mathbf{G}$ and $\mathbf{F}$ is the unipotent radical of $\mathbf{N}$. Consider now the groups $M = N \cap \theta N \theta^{-1}$ and $M_0 = N_0 \cap \theta N_0 \theta^{-1}$. Since $N_0/N_0 \cap \Gamma$ and $N'_0/N'_0 \cap \Gamma$ are both compact, one sees easily that $M_0/M_0 \cap \Gamma$ is compact. Let $\mathbf{M}_1$ be the Zariski-closure of $M_0 \cap \Gamma$ and $M_1 = \mathbf{M}_1 \cap G$. Now $\sigma(M_0 \cap \Gamma)$ is a subgroup of $GL(\mathbf{f})$ which leaves invariant the $\mathbf{Z}$-span $\mathscr{L}$ of $\exp^{-1}(F \cap \Gamma)$. If $GL(\mathfrak{F})$ is given the $\mathbf{Q}$-structure defined by $\mathscr{L}$ ($\mathscr{L}$ is a lattice in the vector space $\mathfrak{f}$), $\sigma(M_0 \cap \Gamma)$ consists of $\mathbf{Q}$-rational points. It follows that $\sigma(\mathbf{M}_1) = \mathbf{M}'_1$ is a $\mathbf{Q}$-subgroup of $GL(\mathfrak{f})$. Since $M_1/M_1 \cap \Gamma$ is compact, $\sigma(M_1 \cap \Gamma)$ is an arithmetic subgroup of $M'_1 = \sigma(M_1)$. Moreover, $F \cap \Gamma$ is maximal unipotent in Γ so that $M_1 \cap \Gamma$ cannot contain any nontrivial unipotent elements. It follows that $\sigma(M_1 \cap \Gamma)$ contains no unipotents; since this group is arithmetic, it follows that $\mathbf{M}'_1$ (= Zariski-closure of $\sigma(M_1 \cap \Gamma)$) is reductive and anisotropic over $\mathbf{Q}$. Since $\sigma \mid_{\mathbf{M}}$ is faithful (Theorem 2.3), $\mathbf{M}_1$ is reductive. Also since M_0/M_1 is compact, M_0 is reductive as well. One now sees easily that **M** is itself reductive and its identity component is of the form **A**. $\mathbf{M}_0^0$ where **A** is a (real) **R**-split toral subgroup of dimension 1 and $\mathbf{M}_0^0$ is the identity component of $\mathbf{M}_0$. Also if we let N_1 denote the group M_1. F, evidently $N_1 / N_1 \cap \Gamma$ is compact. Also, $\mathbf{N}_1$ carries a natural structure of a **Q**-group such that $N_1 \cap \Gamma$ is arithmetic. On the other hand if **D** is the Zariski-closure of $N_0 \cap \Gamma$ in **G** and $D = \mathbf{D} \cap G$, D/N_1 is compact and moreover $\sigma(\mathbf{D})$ is defined over **Q**, since $\sigma(D \cap \Gamma)$ consists of **Q** rational points of $GL(\mathfrak{F})$. It follows that $N_1 \cap \Gamma$ has finite index in $D \cap \Gamma$. In particular N_1 has finite index in **D** and $\mathbf{M}_1^0$ (= identity component of $\mathbf{M}_1$) has finite index in a maximal reductive **Q**-subgroup of $\mathbf{N}_1^0 = \mathbf{D}^0$ (= identity component of **D**). We summarise some of the above remarks in

3.8. Proposition. Let Φ, F, N *and* N_0 *be as in Theorem* 3.5. *Let* **D** *be the Zariski-closure of* $N \cap \Gamma$ *in* **G**. *Then* $\mathbf{D}^0$ *carries a natural* **Q**-*structure such that* $\mathbf{D}^0 \cap \Gamma$ *is an arithmetic subgroup of* $D^0(\subset \mathbf{D}_\mathbf{R})$ *such that* $D^0/D^0 \cap \Gamma$ *is compact. Moreover if* $\theta \in \Gamma$ *is such that* $\theta \notin N$, *than* $\theta \mathbf{N}_0 \theta^{-1} \cap \mathbf{D}^0$ *is a maximal reductive* **Q**-*subgroup of* $\mathbf{D}^0$ *and* $(\theta\, N_0\, \theta^{-1} \cap D^0)/(\theta N_0 \theta^{-1} \cap D^0 \cap \Gamma)$ *is compact.* Also **N** *is a parabolic subgroup of* **G** *and* **F** *is its unipotent radical.*

3.9. From now on we make the following additional hypothesis:

$$\mathbf{R}\text{-}rank\ of\ G > 1.$$

This implies in particular that M_0 is non-compact and that $M_0 \cap \Gamma$ is non-trivial. In fact $\mathbf{M}_1$ contains all **R**-split toral subgroups of $\mathbf{M}_0$ as well as all its **R**-isotropic simple factors. Our immediate aim now is to establish the following.

3.10. Proposition. *The centraliser of* $\mathbf{M}_1$ *in* $\mathfrak{F}$ *is contained in the centre of* $\mathfrak{F}$.

3.11. We consider first the case when G is not simple. Let $\mathfrak{g} = {}_{i\in I}\, \mathfrak{g}^i$ be the decomposition of (the Lie algebra) $\mathfrak{g}$ (of G) into its simple ideals. Let $X \in \mathfrak{g}$ be an element such that ad X is semisimple, has all eigen-values real and with $\mathfrak{f}$ as precisely the sum of the eigen-spaces corresponding to positive eigen-values. Let X_i be the projection of X

on $\mathfrak{g}_i$. We first claim that $X_i \neq 0$ for any i. In fact if $X_{i_0} = 0$ for some $i_0 \in I$, we see that $\mathfrak{f} \subset \prod_{i \neq i_0} \mathfrak{g}^i (= \mathfrak{h}$, say) so that F is contained in the connected normal subgroup $H(\neq G)$ corresponding to the ideal $\mathfrak{h}$. It follows that $F \cap \Gamma = \Phi \subset H$. This contradicts Corollary 1.14 in view of the irreducibility of Γ. We assume, as we may, that X belongs to $\mathfrak{m}$, the Lie algebra of M. Let $\mathfrak{m}_0$ be the Lie algebra of M_0. If we let σ denote the adjoint representation of $\mathfrak{m}$ on $\mathfrak{f}$ as well, we have

$$\mathfrak{m}_0 = \{ Y \in \mathfrak{m} \mid \text{trace } \sigma(Y) = 0 \}.$$

Let i, j be elements of I with $i \neq j$. Then we can find a non-zero linear combination $X' = \lambda X_i + \mu X_j$, $\lambda, \mu \neq 0$, which belongs to $\mathfrak{m}_0$. The one parameter subgroup $\{\exp tX'\}_{-\infty < t < \infty}$ is contained in a $\mathbf{R}$-split-torus of $\mathbf{M}_0$, so that this one parameter subgroup is contained in M_1. We conclude that the centraliser of M_1 in $\mathfrak{f}_i = \mathfrak{f} \cap \mathfrak{g}_i$ is contained in the centraliser of X_i in $\mathfrak{f}_i$; it is hence trivial. Since I was assumed to contain at least two elements, the centraliser of M_1 in $\mathfrak{f}$ is trivial.

3.12. Consider next the case when $\mathbf{G}$ has all simple-components of type G_2. Then G is either the *split* real form of G_2 or G is a *complex* Lie group of type G_2. $\mathfrak{F}$ is the Lie algebra of the unipotent radical of either a Borel subgroup or a maximal parabolic subgroup of $\mathbf{G}$ defined over $\mathbf{R}$.

If $\mathbf{N}$ is a Borel subgroup, one sees easily that to prove Proposition 3.10 in this case we need only show the following: no root of G_2 is proportional to the sum of all the positive roots. This is easily checked directly from an examination of the root system of G_2.

Assume now that $\mathbf{N}$ is a *maximal* $\mathbf{R}$-parabolic subgroup of $\mathbf{G}$. We will treat the situation when $\mathbf{G}$ is absolutely simple (hence isomorphic to the split form of G_2) in detail (entirely analogous arguments apply in the non-absolutely simple case). In this case M_0 is locally isomorphic to $SL(2, \mathbf{R})$. We can assume that the Lie algebra $\mathfrak{M}_0$ of $\mathbf{M}_0$ is generated by the root spaces $\mathfrak{G}^{\pm\alpha}$ of a simple root α of $\mathbf{G}$ w.r.t. a suitable torus $\mathbf{T}$ and an order on $X(\mathbf{T})$. If α is the short root one sees that the descending central series for $\mathfrak{f}$ is of length 2:

$$\mathfrak{f} = \mathfrak{f}_0 \supset \mathfrak{f}_1 \supset \mathfrak{f}_2 \supset \{0\}$$

and that $\mathfrak{f}_i/\mathfrak{f}_{i+1}$ for $i = 0, 2$ are both 2 dimensional while $\mathfrak{f}_1/\mathfrak{f}_2$ is of dimension 1. Since once again the representations of $\mathbf{M}_1$ on $\mathfrak{F}_i / \mathfrak{F}_{i+1}$ are defined over $\mathbf{Q}$ and dim $\mathfrak{F}/\mathfrak{F}_{i+1} \leqslant 2$ while $\mathbf{M}_1$ is an anisotropic $\mathbf{Q}$-form of $SL(2, \mathbf{C})$, we obtain a contradiction. Thus the root α is necessarily the long root and one checks immediately that the centraliser of M_0 is then the centre of $\mathfrak{f}$.

It is easily seen that the following result proved in the appendix completes the proof of Proposition 3.10.

3.13. PROPOSITION. *Let* $\mathbf{G}$ *be a connected algebraic group defined and simple over* $\mathbf{R}$ *and of* $\mathbf{R}$*-rank at least* 2. *Assume that* $\mathbf{G}$ *is not of type* G_2. *Let* $\mathbf{N}$ *be a parabolic subgroup of* $\mathbf{G}$ *defined over* $\mathbf{R}$ *and* $\mathbf{F}$ *its unipotent radical. Let* $\mathbf{M}$ *be a maximal reductive* $\mathbf{R}$*-subgroup of* $\mathbf{N}$ *and* σ *the adjoint representation of* $\mathbf{M}$ *on* $\mathfrak{F}$ (= *Lie algebra of* $\mathbf{F}$). *Let* $\mathbf{M}_0 = \{x \in M \mid \det \sigma(x) = 1\}$ *and* $\mathbf{M}_0^*$ *a minimal connected normal* $\mathbf{R}$*-subgroup* of $\mathbf{M}_0$ *such that* $\mathbf{M}_{0\mathbf{R}}/\mathbf{M}^*_{0\mathbf{R}}$ *is compact. Then the centraliser of* $\mathbf{M}_0^*$ *in* $\mathbf{F}$ *is contained in the centre of* $\mathbf{F}$.

3.14. We choose now an element $\theta_0 \in \Gamma$ not belonging to N_0 and introduce the following notational convention: if $B \subset \mathbf{G}$ is any subgroup $B^- = \theta_0 B \theta_0^{-1}$ and $\mathfrak{b} \subset \mathfrak{G}$ a subspace $\mathfrak{b}^- = \mathrm{Ad}\ \theta_0 (\mathfrak{b})$. Let

$$\mathbf{M} = \mathbf{N} \cap \mathbf{N}^-, \quad M = \mathbf{M} \cap G,$$
$$\mathbf{M}_0 = \mathbf{N}_0 \cap \mathbf{N}_0^- \ (= \{x \in M \mid \det \sigma(x)\} = 1), \quad M_0 = \mathbf{M}_0 \cap G,$$
$$\mathbf{M}_1 = \text{Zariski-closure of } M_0 \cap \Gamma \text{ in } \mathbf{G}, \qquad M_1 = \mathbf{M}_1 \cap G,$$
$$L = \text{normaliser of } M_1^0 (= \text{identity component of } M_1).$$

We fix once for all a maximal $\mathbf{R}$-split torus $\mathbf{T}$ in $\mathbf{M}$; such a torus is also maximal $\mathbf{R}$-split in $\mathbf{G}$. Then, as is easily seen $\mathbf{T}_0 = \mathbf{T} \cap \mathbf{M}_1^0$ is a maximal $\mathbf{R}$-split torus in $\mathbf{M}_1^0$; $\mathbf{T}_0$ is also a maximal $\mathbf{R}$-split torus in $\mathbf{M}_0$ ($\mathbf{M}_{0\mathbf{R}}/\mathbf{M}_{1\mathbf{R}}$ is compact) as well as $\mathbf{L}$. Suppose now $g \in L$ is any element; consider the $\mathbf{R}$-split torus $\mathbf{T}' = g\mathbf{T}g^{-1}$; this torus $\mathbf{T}'$ can be conjugated by an element $g' \in L^0$ into $\mathbf{T}$. Thus we see that each element g^- of L/L_0 has a representative $g'g = g^*$ in L which normalises $\mathbf{T}$ (as well as $\mathbf{M}_1^0$). Next let E be the connected centraliser of M_1^0 in G; since M_1^0 is reductive, so is E. Let $\mathfrak{e}$ be the Lie algebra of E. Then it is easily seen that $\mathfrak{e}$ decomposes into a direct sum:

$$\mathfrak{e} = \mathfrak{e} \cap \mathfrak{f}^- \oplus \mathfrak{e} \cap \mathfrak{m} \oplus \mathfrak{e} \cap \mathfrak{f}.$$

Moreover the group $\mathbf{E}' = [\mathbf{E}, \mathbf{E}]$ has $\mathbf{R}$-rank 0 or 1 and in the latter case $\mathbf{E} \cap \mathbf{F}$ (resp. $\mathbf{E} \cap \mathbf{F}^-$) is a maximal unipotent subgroup of $\mathbf{E}$ (as also $\mathbf{E}'$) defined over $\mathbf{R}$. The element g^* chosen above normalises $\mathbf{T}$ as well as $\mathbf{E}$ hence also E', the identity component of $\mathbf{E}'_{\mathbf{R}}$. It normalises therefore the $\mathbf{R}$-split torus $\mathbf{S} = (\mathbf{T} \cap \mathbf{E}')^0$ of $\mathbf{E}'$. When $\mathbf{R}$-rank $\mathbf{E}' > 0$, $\mathbf{E} \cap \mathbf{F}$ and $\mathbf{E} \cap \mathbf{F}^-$ are the only maximal unipotent $\mathbf{R}$-subgroups of $\mathbf{E}'$ normalised by the torus $\mathbf{S}$; it follows that inner conjugation by g^* must permute these two groups. Now there exists an element $s \in E'$ normalising $\mathbf{S}$ and hence $\mathbf{T}$ such that $s(\mathbf{E} \cap \mathbf{F})\, s^{-1} = \mathbf{E} \cap \mathbf{F}^-$. We conclude therefore that we can find g'' such that $g''\, g^*$ normalises $\mathbf{T}$ and maps $\mathbf{F} \cap \mathbf{E}$ and $\mathbf{F}^- \cap \mathbf{E}$ into themselves. We have thus proved the following result:

3.15. PROPOSITION. *There exists a finite subset $W_1 \subset L$ such that $W_1 L^0 = L$ and each element $w \in W_1$ normalises a given maximal split torus* $\mathbf{T}$ *in* $\mathbf{M}_1^0$ *as well as the groups* $\mathbf{E} \cap \mathbf{F}$ *and* $\mathbf{E} \cap \mathbf{F}^-$ (*where* $\mathbf{E}$ *is the connected centraliser of* $\mathbf{M}_1$ *in* $\mathbf{G}$).

3.16. We denote by E(resp. E') the identity component of the group $\mathbf{E}_{\mathbf{R}}$ (resp. $\mathbf{E}'_{\mathbf{R}}$, $\mathbf{E}' = [\mathbf{E}, \mathbf{E}]$). As was remarked earlier, $\mathbf{E}'$ is a $\mathbf{R}$-group of $\mathbf{R}$-rank 1 or 0. If $\mathbf{E}'$ has $\mathbf{R}$-rank 0, then $\mathbf{E} \cap \mathbf{F}$ and $\mathbf{E} \cap \mathbf{F}^-$ are trivial so that $\mathbf{E} \subset \mathbf{M}$. It follows that in this case $\mathbf{L}^0 = \mathbf{E}.\ \mathbf{M}_1^0 \subset \mathbf{M}$ and $L^0 = E.\ M_1^0 \subset M_0$. If on the other hand $\mathbf{E}'$ has $\mathbf{R}$-rank 1, the Bruhat-decomposition for the reductive $\mathbf{R}$-group $\mathbf{E}$ says the following: there exists an element $s \in E'$ normalising the torus $\mathbf{T}$ as well as the identity component $\mathbf{S}$ of $\mathbf{T} \cap \mathbf{E}'$ such that any element $1 \in E$ can be written in the form

$$1 = u\, s\, z\, v$$

or

$$1 = z.v$$

where $u, v \in F \cap E$, $z \in M \cap E$.

3.17. PROPOSITION. *Any element* $1 \in L^0$ *can be written in the form*

$$1 = u\, s\, z\, v$$

or

$$1 = z v$$

where $u, v \in F \cap L^0$, $z \in L^0 \cap M$ *and* $s \in E'$ *is an element (chosen once for all) which normalises a given* **R***-split torus* **T** *in* **M** *and is such that* $s(\mathbf{F} \cap \mathbf{E}') = \mathbf{F}^- \cap \mathbf{E}'$.

The proposition follows from the considerations of 3.16, once one remarks that M_1^0 normalises F.

3.18. Consider now the set W_1 of Proposition 3.15. If $w \in W_1$, then w normalises **T** as well as $\mathbf{M}_1^0$. It follows that Ad w leaves the corresponding Lie algebras $\mathfrak{T}$ and $\mathfrak{M}_1$ as also $\mathfrak{t} = \mathfrak{g} \cap \mathfrak{T}$ and $\mathfrak{m}_1 = \mathfrak{g} \cap \mathfrak{M}_1$ stable. We conclude then that w leaves invariant the orthogonal complements (if $L \cap F \neq \{e\}$, $\mathfrak{S}$ is indeed the Lie algebra of **S** of 3.16) of $\mathfrak{m}_1 \cap \mathfrak{t}$ in $\mathfrak{t}$; this last space is easily seen to be of dimension 1 — in fact it is of the form $\mathbf{R}.X$ with $X \neq 0$ in $\mathfrak{g}$ having the following properties:

(i) ad X is semisimple and has all eigen-values real,

(ii) $\mathfrak{m}$ is the centraliser of X in $\mathfrak{g}$,

(iii) $\mathfrak{f}$ (resp. $\mathfrak{f}^-$) is the sum of eigen-spaces of ad X corresponding to positive (resp. negative) eigen-values.

Since Ad w maps $\mathbf{R}X$ isomorphically onto itself and preserves the restriction of the Killing form to $\mathbf{R}X$, one sees that

$$\mathrm{Ad}\, w(X) = \pm X \text{ for all } w \in W_1.$$

This means that any element $w \in W_1$ either $w\,\mathbf{F}w^{-1} = \mathbf{F}$ or $w\,\mathbf{F}\,w^{-1} = \mathbf{F}^-$; in the former case w centralises X and belongs to $M = N \cap N^-$. Moreover if $\mathbf{E}' = [\mathbf{E}, \mathbf{E}]$ (**E** is the connected centraliser of $\mathbf{M}_1^0$) is of **R**-rank 1, $\mathbf{F} \cap \mathbf{E} \neq \{e\}$ and from our choice of W_1, $w(\mathbf{F} \cap \mathbf{E})\, w^{-1} = \mathbf{F} \cap \mathbf{E}$ for all $w \in W_1$; consequently in this case we find that $w\,\mathbf{F}\,w^{-1} = \mathbf{F}$ for all $w \in W_1$ i.e., $w \in M \subset N \cap N^-$ for all $w \in W_1$. If $\mathbf{E}'$ is of **R**-rank 0, either $W_1 \subset M$ or there exists an element $w_0 \in W_1$ such that $w_0 \mathbf{F} w_0^{-1} = \mathbf{F}^-$ and for $w \in W_1$, either $w \in W_1 \cap M$ or $w = w_0\, w'$ with $w' \in M \cap L$. Combining now Propositions 3.15 and 3.17 we obtain the following result.

3.19. PROPOSITION. *Let* **T** *be a maximal* **R***-split torus in* **M** *and* $\mathfrak{t}$ *the Lie algebra of* $\mathbf{T}_{\mathbf{R}}$. *Let* $\mathfrak{s}$ *be the orthogonal complement of* $\mathfrak{m}_1$(= *Lie algebra of* M_1^0) *in* $\mathfrak{t}$. *Let* **S** *be the torus in* **G** *corresponding to* $\mathfrak{s}$. *Let* $Z(\mathbf{S})$

be the centraliser of $\mathbf{S}$ *in* G *and* $N(\mathbf{S})$ *the normaliser of* $\mathbf{S}$ *in* $\mathbf{G}$. *Then the group* $N(\mathbf{S}) \cap L/Z(\mathbf{S}) \cap L$ *is of order at most* 2. *Further if* $W \subset N(\mathbf{S}) \cap L$ *is a set of representatives for this group containing the identity element then every element* $g \in L$ *can be written in the form*

$$g = u\, w\, z\, v$$

where $u, v \in F \cap L$, $w \in W$ *and* $z \in Z(\mathbf{S}) \cap L = M \cap L$.

3.20. We denote by σ the adjoint representation of $\mathbf{N}$ on the space $\mathfrak{F}$. As in 3.7 let $\mathbf{D}$ be the Zariski closure of $N_0 \cap \Gamma$ in $\mathbf{G}$. Then we can write $\mathbf{D} = \mathbf{N}_0^- \cap \mathbf{D}.\mathbf{F}$. We claim that $\mathbf{N}_0^- \cap \mathbf{D}^0 = \mathbf{M}_1^0$. Consider the $\mathbf{Q}$-structure on $GL(\mathfrak{F})$ defined by the lattice $\mathscr{L}$ spanned by $\exp^{-1}(F \cap \Gamma)$ in $\mathfrak{F}$. Since $\sigma(D \cap \Gamma)$ stabilises $\mathscr{L}$, $\sigma(D \cap \Gamma)$ consists of $\mathbf{Q}$-rational points; as $\sigma(D \cap \Gamma)$ is Zariski-dense in $\sigma(\mathbf{D})$, $\sigma(\mathbf{D})$ is defined over $\mathbf{Q}$. For similar reasons $\sigma(\mathbf{M}_1^0)$ and hence $\sigma(\mathbf{M}_1^0.\,\mathbf{F})$ are also defined over $\mathbf{Q}$. Evidently $\sigma(\mathbf{M}_1^0.\,\mathbf{F}) \subset \sigma(\mathbf{D}) = \mathbf{D}'$. The group $\sigma(D \cap \Gamma)$ is an arithmetic subgroup of $\mathbf{D}'_{\mathbf{R}}$ such that $\mathbf{D}'_{\mathbf{R}} / \sigma(D \cap \Gamma)$ is compact. On the other hand if $\mathbf{D}'' = \sigma(\mathbf{M}_1^0.\,\mathbf{F})$, $\mathbf{D}'_{\mathbf{R}}/\mathbf{D}''_{\mathbf{R}}$ is compact and $\mathbf{D}''_{\mathbf{R}}/\sigma\,(D \cap \Gamma) \cap \mathbf{D}''_{\mathbf{R}}$ is also compact. It follows that the projection of $\sigma(D \cap \Gamma)$ in $\mathbf{D}'/\mathbf{D}''$ is finite. Since $\sigma(D \cap \Gamma)$ is Zariski-dense in $\mathbf{D}'$, this means that $\mathbf{D}''$ has finite index in $\mathbf{D}'$. This proves our contention that $\mathbf{D}^0 = \mathbf{M}_1^0.\mathbf{F}$. The map σ restricted to $\mathbf{M}_1^0$ is an isomorphism onto a $\mathbf{Q}$ subgroup $GL(\mathfrak{F})$. Transporting the $\mathbf{Q}$-structure to $\mathbf{M}_1^0$, we see that for this $\mathbf{Q}$-structure the adjoint action of $\mathbf{M}_1^0$ on $\mathbf{F}$ is defined over $\mathbf{Q}$. We obtain thus a $\mathbf{Q}$-structure on $\mathbf{D}^0$ such that $\mathbf{M}_1^0$ is a maximal connected reductive $\mathbf{Q}$-subgroup of $\mathbf{D}^0$. Further $\mathbf{D}_{\mathbf{R}}^0 \cap \Gamma$ is an arithmetic subgroup of $\mathbf{D}_{\mathbf{R}}^0$ for this $\mathbf{Q}$-structure. Suppose now $\gamma \in \Gamma$ is any element not contained in N. Let $\mathbf{M}_\gamma$ be the Zariski-closure of $\gamma\, N\, \gamma^{-1} \cap N$ in $\mathbf{G}$. Then the foregoing considerations show that $\mathbf{M}_\gamma^0 \subset \mathbf{D}^0$ and that it is a maximal (connected) reductive subgroup of $\mathbf{D}^0$. Since $\mathbf{M}_\gamma^0 \cap \Gamma$ is Zariski-dense in $\mathbf{M}_\gamma^0$ and $\mathbf{D}^0 \cap \Gamma$ is arithmetic, we see that $\mathbf{M}_\gamma^0$ is defined over $\mathbf{Q}$ as well. We can therefore find $\alpha \in \mathbf{F}_{\mathbf{Q}}$ such that $\mathbf{M}_\gamma^0 = \alpha\, \mathbf{M}_1^0\, \alpha^{-1}$. On the other hand $\mathbf{M}_\gamma$ is the Zariski-closure of $\gamma(N \cap \gamma^{-1} N\, \gamma)\gamma^{-1}$ so that $\mathbf{M}_\gamma^0 = \gamma\, \mathbf{M}_{\gamma-1}^0\, \gamma^{-1}$. There exists $\beta \in \mathbf{F}_{\mathbf{Q}}$ such that $\mathbf{M}_{\gamma-1}^0 = \beta^{-1}\mathbf{M}_1^0$. We find thus that there exist $\alpha, \beta \in \mathbf{F}_{\mathbf{Q}}$ such that

$$\alpha^{-1}\gamma\beta^{-1}.\mathbf{M}_1^0.\beta\gamma^{-1}\alpha=\mathbf{M}_1^0.$$

In other words if $\gamma\in\Gamma$ is any element not belonging to N, we have

$$\gamma=\alpha g\beta$$

where $\alpha,\beta\in F_Q$ and $g\in L$ (= normaliser of M_1^0). In view of Proposition 3.19 and Proposition 3.10 we obtain the following:

3.21. **Proposition.** *Every element $\gamma\in\Gamma$ can be written in the form*

$$\gamma=\alpha u w z v\beta \qquad (*)$$

where $\alpha,\beta\in F_Q$, u v, belong to the centre of F as well as the centraliser of M_1^0, $z\in M\cap L$ and $w\in W$.

3.22. **Remarks.** (a) When $\gamma\in N$, the assertion in Proposition 3.21 is trivial.

(b) Since Γ is Zariski-dense in $\mathbf{G}$ Proposition 3.21 guarantees that W cannot reduce to a single element.

3.23. **Corollary.** $M\subset L$.

3.24. **Proof.** Consider the map $\Psi:\mathbf{F}\times\mathbf{M}\times\mathbf{F}\to\mathbf{G}$ given by $\Psi(f_1,m,f_2)=f_1w_0mf_2$, f_1,f_2 F, $m\in\mathbf{M}$ and $w_0\in\mathbf{W}$ is the unique element different from identity (cf. Remark 3.22 (b)). Since $w_0Fw_0^{-1}\cap F=\{e\}$, it is easily seen that Ψ is an isomorphism onto an open subset $\mathbf{U}$ of $\mathbf{G}$. If $\pi:\mathbf{F}\times\mathbf{M}\times\mathbf{F}\to\mathbf{M}$ is the natural projection, clearly $\pi\Psi^{-1}(\Gamma)$ must be Zariski-dense in $\mathbf{M}$. On the other hand for $\gamma\in\Gamma-N$, $\gamma=\alpha u w_0 z v\beta$, $\alpha,\beta\in U_Q$, $\alpha,\beta\in$ Centre of F and $z\in M\cap L$ so that $\pi\Psi^{-1}(\gamma)\in M\cap L$ for all $\gamma\in$ Image $\Psi\cap\Gamma$. Thus $M\cap L$ must be Zariski-dense in $\mathbf{M}$. We conclude from this that $\mathbf{M}$ normalises M_1^0. It follows that M normalises M_1^0, i.e. $M\subset L$.

3.25. **Corollary.** *$\theta_0=w_0z_0v_0\beta_0$ where (θ_0 was the element chosen at the beginning of* 3.14) *$w_0\in W$ is the element different from the identity, $z_0\in M$, $v_0\in$ Centre $F\cap$ Centraliser M_1^0 and $\beta_0\in\mathbf{F}_Q$.*

This follows from Proposition 3.21, once one observes that for $x\in\mathbf{F}$, $x\mathbf{F}^-x^{-1}\neq\mathbf{F}^-$ unless x is the identity.

3.26. PROPOSITION. *For* $\gamma \in \Gamma$ *either* $\gamma \in N$ *or* γ *can be written in the form*

$$\gamma = \alpha\, u\, z\, \tau\, v\, \beta$$

where $\alpha, \beta \in \mathbf{F}_{\mathbf{Q}}$, u, v *(Centraliser* $M \cap$ *Centre* F*)* $z \in M$ *and* $\tau = w_0 z_0$ *(*w_0, z_0 *being defined by Corollary* 3.25*). Then* $\sigma(z) \in GL(\mathfrak{f})_{\mathbf{Q}}$.

The first assertion is just a reformulation of Proposition 3.21. To prove the second assertion, we argue as follows. We have $\gamma = \alpha\, u\, z\, \theta_0\, \beta_0^{-1}\, v_0^{-1}\, v\, \beta = \alpha\, u\, z\, \theta_0\, v'\, \beta'$ where $\beta' \in \mathbf{F}_{\mathbf{Q}}$ and v' is in the centre of F. Now since $\beta' \in \mathbf{F}_{\mathbf{Q}}$, we can find a subgroup $\Phi' \subset \Phi$ of finite index such that $\beta'\, \Phi'\, \beta'^{-1} \subset \Phi$. Let $\varphi' \in \Phi'$; then we have $\varphi = v'\, \beta'\, \varphi'\, \beta'^{-1}\, v' = \beta'\, \varphi'\, \beta'^{-1} \in \Phi$. We find therefore that $\gamma\, \varphi' \gamma^{-1} = \alpha\, u\, z\, (\theta_0\, \varphi\, \theta_0^{-1}).\, z^{-1}\, u^{-1}\, \alpha^{-1}$. Now $\theta_0\, \varphi\, \theta_0^{-1} \in \Gamma$; in fact it belongs to Φ^- so that

$$\theta_0\, \varphi\, \theta_0^{-1} = a\, \xi\, t\, \tau\, \eta\, b$$

with $a, b \in \mathbf{F}_{\mathbf{Q}}$, $\xi, \eta \in$ Centre of F, $t \in M$. We obtain from this

$$\gamma\, \varphi'\, \gamma^{-1} = \alpha.\, u.\, (z a z^{-1})\, (z \xi z^{-1})\, z t \tau z^{-1}\, (z \eta z^{-1})\, (z b z^{-1}).\, u^{-1}\, \alpha^{-1}.$$

Since the map Ψ of 3.24 is injective, we conclude that

$$za\, \xi\, z^{-1} \in \mathbf{F}_{\mathbf{Q}}.\ (\text{Centre of } F \cap \text{Centraliser } M_1^0). \qquad (*)$$

Now as φ' varies over Φ', φ varies over a Zariski-dense subgroup Ψ of $\mathbf{F}$ and $\theta_0 \varphi\, \theta_0^{-1}$ varies over a Zariski-dense subgroup Ψ^- of F^-

3.27. Consider now $\mathbf{F}^- \cap \text{Image}\, \Psi = \mathbf{U}^F$, say. Each element x of $\mathbf{U}^F$ can be written in the form

$$x = \mathbf{u}(x)\tau.\ \mathbf{z}(x).\ \mathbf{v}(x)$$

where $\mathbf{u} : \mathbf{U}^F \to \mathbf{F}$, $\mathbf{v} : \mathbf{U}^F \to \mathbf{F}$, $\mathbf{z} : \mathbf{U}^F \to \mathbf{M}$ are morphisms. We claim that $\mathbf{u}$ is dominant.

3.28. Assume this claim for the moment. Then $\mathbf{u}(\Psi^-)$ is Zariski-dense in $\mathbf{F}$. It follows that the group Δ generated by $\{a, \xi\}$ as $\varphi' \in \Phi'$ varies is Zariski-dense in $\mathbf{F}$. Let $H =$ Centraliser $M_1^0 \cap$ Centre of F. Evidently $\Delta \subset \mathbf{F}_{\mathbf{Q}}.\, H$. On the other hand, in view of the Zariski density of Δ, one sees easliy that if $\alpha \in \mathbf{F}_{\mathbf{Q}}$, there is an integer $p \geqslant 1$ such that $\alpha^p \in \Delta$. It follows that z normalises $\mathbf{F}_{\mathbf{Q}}.H$. If H is trivial this proves that $\sigma(z) \in GL(\mathfrak{f})_{\mathbf{Q}}$. If H is non-trivial, we assert that $H =$ centre of F.

The Lie algebra $\mathfrak{h}$ of H can be characterised as the Ad M_1^0 invariants in the centre of $\mathfrak{f}$. Since $H/H \cap \Gamma$ is compact, from the irreducibility of the lattice, it is not difficult to see that $\mathfrak{H}$ has a non-trivial projection into every simple factor of $\mathfrak{G}$. On the other hand, $\mathbf{M}$ normalises $\mathbf{M}_1^0$ (Corollary 3:23) so that $\mathfrak{H}$ must be $\mathbf{M}$-stable; and it is known (and easy to prove) that, the projection of the centre of $\mathfrak{F}$ on each simple component of $\mathfrak{G}$ is an irreducible $\mathbf{M}$-module. We conclude from these considerations that $\mathfrak{H}$ = Centre of $\mathfrak{F}$ and hence H = centre of F. The Lie algebra $\mathfrak{F}$ then decomposes into a direct sum

$$\mathfrak{F} = \mathfrak{F}_1 \oplus \text{Centre of } \mathfrak{F},$$

where $\mathfrak{F}_1$ is $\mathbf{M}_0$-stable and defined over $\mathbf{Q}$. The space $\mathfrak{F}_1$ is in fact uniquely determined as the sum of *all* irreducible non-trivial $\mathbf{M}_1^0$-submodules of $\mathfrak{F}$. It follows that $\mathfrak{F}$ is indeed $\mathbf{M}$-stable. We can then write any element $x \in \mathbf{F}$ in the form

$$x = x_1 . x_2$$

where $x_1 \in \mathbf{F}_1 = \exp \mathfrak{F}_1$ and $x_2 \in$ Centre of $\mathbf{F}$. Moreover if $x \in \mathbf{F}_{\mathbf{Q}}.H$, $x_1 \in \mathbf{F}_{\mathbf{Q}} \cap \mathbf{F}_1 = \mathbf{F}_{1\mathbf{Q}}$; also if x varies over a Zariski-dense subset of $\mathbf{F}$, x_1 varies over a Zariski-dense subset of $\mathbf{F}_1$. Writing now $a.\xi$ as $a_1.\xi_1$ with $a_1 \in \mathbf{F}_{1\mathbf{Q}}$, $\xi_1 \in H$ and observing that $z\,\mathbf{F}_1 z^{-1} = \mathbf{F}_1$, we find that $z\,\mathbf{F}_{1\mathbf{Q}} z^{-1} \subset \mathbf{F}_{1\mathbf{Q}}$. It is moreover easily seen that $\mathbf{F}_1$ generates all of $\mathbf{F}$ (this follows from the structure of parabolic subgroups of $\mathbf{G}$). Hence the group Δ' generated by $\mathbf{F}_{1\mathbf{Q}}$ is Zariski-dense in $\mathbf{F}$ and contained in $\mathbf{F}_{\mathbf{Q}}$. We conclude from this that $\sigma(z) \in GL\,(\mathfrak{f})_{\mathbf{Q}}$.

3.29. We will now establish the claim made in 3.27. Let $n = \dim \mathbf{F}$ and $\mathbf{E} = \wedge^n \mathfrak{G}$. Let $P(\mathbf{E})$ be the associated projective space of $\mathbf{E}$ and $p \in P(\mathbf{E})$ the point defined by the (n-dimensional) subspace $\mathfrak{F}$ of $\mathfrak{G}$. Consider the orbit map

$$\lambda : \mathbf{F}^- \to P(\mathbf{E})$$

given by $\lambda(x) = x.\,p$. For $x \in \mathbf{U}$ evidently

$$\lambda(x) = u(x).\,\tau.\,p$$

($z(x)$ and $v(x)$ normalise $\mathfrak{F}$.) On the other hand, since the isotropy at p for the action of G is $\mathbf{N}$ and $\mathbf{N} \cap \mathbf{F}^- = \{e\}$, λ is a 1–1 map. It follows that $\mathbf{u}: \mathbf{U}^F \to \mathbf{F}$ is a 1–1 map. Since $\Gamma \cap \mathbf{U}^F = F^- \cap \Gamma$ is Zariski-dense in $\mathbf{F}^-$, $\mathbf{U}^F$ is non-empty, open and dim $\mathbf{U}^F = \dim \mathbf{F}^- = \dim \mathbf{F}$. Thus the image of $\mathbf{U}^F$ under $\mathbf{u}$ is Zariski-dense in $\mathbf{F}$, i. e. $\mathbf{u}$ is dominant.

3.30. As was observed in 3.24, as γ varies over $\Gamma - N$ we have $\gamma = \alpha u z \tau v \rho$ with z running over a Zariski-dense subset of $\mathbf{M}$. Since $\sigma(z) \in GL(\mathfrak{F})_{\mathbf{Q}}$, for such $z \in \mathbf{M}$, we conclude that $\sigma(M)$ is a $\mathbf{Q}$-subgroup of $GL(\mathfrak{f})$. It is now immediate that $\mathbf{N} = \mathbf{M}.\mathbf{F}$ carries a natural $\mathbf{Q}$-structure such that

(i) $\mathbf{M}$ is defined over $\mathbf{Q}$

and

(ii) $\mathbf{N} \cap \Gamma$ is an arithmetic subgroup of N for this $\mathbf{Q}$-structure.

We now define a $\mathbf{Q}$-structure on all of $\mathbf{G}$. Since $\mathbf{G}$ was assumed to have trivial centre, it is isomorphic to its adjoint group. It suffices to define a $\mathbf{Q}$-structure on the Lie algebra. We have in fact the following general result.

3.31. Theorem. *Let $\mathfrak{G}$ be a complex semisimple algebra and $\mathfrak{N}$ the Lie algebra corresponding to a parabolic subgroup* $\mathbf{N}$ *of* $\mathbf{G}$ *(= a complex Lie group with $\mathfrak{G}$ as Lie algebra). Let $\mathfrak{n}_{\mathbf{Q}}$ be a Lie algebra over* $\mathbf{Q}$ *and $i : \mathfrak{n}_{\mathbf{Q}} \to \mathfrak{N}$ an injective homomorphism such that $i \otimes 1 : \mathfrak{n}_{\mathbf{Q}} \otimes_{\mathbf{Q}} \mathbf{C} \to \mathfrak{N}$ is an isomorphism. Then there exists a* $\mathbf{Q}$*-Lie algebra $\mathfrak{g}_{\mathbf{Q}}$ and injective homomorphisms $j : \mathfrak{g}_{\mathbf{Q}} \to \mathfrak{G}$ and $\alpha : \mathfrak{n}_{\mathbf{Q}} \to \mathfrak{g}_{\mathbf{Q}}$ such that*

(i) *$j \otimes 1 : \mathfrak{g}_{\mathbf{Q}} \otimes_{\mathbf{Q}} \mathbf{C} \to \mathfrak{G}$ is an isomorphism and*

(ii) *the diagram*

$$\begin{array}{ccc} \mathfrak{g}_{\mathbf{Q}} & \xrightarrow{\quad j \quad} & \mathfrak{G} \\ {\scriptstyle\alpha}\uparrow & & \uparrow{\scriptstyle\beta} \\ \mathfrak{n}_{\mathbf{Q}} & \xrightarrow{\quad i \quad} & \mathfrak{N} \end{array}$$

is commutative, β being the natural inclusion.

3.32. We first show that $\mathfrak{n}_{\mathbf{Q}}$ admits a decomposition $\mathfrak{n}_{\mathbf{Q}} = \mathfrak{m}_{\mathbf{Q}} \oplus \mathfrak{f}_{\mathbf{Q}}$ where $\mathfrak{m}_{\mathbf{Q}}$ is reductive and $\mathfrak{f}_{\mathbf{Q}}$ is the maximum nilpotent ideal. (If the $\mathbf{Q}$-structure is obtained from a $\mathbf{Q}$-structure on $\mathbf{N}$, this assertion is immediate). Let $\mathfrak{r}_Q$ be the radical of $\mathfrak{n}_Q$ and $\mathfrak{f}_Q$ the maximal nilpotent ideal. Then $\mathfrak{n}_Q$ is a semi-direct product of the form $\mathfrak{n}_Q = \mathfrak{m}'_Q \oplus \mathfrak{r}_Q$ where $\mathfrak{m}'_Q$ is a semisimple Lie subalgebra of $\mathfrak{n}_Q$. From the structure of parabolic groups, one knows that $\mathfrak{N}$ is a semi-direct product of the form $\mathfrak{N} = \mathfrak{M} \oplus \mathfrak{F}$ where $\mathfrak{F}$($=\mathbf{C}$-Linear span of $i(\mathfrak{f}_Q)$) is the maximal nilpotent ideal in $\mathfrak{N}$ and $\mathfrak{M}$ is reductive; moreover $\mathfrak{M} = \mathfrak{M}' \oplus \mathfrak{A}$ with $\mathfrak{A}$ = Centre of $\mathfrak{M}$ and $\mathfrak{M}' = [\mathfrak{M}, \mathfrak{M}]$ is the $\mathbf{C}$-linear span of $i(\mathfrak{m}'_Q)$. Further the centre $\mathfrak{A}$ of $\mathfrak{M}$ contains an element X such that the eigenvalues of ad X acting on $\mathfrak{F}$ are all positive integers; also if for integer $i > 0$, we let

$$\mathfrak{F}(i) = \{v \in \mathfrak{F} \mid [X, v] = i.\, v\},$$

then $\mathfrak{F}^{(i)} = \sum_{k>i} \mathfrak{F}(k)$, $\mathfrak{F}(m) = 0$, $\mathfrak{F}(m-1) \neq 0$, $0 \leqslant k \leqslant m$, ($m$ a suitable integer) is the descending central series for $\mathfrak{F}$. On $\mathfrak{F}^{(i)}/\mathfrak{F}^{(i+1)}$ ad X acts as the scalar i. Now if $\mathfrak{f}_Q^{(i)}$ is the descending central series for $\mathfrak{f}_Q$, evidently $\mathfrak{F}^{(i)}$ is the $\mathbf{C}$-linear span of $\mathfrak{f}_Q^{(i)}$ and the natural action of $\mathfrak{r}_Q/\mathfrak{f}_Q$ on $\mathfrak{f}_Q^{(i)}/\mathfrak{f}_Q^{(i+1)}$ is completely reducible and none of the simple components are trivial. It follows from the main result in the Appendix II that $H^p(\mathfrak{r}_Q/\mathfrak{f}_Q, \mathfrak{f}_Q/\mathfrak{f}_Q^{i+1}) = 0$ for all $p \geqslant 0$. A standard inductive argument now shows that we have a semi-direct product decomposition $\mathfrak{r}_Q = \mathfrak{a}_Q \oplus \mathfrak{f}_Q$ and we can find $X \in \mathfrak{a}_Q$ such that the eigenvalues of ad X in $\mathfrak{f}_Q$ are positive integers, $\mathfrak{a}_Q$ is abelian and centralises $\mathfrak{m}'_Q$. We fix now a semi-direct product decomposition $\mathfrak{n}_Q = (\mathfrak{m}'_Q \oplus \mathfrak{a}_Q) \oplus \mathfrak{f}_Q$ and fix also an element $X \in \mathfrak{a}_Q$ as above. Let $\mathfrak{m}_Q = \mathfrak{m}'_Q \oplus \mathfrak{a}_Q$. With this notation, we have the following sharper version of Theorem 3.31.

3.33. THEOREM. *Let $\mathfrak{F}^-$ be the sum of the eigen-spaces of* ad$(i(X))$ *corresponding to negative eigenvalues. Then $\mathfrak{g}_Q$ and j of Theorem* 3.31 *can be chosen to have the following property in addition to those in Theorem* 3.31. *Let $\mathfrak{f}_Q^- = j^{-1}(\mathfrak{f}^-)$; then $j' = j|_{\mathfrak{f}_Q} : \mathfrak{f}_Q^- \to \mathfrak{F}^-$ induces an isomorphism $j' \otimes 1 : \mathfrak{f}_Q^- \otimes \mathbf{C} \to \mathfrak{F}^-$.*

In the sequel we identify $\mathfrak{n}_Q$ with the $\mathbf{Q}$-subspace $i(\mathfrak{n}_Q)$ of $\mathfrak{g}$. In particular X will be regarded as an element of $\mathfrak{G}$. For a positive integer q, let

$$\mathfrak{k}_{\mathbf{Q}}(q) = \{v \in \mathfrak{k}_{\mathbf{Q}} \mid [X,v] = qv\}$$
$$\mathfrak{F}^{-}(q) = \{v \in \mathfrak{F}^{-} \mid [X,v] = -qv\}$$
$$\mathfrak{F}^{-} = \sum_{q>0} \mathfrak{F}^{-}(q).$$

Then $\mathfrak{F}^{-}$(resp. $\mathfrak{k}_{\mathbf{Q}}$) is generated by $\mathfrak{F}^{-}(1)$ (resp. $\mathfrak{k}_{\mathbf{Q}}(1)$) as a Lie algebra. Let

$$\mathfrak{k}_{\mathbf{Q}}^{-}(1) = \{v \in \mathfrak{F}^{-}(1) \mid \langle v,w \rangle \in \mathbf{Q} \text{ for } w \in \mathfrak{k}_{\mathbf{Q}}\}$$

where $\langle\ ,\ \rangle$ denotes the Killing form. Let

$$\mathfrak{k}_{\mathbf{Q}}^{-} = \{v \in \mathfrak{F}^{-} \mid \langle v,w \rangle \in \mathbf{Q} \text{ for } w \in \mathfrak{k}_{\mathbf{Q}}\}.$$

Evidently $\mathfrak{k}_{\mathbf{Q}}^{-}(1) \subset \mathfrak{k}_{\mathbf{Q}}^{-}$ and $\mathfrak{k}_{\mathbf{Q}}^{-}(1)$ (resp. $\mathfrak{k}_{\mathbf{Q}}^{-}$) spans $\mathfrak{F}^{-}(1)$ (resp. $\mathfrak{F}^{-}$) as a vector space over $\mathbf{C}$. Let $\mathfrak{k}'_{\mathbf{Q}}$ be the Lie subalgebra over $\mathbf{Q}$ generated by $\mathfrak{k}_{\mathbf{Q}}^{-}(1)$.

3.34. CLAIM. $\mathfrak{k}'_{\mathbf{Q}} = \mathfrak{k}_{\mathbf{Q}}^{-}$.

To prove the claim, we begin with the observation that for $v, w \in \mathfrak{m}_{\mathbf{Q}}$, $\langle v, w\rangle \in \mathbf{Q}$. Let σ (resp. ρ) denote the adjoint representation of $\mathfrak{m}_{\mathbf{Q}}$ on $\mathfrak{k}_{\mathbf{Q}}$ (resp. $\mathfrak{m}_{\mathbf{Q}}$). Then for $v, w \in \mathfrak{m}_{\mathbf{Q}}$, we have

$$\langle v, w\rangle = \text{trace } \rho(v)\rho(w) + 2 \text{ trace } \sigma(v)\sigma(w)$$

and evidently the right-hand side belongs to $\mathbf{Q}$. Since the restriction of the Killing form to $\mathfrak{M}$ is non-degenerate and element $v \in \mathfrak{M}$ belongs to $\mathfrak{m}_{\mathbf{Q}}$ if and only if $\langle v, w\rangle \in \mathbf{Q}$ for all $w \in \mathfrak{m}_{\mathbf{Q}}$. Using this criterion, we now show that $[\mathfrak{k}_{\mathbf{Q}}^{-}(1), \mathfrak{k}_{\mathbf{Q}}(1)] \subset \mathfrak{m}_{\mathbf{Q}}$. In fact if $A \in \mathfrak{k}_{\mathbf{Q}}^{-}(1)$, $Y \in \mathfrak{k}_{\mathbf{Q}}(1)$ and $V \in \mathfrak{m}_{\mathbf{Q}}$, we have

$$\langle [A, Y], V\rangle = -\langle A, [V, Y]\rangle \in \mathbf{Q}$$

since $[V, Y] \in \mathfrak{k}_{\mathbf{Q}}$. We assert now that for $i > 1$, we have

$$[\mathfrak{k}_{\mathbf{Q}}^{-}(1), \mathfrak{k}_{\mathbf{Q}}(i)] \subset \mathfrak{k}_{\mathbf{Q}}(i-1). \tag{*}$$

We argue by induction on i. Assume the assertion proved for all $j < i$, $j \geqslant 2$. Then we need only prove that for $Y \in \mathfrak{k}_{\mathbf{Q}}(1)$, $Z \in \mathfrak{k}_{\mathbf{Q}}(i-1)$ and $A \in \mathfrak{k}_{\mathbf{Q}}^{-}(1)$ we have

$$[A[Y,Z]] \in \mathfrak{k}_{\mathbf{Q}} \text{ (hence } \mathfrak{k}_{\mathbf{Q}}(i-1)).$$

In fact, $[A, [Y, Z]] = [[A, Y], Z] + [Y,[A, Z]]$. Since $[A, Y] \in \mathfrak{m}_{\mathbf{Q}}$, $[[A,Y], Z] \in \mathfrak{k}_{\mathbf{Q}}(i-1)$; on the other hand, by induction hypothesis

$[A, Z] \in \mathfrak{k}_{\mathbf{Q}}(i-2)$ (note that $i > 2$) so that $[Y, [A, Z]] \in \mathfrak{k}_{\mathbf{Q}}$. At the start of the induction, when $i = 2$, we have $Y, Z \in \mathfrak{k}_{\mathbf{Q}}^{-}(1)$ so that $[A, Y]$ and $[A, Z]$ are in $\mathfrak{m}_{\mathbf{Q}}$ and the result follows.

3.35. Now let $\mathfrak{k}'_{\mathbf{Q}}(i)$ be defined inductively as follows: $\mathfrak{k}'_{\mathbf{Q}}(1) = \mathfrak{k}_{\mathbf{Q}}^{-}(1)$ and for $i > 1$, $\mathfrak{k}'^{(i)}_{\mathbf{Q}} = [\mathfrak{k}_{\mathbf{Q}}^{-}(1), \mathfrak{k}'_{\mathbf{Q}}(i-1)]$. We will show inductively that for all $i \geqslant 1$,

$$\mathfrak{k}'_{\mathbf{Q}}(i) \subset \mathfrak{k}_{\mathbf{Q}}^{-}(i) = \{v \in \mathfrak{k}_{\mathbf{Q}}^{-} \mid [X, v] = i.v\} \tag{*}$$

When $i = 1$, this follows from the definitions. Assume that (*) is proved for $1 \leqslant j < i$. Let $A \in \mathfrak{k}_{\mathbf{Q}}^{-}(1)$, $B \in \mathfrak{k}'_{\mathbf{Q}}(i-1)$ $Y \in \mathfrak{k}_{\mathbf{Q}}(1)$, $Z \in \mathfrak{k}_{\mathbf{Q}}(i-1)$. Then we have

$$\begin{aligned}\langle [A, B], [Y, Z] \rangle &= -\langle B, [A, [Y, Z]] \rangle \\ &= -\langle B, [[A, Y], Z] \rangle - \langle B, [Y[A, Z]] \rangle.\end{aligned}$$

Now $[A, Y] \in \mathfrak{m}_{\mathbf{Q}}$ and $[A, Z] \in \mathfrak{k}_{\mathbf{Q}}(i-2)$ if $i > 2$ and $[A, Z] \in \mathfrak{m}_{\mathbf{Q}}$ if $i = 2$, so that from the induction hypothesis we conclude that the right-hand side belongs to $\mathbf{Q}$. Since the elements $[Y, Z]$, $Y \in \mathfrak{k}_{\mathbf{Q}}^{-}(1)$, $Z \in \mathfrak{k}'_{\mathbf{Q}}(i-1)$ span all of $\mathfrak{k}'_{\mathbf{Q}}(i)$ the result (*) follows. We see further that $\mathfrak{k}'_{\mathbf{Q}} = \sum_{i \geqslant 1} \mathfrak{k}'_{\mathbf{Q}}(i)$ is precisely the subalgebra generated by $\mathfrak{k}_{\mathbf{Q}}^{-}(1)$. Since the $\mathbf{C}$-span of $\mathfrak{k}_{\mathbf{Q}}^{-}(1)$ is easily seen to be $\mathfrak{k}^{-}(1)$ ($= \{v \in \mathfrak{k}_{1}^{-} [X, v] = -v\}$) one sees immediately that $\mathfrak{k}_{\mathbf{Q}}^{-}$ spans all of $\mathfrak{F}^{-}$. Since $\mathfrak{k}'_{\mathbf{Q}} \subset \mathfrak{k}_{\mathbf{Q}}^{-}$ (as follows from (*)) and $\mathfrak{k}_{\mathbf{Q}}^{-}$ is a $\mathbf{Q}$-vector space of the same dimension as $\mathfrak{F}^{-}$ over $\mathbf{C}$, Claim 3.34 is established.

3.36. Consider now the subspace $\mathfrak{g}_{\mathbf{Q}}^{-} = \mathfrak{k}_{\mathbf{Q}}^{-} \oplus \mathfrak{m}_{\mathbf{Q}} \oplus \mathfrak{k}_{\mathbf{Q}}$. This is a $\mathbf{Q}$-vector subspace of $\mathfrak{G}$ of the same dimension as $\mathfrak{G}$ over $\mathbf{C}$. We claim that $\mathfrak{g}_{\mathbf{Q}}$ is a Lie subalgebra. The relations $[\mathfrak{m}_{\mathbf{Q}}, \mathfrak{m}_{\mathbf{Q}}] \subset \mathfrak{m}_{\mathbf{Q}}$, $[\mathfrak{m}_{\mathbf{Q}}, \mathfrak{k}_{\mathbf{Q}}] \subset \mathfrak{k}_{\mathbf{Q}}$, $[\mathfrak{k}_{\mathbf{Q}}, \mathfrak{k}_{\mathbf{Q}}] \subset \mathfrak{k}_{\mathbf{Q}}$ follow from the hypothesis. From the invariance of the Killing form, one sees easily that $[\mathfrak{m}_{\mathbf{Q}}, \mathfrak{k}_{\mathbf{Q}}^{-}] \subset \mathfrak{k}_{\mathbf{Q}}^{-}$. That $\mathfrak{k}_{\mathbf{Q}}^{-}$ is a Lie subalgebra follows from Claim 3.34 established above. We need only prove now the following assertions:

(a) $[\mathfrak{k}_{\mathbf{Q}}^{-}(i), \mathfrak{k}_{\mathbf{Q}}(i)] \subset \mathfrak{m}_{\mathbf{Q}}$,

(b) $[\mathfrak{k}_{\mathbf{Q}}^{-}(i), \mathfrak{k}_{\mathbf{Q}}(j)] \subset \mathfrak{k}_{\mathbf{Q}}(j-i)$, if $j > i$,

(c) $[\mathfrak{k}_{\mathbf{Q}}^{-}(i), \mathfrak{k}_{\mathbf{Q}}(j)] \subset \mathfrak{k}_{\mathbf{Q}}(i-j)$.

For $A \in \mathfrak{k}_{\mathbf{Q}}^{-}(i)$, $Y \in \mathfrak{k}_{\mathbf{Q}}(i)$ and $V \in \mathfrak{m}_{\mathbf{Q}}$, we have

$$\langle [A, Y], V\rangle = \langle A, [Y, V]\rangle \in \mathbf{Q}.$$

This proves (a). To prove (b) we argue by induction on i. When $i=1$, this follows from (*) of 3.34. Let $A \in \mathfrak{f}_{\mathbf{Q}}^{-}(1)$ and $B \in \mathfrak{f}_{\mathbf{Q}}^{-}(i-1)$. Let $Y \in \mathfrak{f}_{\mathbf{Q}}(j)$, $j > i > 1$ be any element. Then we have

$$[[A, B], Y] = [A, [B, Y]] + [B, [A, Y]].$$

By induction hypothesis $[B, Y] \in \mathfrak{f}_{\mathbf{Q}}(j-i+1)$ and $[A, Y] \in \mathfrak{f}_{\mathbf{Q}}(j-1)$ where we have set $\mathfrak{m}_{\mathbf{Q}} = \mathfrak{f}_{\mathbf{Q}}(0)$. Induction hypothesis applied once again gives us the desired result. One argues analogously to prove (c); the start of the induction is the analogue of (*) of 3.34 with the roles of $\mathfrak{F}$ and $\mathfrak{F}^{-}$ reversed; and this analogue can be established since we now have $[\mathfrak{m}_{\mathbf{Q}}, \mathfrak{f}_{\mathbf{Q}}^{-}] \subset \mathfrak{f}_{\mathbf{Q}}^{-}$ and $\mathfrak{f}_{\mathbf{Q}}^{-}$ is a Lie subalgebra of $\mathfrak{G}$. This completes the proof of Theorem 3.33. As a corollary, we obtain the following (cf. Remarks in 3.27).

3.37. THEOREM. *$\mathbf{G}$ has a structure of an algebraic $\mathbf{Q}$-group with the following properties*: (a) *G is contained in $\mathbf{R}$-rational points of $\mathbf{G}$: in other words, the $\mathbf{Q}$-structure is compatible with the $\mathbf{R}$-structure*; (b) $\mathbf{N}$ *and* $\mathbf{N}^{-}$ *are $\mathbf{Q}$-subgroups of* $\mathbf{G}$; (c) $\mathbf{N} \cap \Gamma$ *is an arithmetic subgroup of* $\mathbf{N}_{\mathbf{R}}$; *and* (d) *the $\mathbf{Q}$-structure has $\mathbf{Q}$-rank* 1.

Only (d) requires some elaboration. $\mathbf{M} = \mathbf{N}^{-} \cap \mathbf{N}$ is a $\mathbf{Q}$-subgroup. Since $\mathbf{M}_{0\mathbf{R}}/M_1^0$ is compact and $M_1^0/M_1^0 \cap \Gamma$ is compact, the group $\mathbf{M}_0$ is anisotropic over $\mathbf{Q}$. Since $\mathbf{M}/\mathbf{M}_0$ has dimension 1 and $\mathbf{M}$ contains a maximal $\mathbf{Q}$-split torus, $\mathbf{Q}$-rank $\mathbf{G} \leqslant 1$. Since $\mathbf{G}$ contains $\mathbf{Q}$-rational unipotents, $\mathbf{Q}$-rank $G = 1$.

3.38. If $\mathbf{S}$ is a $\mathbf{Q}$-split torus contained in $\mathbf{M}$, $\mathbf{M}$ is precisely the Centraliser of $\mathbf{S}$. Also, $\mathbf{N}(\mathbf{S})/Z(\mathbf{S})$ is a group of order 2 ($\mathbf{N}(\mathbf{S})=$ normalizer of $\mathbf{S}$; $Z(\mathbf{S})=$ Centraliser of $\mathbf{S}$). Let $k \in \mathbf{N}(\mathbf{S})_{\mathbf{Q}}$ be a representative of the non-trivial element of $\mathbf{N}(\mathbf{S})/Z(\mathbf{S})$. One sees now easily from the fact, that $\theta_0 \mathbf{N} \theta_0^{-1} = \mathbf{N}^{-}$ and $\theta_0 = \tau v_0 \beta_0$ where $v_0 \in$ (Centre of F) $\cap$ Centraliser of M^0 and $\beta_0 \in \mathbf{F}_{\mathbf{Q}}$, that we have $\tau = z_1 k$ with $z_1 \in M$. Thus we have

$$\theta_0 \equiv z_1 k . v_0 \beta_0$$

with $\beta_0 \in \mathbf{F}_{\mathbf{Q}}$, and $v_0 \in$ Centre of $F \cap$ Centraliser M_1^0. Let $\varphi \in F \cap \Gamma = \Phi$. Then we have

$$\theta_0\varphi\theta_0^{-1} = a\xi t\tau\eta = a\xi t z_1 kb\eta$$

where a, $b \in \mathbf{F_Q}$, $\xi, \eta \in$ Centre of $F \cap$ Centraliser M_1^0 and $t \in M$. On the other hand we have

$$\theta_0 \varphi \theta_0^{-1} = z_1 x z_1^{-1}$$

with $x \in \mathbf{G_Q} \cap F^-$: this follows from the assumption that $k \in \mathbf{N(S)_Q}$. Now x has the Bruhat-decomposition

$$x = \lambda \xi k \mu$$

with $\lambda, \mu \in \mathbf{F_Q}$ and $\xi \in \mathbf{M_Q}$. We conclude from this that $z_1 \lambda z_1^{-1} \in \mathbf{F_Q}.H$ (H was defined in 3.28). The arguments of 3.26–3.29 used to prove Proposition 3.26 can be imitated to conclude analogously that $\sigma(z_1) \in GL(\mathfrak{F})_\mathbf{Q}$. This is equivalent to saying that $z_1 \in \mathbf{M_Q}$. It follows immediately that we have

3.39. PROPOSITION. $\theta_0 \Phi \theta_0^{-1} \subset \mathbf{G_Q}$.

3.40. COROLLARY. *Let* $0 \neq Z \in \mathfrak{F}$ *be any element such that* $\exp Z \in \Phi$. *Then for any* $\theta_0 \in \Gamma$, $\langle \mathrm{ad}\ \theta_0(Z), Z\rangle \in \mathbf{Q}$. *More generally for* $\theta, \theta' \in \Gamma$, $\langle \mathrm{ad}\ \theta(Z), \mathrm{ad}\ \theta'(Z)\rangle \in \mathbf{Q}$.

The first statement is an immediate consequence of Proposition 3.36. The second now follows from the invariance of the Killing form.

3.41. Fix $Z \neq 0$ in $\mathfrak{F}$ as in Corollary 3.40. Let $V_\mathbf{Q}$ be the $\mathbf{Q}$-linear span of $\{\mathrm{ad}\ \theta(Z) \mid \theta \in \Gamma\}$. The $\mathbf{R}$-span V of $V_\mathbf{Q}$ is a Γ-stable subspace of $\mathfrak{g}$ and is hence an ideal. The corresponding Lie group evidently contains non-trivial unipotent elements of Γ so that it follows that $\mathfrak{g} = V$. The Killing form is non-degenerate on $V = \mathfrak{g}$ and takes rational values on $V_\mathbf{Q}$, a $\mathbf{Q}$-subspace which spans $V = \mathfrak{g}$. It follows that $\dim_\mathbf{Q} V_\mathbf{Q} = \dim_\mathbf{R} \mathfrak{g}$. Since $\mathrm{ad}\ \theta(V_\mathbf{Q}) \subset V_\mathbf{Q}$ for all $\theta \in \Gamma$, $\mathrm{ad}\ \Gamma \subset GL(\mathfrak{G})_\mathbf{Q}$ for the $\mathbf{Q}$-vector space structure defined on $\mathfrak{G}$ by $V_\mathbf{Q}$. Taking Zariski-closures, one sees that $\mathbf{G} \simeq \mathrm{ad}\ \mathbf{G}$ *carries a* $\mathbf{Q}$*-structure such that* $\Gamma \subset \mathbf{G_Q}$. Moreover it is easily seen that this $\mathbf{Q}$-structure is the same as that introduced in Theorem 3.37. Since $\Gamma \subset \mathbf{G_Q}$, we have the following: if $\Phi \subset \Gamma$ is any maximal unipotent subgroup of Γ and $\mathbf{F}$ is the Zariski-closure of Φ in $\mathbf{G}$, $\mathbf{F}$ is the unipotent radical of a

Q-parabolic subgroup **N** (= normaliser of **F**). **N** ∩ Γ is an arithmetic subgroup of **N** and $N_0/N_0 \cap \Gamma$ is compact where

$$N_0 = \{x \in G \cap \mathbf{N} \mid \det \sigma(x) = 1\},$$

σ being the adjoint representation of **N** on the Lie algebra $\mathfrak{F}$ of **F**.

3.42. Suppose now that $\mathbf{P} \subset \mathbf{G}$ is any parabolic subgroup defined over **Q**. Let Φ be any maximal unipotent subgroup of Γ with Zariski-closure **F** in **G** and **N** the normaliser of **F**. Then **N** is also a parabolic subgroup defined over **Q** and since **G** has **Q**-rank 1, $\mathbf{P} \cap \mathbf{N} = \mathbf{M}$ is a maximal reductive subgroup of **P** as well as **N**. Since $\mathbf{N} \cap \Gamma$ is arithmetic $\mathbf{M} \cap \Gamma$ is arithmetic in **M** and $\mathbf{M}_{0\mathbf{R}}/\mathbf{M}_{0\mathbf{R}} \cap \Gamma$ is compact ($\mathbf{M}_0 = \{x \in \mathbf{M} \mid \det \sigma(x) = 1\}$). Further since $\mathbf{M}_{0\mathbf{R}}$ is non-compact $\mathbf{M}_{0\mathbf{R}} \cap \Gamma$ is non-trivial. Consider now a conjugate $\mathbf{N}' = \theta \mathbf{N} \theta^{-1}$ of **N** by an element $\theta \in \Gamma - \mathbf{N}$. Then the discussion above is valid for **N′** as well so that if $\mathbf{M}' = \mathbf{P} \cap \mathbf{N}'$ and $\mathbf{M}'_0 = \{x \in \mathbf{M}' \mid \det \sigma'(x) = 1\}$, σ′ denoting the adjoint representation of **M′** on $\mathbf{F}'(= \theta F \theta^{-1})$, then $\mathbf{M}' \cap \Gamma$ is arithmetic in **M′** as well. Now **M** and **M′** being maximal reductive subgroups of **P**, there exists an element $u \in \mathbf{U_Q}$, **U** the unipotent radical of **P**, such that $u\mathbf{M}u^{-1} = \mathbf{M}'$; the groups $u(\mathbf{M} \cap \Gamma)u^{-1}$ and $\mathbf{M}' \cap \Gamma$ are then both arithmetic in **M′** and are hence commensurable. Let $\Delta \subset \mathbf{M} \cap \Gamma$ be a subgroup of finite index such that $u \Delta u^{-1} \subset \mathbf{M}' \cap \Gamma$. We assert that for a suitable choice of θ, all of Δ cannot commute with θ. The Zariski-closure of Δ contains the connected component of the identity in $\mathbf{M}_1$ = Zariski-closure of $\Gamma \cap \mathbf{M}$ so that it suffices to show that θ can be chosen such that $\mathbf{M}_1 \not\subset \theta \mathbf{N} \theta^{-1}$; and this follows from the Zariski density of Γ in **G**. We can thus assume that there exists $\delta \in \Delta$ such that $\delta' = u \delta u^{-1} \neq \delta$. Consider now the element $\delta' \delta^{-1} \in \Gamma \cap P$. Clearly

$$\alpha = \delta' \delta^{-1} = u \delta u^{-1} \delta^{-1} = u(\delta u^{-1} \delta^{-1}) \in U$$

since **P** normalises **U**, it follows that $\mathbf{U} \cap \Gamma \ni \{\alpha\}$ and $\alpha \neq$ identity. Let Θ be the unique maximal unipotent subgroup of Γ containing α. Then since $\Theta \subset \mathbf{G_Q}$, and **U** is the *unique* maximal unipotent **Q**-subgroup containing α, $\mathbf{U} \supset \Theta$ (the uniqueness of the maximal unipotent **Q**-group containing α follows from the fact that **G** has *Q*-rank 1 (see for instance Raghunathan [1], § 12.15)). On the other hand the

Zariski-closure $\mathbf{U}'$ of Θ which is necessarily a $\mathbf{Q}$-group is the unipotent radical of a parabolic group. It follows that $\mathbf{U}' = \mathbf{U}$ and $(\mathbf{U} \cap G)/\mathbf{U} \cap \Gamma$ is compact and that $\mathbf{P} \cap \Gamma$ is an arithmetic subgroup of $\mathbf{P}$.

4. L-subgroups of rank $\geqslant 2$. As hitherto G will denote a connected semisimple Lie group with trivial centre and no compact factors and Γ an irreducible non-uniform L-subgroup. We assume throughout this section that Γ *has rank* $\geqslant 2$.

4.1. NOTATION. We fix a connected $\mathbf{R}$-algebraic group $\mathbf{G}$ with trivial centre such that G is isomorphic to the identity component of $\mathbf{G}_R$, the group of $\mathbf{R}$-rational points of $\mathbf{G}$. We choose once and for all one such isomorphism and identify G with a subgroup of $\mathbf{G}$. For a subgroup $H \subset G$, we denote by $\mathbf{H}$ the Zariski-closure of H in $\mathbf{G}$. As usual for a Lie subgroup $H \subset \mathbf{G}$, H^0 will denote the identity component of H. (We remark that if $H \subset G$, $(\mathbf{H}^0) = (\mathbf{H})^0$ so that, this group can be denoted $\mathbf{H}^0$ without any cause for confusion.) The Lie algebra of G (resp. $\mathbf{G}$) is denoted $\mathfrak{g}$ (resp. $\mathfrak{G}$) and for a linear subspace $E \subset \mathfrak{g}$, $\mathbf{E}$ denotes its $\mathbf{C}$-linear span. When a subspace of $\mathfrak{g}$ is denoted by a gothic lower case letter, its $\mathbf{C}$-linear span is denoted by the corresponding capital gothic letter. The following theorem summarises the results of § 2.

4.2. THEOREM. *There is a parabolic subgroup* $\mathbf{N} \subset \mathbf{G}$ *defined over* $\mathbf{R}$ *with the following properties:*

(i) *Let* $\mathbf{F}$ *be the unipotent radical and* $F = \mathbf{F} \cap G^{(*)}$; *then* $F \cap \Gamma = \Phi$ *is a lattice in* F. *Consequently the* $\mathbf{Z}$*-linear span* $\mathscr{L}$ *of* $\exp p^{-1}(\Phi)$ *in* $\mathfrak{f}$ (= *Lie algebra of* F) *is a lattice in* $\mathfrak{f}$. $GL\,(\mathfrak{F})$ *has a natural* $\mathbf{Q}$-*structure defined by* $\mathscr{L}$.

(ii) *Let* σ *denote the adjoint representation of* $\mathbf{N}$ *on* $\mathfrak{F}$. *Let* $\mathbf{D}$ *be the Zariski-closure of* $N \cap \Gamma$ *in* $\mathbf{G}$. *Then* $\mathbf{D}' = \sigma(\mathbf{D})$ *is a* $\mathbf{Q}$-*subgroup of* $GL\,(\mathfrak{f})$ *and* $\sigma(\mathbf{D} \cap \Gamma)$ *is an arithmetic subgroup of* $\mathbf{D}'_{\mathbf{R}}$.

(iii) *Let* $N = \mathbf{N} \cap G^*$ *and* $N_0 = \{x \in N \mid det_{\cdot}\sigma\,(x) = \pm 1\}$. *Let* $D = \mathbf{D} \cap G^{\dagger}$; *then* $D \subset N_0$ *and* N_0/D *is compact. If* $\mathbf{B} \subset \mathbf{N}_0$ *is any* $\mathbf{R}$-*split torus; then* $\mathbf{B} \subset \mathbf{D}$. *Moreover if* $\mathbf{B} \subset \mathbf{N}$ *is any* $\mathbf{R}$-*simple connected subgroup*

† $\mathbf{F} \cap G$, $\mathbf{N} \cap G$ and $\mathbf{D} \cap G$ are easily seen to be Zariski-dense in $\mathbf{F}$, $\mathbf{N}$ and $\mathbf{D}$ respectively.

with $\mathbf{B_R}$ *non-compact, then* $\mathbf{B} \subset \mathbf{D}$. *The natural map* $N_0/N_0 \cap \Gamma \to G/\Gamma$ *is proper.*

(iv) *The group* $\mathbf{D}/\mathbf{F}$ *is reductive and rank* $\Gamma = \mathbf{Q}$*-rank* $(\mathbf{D}/\mathbf{F}) + 1$. ($\mathbf{D}/\mathbf{F}$ *has a natural* $\mathbf{Q}$*-structure*).

4.3. DEFINITION. *A parabolic subgroup* $\mathbf{N}$ *of* $\mathbf{G}$ *as in Theorem* 4.2 *will be called* Γ*-adapted. The group* $\mathbf{N}$ *is* Γ*-rational if in addition it satisfies the following condition: there exists a connected reductive subgroup* $\mathbf{M}_1$ *such that* $\mathbf{M}_1 \cap \Gamma$ *is Zariski-dense in* $\mathbf{M}_1$, $\sigma(\mathbf{M}_1)$ *is a maximal connected reductive subgroup of* $\mathbf{D}'$ *defined over* $\mathbf{Q}$ *and* $\sigma(\mathbf{M}_1) \cap \Gamma$ *is an arithmetic subgroup of* $\sigma(\mathbf{M}_1)_{\mathbf{R}}$.

We will prove later in §5 that there always exist Γ-rational parabolic subgroups of $\mathbf{G}$. The proof of this result is however somewhat involved. We will therefore first obtain further properties of Γ under the assumption that $\mathbf{G}$ admits Γ-rational parabolic subgroups.

4.4. THEOREM. *Assume that* $\mathbf{G}$ *admits a* Γ*-rational parabolic subgroup. Then there exists a* $\mathbf{Q}$*-group* $\mathbf{G}^*$ *and an* $\mathbf{R}$*-isomorphism* $u: \mathbf{G}^* \to \mathbf{G}$ *such that*

(i) $\Gamma' = u^{-1}(\Gamma) \subset \mathbf{G}^*_{\mathbf{Q}}$, *and*

(ii) *for every parabolic sugbroup* $\mathbf{P}$ *of* $\mathbf{G}^*$ *defined over* $\mathbf{Q}$,

$$\mathbf{P} \cap \Gamma' \text{ is an arithmetic subgroup } \mathbf{P}_{\mathbf{R}}.$$

4.5. We fix a Γ-rational parabolic subgroup $\mathbf{N}$ and choose also a subgroup $\mathbf{M}_1$ as in Definition 4.3. We will use the notation introduced in Theorem 4.2. Since $\mathbf{D}/\mathbf{F}$ is reductive, it follows that $\mathbf{D}^0 = \mathbf{M}_1 . \mathbf{F}$. The lattice $\Phi = F \cap \Gamma$ in F defines a natural $\mathbf{Q}$-structure on $\mathbf{F}$. Since $\mathbf{M}'_1 = \sigma(\mathbf{M}_1)$ is a $\mathbf{Q}$-subgroup of $GL(\mathfrak{F})$ and σ maps $\mathbf{M}_1$ isomorphically onto $\mathbf{M}'_1$, we see that $\mathbf{M}_1$ carries a $\mathbf{Q}$-structure such that the action of $\mathbf{M}_1$ on $\mathbf{F}$ is defined over $\mathbf{Q}$. We obtain then a $\mathbf{Q}$-structure on $\mathbf{D}^0$ such that $\mathbf{M}_1$ is a (maximal reductive) $\mathbf{Q}$-subgroup of $\mathbf{D}^0$. Further, $\mathbf{M}_1 \cap \Gamma$ is an arithmetic subgroup of $\mathbf{M}_1$ for this $\mathbf{Q}$-structure. Since $\mathbf{F} \cap \Gamma$ is an arithmetic subgroup of $\mathbf{F}$, $(M_1 \cap \Gamma).(F \cap \Gamma)$ is an arithmetic subgroup $\mathbf{D}^0$. Consequently $\mathbf{D}^0 \cap \Gamma$ is an arithmetic subgroup of $\mathbf{D}^0$.

4.6. Now let $\mathbf{T}$ be a maximal $\mathbf{Q}$-split torus in $\mathbf{M}_1$ and $X(T)$ the group of rational characters on T. Fix an ordering of $X(T)$ and let

Σ (resp. Σ^+, Δ) denote the system of **Q**-roots (resp. positive **Q**-roots, simple **Q**-roots) of $\mathbf{M}_1$ with respect to **T**. For $\alpha \in \Sigma$, let

$$\mathfrak{M}(\alpha) = \{v \in \mathfrak{M}_1 \mid \mathrm{ad}\, t(v) = \alpha(t) . v \text{ for } t \in \mathbf{T}\}$$

where $\mathfrak{M}_1$ is the Lie algebra of $\mathbf{M}_1$. Let $\mathfrak{U}_1 = \sum_{\alpha \in \Sigma^+} \mathfrak{M}(\alpha)$ and $\mathbf{U}_1$ the corresponding Lie subgroup of $\mathbf{M}_1$. Then $\mathbf{U}_1$ is a maximal unipotent **Q**-subgroup of $\mathbf{M}_1$. Let $\mathbf{U} = \mathbf{U}_1.\mathbf{F}$ and $U = \mathbf{U} \cap G$. We will now prove the following

4.7. Lemma. $\mathbf{U} \cap \Gamma$ *is a maximal unipotent subgroup of* Γ.

Let $\Psi \supset \mathbf{U} \cap \Gamma$ be any unipotent subgroup of Γ. Let **S** be the Zariski-closure of Ψ and $\mathfrak{S}$ the Lie algebra of **S**. Then **S** is unipotent so that $\mathfrak{S}$ is orthogonal to itself hence a fortiori to $\mathfrak{F}$ with respect to the Killing form. On the other hand $\mathfrak{N}$ is precisely the orthogonal complement of $\mathfrak{F}$ with respect to the Killing form. It follows that $\mathfrak{S}$ is contained in $\mathfrak{N}$. Hence $\mathbf{S} \subset \mathbf{N}$. On the other hand, as is easily seen, $U/U \cap \Gamma$ is compact where $U = \mathbf{U} \cap G$ so that **U** is contained in the Zariski-closure of Ψ, i.e. $\mathbf{U} \subset \mathbf{S}$. It suffices therefore to prove that $\mathbf{U} \cap \Gamma$ is a maximal unipotent subgroup of $\mathbf{N} \cap \Gamma$; and this follows easily from the arithmeticity of the group σ $(\mathbf{N} \cap \Gamma)$.

4.8. Lemma. **U** *is the unipotent radical of a parabolic group in* **G**.

Let **M** be a maximal reductive **R**-algebraic subgroup of **N** containing $\mathbf{M}_1$. Let $\mathbf{M}_0 = \mathbf{M} \cap N_0$. The group **M** is an almost direct product $\mathbf{A}.\mathbf{M}_0$ where **A** is the central **R**-split torus whose Lie algebra $\mathfrak{A}$ is the orthogonal complement of $\mathfrak{M}_0$ (= Lie algebra of $\mathbf{M}_0$) in $\mathfrak{M}$ (= Lie algebra of **M**) with respect to the Killing form. Now let **P** be the normaliser of **U**; $\mathbf{P} \cap \mathbf{M}_1$ is evidently a parabolic subgroup of $\mathbf{M}_1$ defined over **R**. Since **M** contains all unipotent **R**-subgroups of $\mathbf{M}_0$, $\mathbf{P} \cap \mathbf{M}_0$ is necessarily a parabolic subgroup of $\mathbf{M}_0$. Finally since $\mathbf{A} \subset \mathbf{P}$, $\mathbf{P} \cap \mathbf{M}$ is a parabolic subgroup of **M**. Since **MF** is a parabolic subgroup of **G**, **P** is a parabolic subgroup of **G**. It is not difficult to see that the Lie algebra $\mathfrak{U}$ of **U** is the orthogonal complement of the Lie algebra $\mathfrak{P}$ of **P**. This shows that **U** is the unipotent radical of **P**. Hence the lemma.

4.9. Now consider any element $\alpha \in \Delta$. Let $\mathfrak{U}(\alpha)$ be the ideal in $\mathfrak{U}$ generated by $\mathfrak{M}(\alpha)$. Now since $U/U \cap \Gamma$ is compact, **U** is a **Q**-subgroup of $\mathbf{D}^0$. As $\mathfrak{M}(\alpha)$ is a **Q**-subspace, $\mathfrak{U}(\alpha)$ is a **Q**-subalgebra of $\mathfrak{U}$. Consequently the corresponding Lie subgroup $\mathbf{U}(\alpha)$ of $\mathbf{D}^0$ is defined over **Q** and $U(\alpha)/U(\alpha) \cap \Gamma$ is compact where $U(\alpha) = \mathbf{U}(\alpha) \cap G$. We will now prove the following

4.10. LEMMA. *For $\alpha \in \Delta$, $\mathbf{U}(\alpha)$ is the unipotent radical of a parabolic subgroup $\mathbf{N}(\alpha)$ of* **G** *defined over* **R**.

4.11. Let $\mathbf{T}^* = \mathbf{A}.\mathbf{T}$ and **B** be a maximal torus of **G** containing $\mathbf{T}^*$ and contained in **M**. Let $X(\mathbf{T}^*)$, $X(\mathbf{A})$ and $X(\mathbf{B})$ denote the groups of characters on $\mathbf{T}^*$, **A** and **B** respectively. Now it is known (see for instance Raghunathan [1 , Chapter XII]) that there is a generating character $\chi_0 \in X(\mathbf{A})$ such that $\mathfrak{F}$ is precisely the sum of all those eigen-spaces of **A** corresponding to positive multiples of χ_0. We introduce an ordering on $X(\mathbf{T}^*)$ as follows: an element $\theta \in X(\mathbf{T}^*)$ is positive if $\theta|_{\mathbf{A}} = m\chi_0$ with $m \geqslant 0$ and if $m = 0$, $\theta|_{\mathbf{T}}$ is positive in the order fixed on $X(\mathbf{T})$ (cf. 4.6). Next we introduce an order on $X(\mathbf{B})$ compatible with that on $X(\mathbf{T}^*)$. Let $\boldsymbol{\Delta}$ be the system of simple roots of **G** w.r.t. **B** and the order chosen above. Let r: $X(\mathbf{B}) \to X(\mathbf{T})$ and $r^* : X(\mathbf{B}) \to X(T^*)$ be the natural restriction maps. We claim that $r(\boldsymbol{\Delta})$ contains Δ. To see this we first observe that $\{\mathfrak{M}(\theta) \mid \theta \in \Delta\}$ generate $\mathfrak{U}_1$ as a Lie algebra. Each $\mathfrak{M}(\theta)$ is evidently the span of the root-spaces $\mathfrak{G}(\boldsymbol{\theta})$ $(= \{v \in \mathfrak{G} \mid \operatorname{ad} t(v) = \boldsymbol{\theta}(t)\, v \text{ for } t \in \mathbf{B}\})$ as $\boldsymbol{\theta}$ runs through those roots (w.r.t. **B**) which are trivial on **A** and restricted **T** are equal to θ. Let θ^* be the lowest of these roots. We claim that θ^* is simple. In fact if $\theta^* = \beta^* + \gamma^*$ with β^*, γ^* positive roots, we necessarily have $\beta^*|_{\mathbf{A}} = \gamma^*|_{\mathbf{A}} = 0$. If β^* and γ^* are both nontrivial on **T**, then $\mathfrak{G}(\beta^*)$ and $\mathfrak{G}(\gamma^*)$ are both contained in $\mathfrak{M}_1$. This follows from the fact that $\mathbf{M}_1$ contains all unipotent subgroups of **M** defined over **R**. But then $\beta = \beta^*|_{\mathbf{T}}$ and $\gamma = \gamma^*|_{\mathbf{T}}$ are both positive **Q**-roots of $\mathbf{M}_1$ w.r.t. **T** and $\theta = \beta + \gamma$, a contradiction. Thus one of β^* or γ^* must be trivial on **T**. But then if $\gamma^* \neq 0$, β^* has the same restriction to **T** as θ^*, vanishes on **A** and $\beta^* < \theta^*$, a contradiction. Thus $r(\boldsymbol{\Delta})$ contains Δ. Moreover if $\theta^* \in \boldsymbol{\Delta}$ is such that θ^* is trivial on **A** and non-trivial on **T**, then $r(\theta^*) \in \Delta$. This follows from our choice of ordering on **B** and the fact that the set $\{\theta^* \in \mathbf{\Delta} \mid \theta^* \text{ trivial on } \mathbf{A}\}$ is precisely the system of simple roots of the reductive group $\mathbf{M}$ (= centraliser of $\mathbf{A}$). We can now describe the Lie algebras of groups $\mathbf{F}$, $\mathbf{U}$ and $\mathbf{U}(\alpha)$ as follows. For a subset $\mathbf{\Delta}' \subset \mathbf{\Delta}$, let

$$\mathbf{\Sigma}(\mathbf{\Delta}') = \{\beta = \sum_{\theta \in \mathbf{\Delta}} m_\theta . \theta \in \Sigma^+ \mid m_\theta > 0 \text{ for some } \theta \in \mathbf{\Delta}'\}.$$

Let

$$\mathbf{\Delta}(\mathbf{F}) = \{\theta \in \mathbf{\Delta} \mid \theta \text{ is non-trivial on } A\}$$

$$\mathbf{\Delta}(\mathbf{U}) = \{\theta \in \mathbf{\Delta} \mid \theta \text{ is non-trivial on } \mathbf{T}^*\} \text{ and}$$

$$\mathbf{\Delta}(\alpha) = \{\theta \in \mathbf{\Delta} \mid r(\theta) = \alpha\}.$$

We set also

$$\mathbf{\Sigma}(\mathbf{F}) = \mathbf{\Sigma}(\mathbf{\Delta}(\mathbf{F}))$$

$$\mathbf{\Sigma}(\mathbf{U}) = \mathbf{\Sigma}(\mathbf{\Delta}(\mathbf{U})) \text{ and}$$

$$\mathbf{\Sigma}(\alpha) = \mathbf{\Sigma}(\mathbf{\Delta}(\alpha)).$$

With this notation, one sees easily that we have

$$\mathfrak{F} = \sum_{\theta \in \mathbf{\Sigma}(\mathbf{F})} \mathfrak{G}(\theta)$$

$$\mathfrak{U} = \sum_{\theta \in \mathbf{\Sigma}(\mathbf{U})} \mathfrak{G}(\theta) \text{ and}$$

$$\mathfrak{U}(\alpha) = \sum_{\theta \in \mathbf{\Sigma}(\alpha)} \mathfrak{G}(\theta).$$

This proves in particular that $\mathbf{U}(\alpha)$ is the unipotent radical of a parabolic subgroup $\mathbf{N}(\alpha)$ (= normaliser of $\mathbf{U}(\alpha)$). This proves Lemma 4.10.

4.12. Lemma. *Let* $\mathbf{V}$ *be one of the unipotent groups* $\{\mathbf{U}(\alpha) \mid \alpha \in \Delta\}$ *or* $\mathbf{F}$. *Suppose* $\mathbf{V}_1 \underset{\neq}{\subset} \mathbf{V}$ *is subgroup with the following properties : if* $V_1 = \mathbf{V}_1 \cap G$, $V_1/V_1 \cap \Gamma$ *is compact and* $\mathbf{V}_1$ *is the unipotent radical of an* $\mathbf{R}$*-parabolic subgroup of* G. *Then* $\mathbf{V}_1 = \{e\}$.

4.13. Consider first the case $\mathbf{V} = \mathbf{F}$. Let $\mathbf{P}$ be the normaliser of $\mathbf{V}_1$. Then $\mathbf{P}$ contains $\mathbf{N}$. Let $\pi \colon P \to P/V_1$ be the natural projection where $P = \mathbf{P} \cap G$. Let $\mathfrak{V}_1$ denote the Lie algebra of $\mathbf{V}_1$ and let ρ denote the adjoint representation of $\mathbf{N}/\mathbf{V}_1$ on $\mathfrak{F}/\mathfrak{V}_1$ and let τ be the corresponding representation of $\mathbf{N}$. Now the group $\tau(\mathbf{D} \cap \Gamma)$ leaves a lattice in $\mathfrak{f}/\mathfrak{v}_1$ ($\mathfrak{v}_1 = \mathfrak{V}_1 \cap \mathfrak{g}$) stable so that det $\tau(x) = \pm 1$ for $x \in \mathbf{D}$. Since N_0/D is compact, det $\tau(N_0)$ is a compact subgroup of $\mathbf{R}^*$ hence det $\tau(N_0) \subset \{\pm 1\}$. Thus both the characters det τ and det σ on $\mathbf{N}$ are trivial on a codimension 1 group. On the other hand it is easy to check that if $\mathbf{V}_1$ is not trivial nor the whole of $\mathbf{F}$, then det τ and det σ are linearly independent characters on $\mathbf{N}$, a contradiction.

NOTE. The argument above has used only the fact that $\mathbf{N}$ is Γ-adapted—we have not used the stronger condition of Γ-rationality.

4.14. Consider now the case when $\mathbf{V} = \mathbf{U}(\alpha)$ for some $\alpha \in \Delta$. Consider the subgroup $\mathbf{U}_1 \cap \mathbf{V}_1 = \mathbf{V}_2$, say. It is easily seen that $\mathbf{V}_2$ is the unipotent radical of a parabolic subgroup of $\mathbf{M}_1$; since $V_2/V_2 \cap \Gamma$ ($V_2 = \mathbf{V}_2 \cap G$) is compact, $\mathbf{V}_2$ is defined over $\mathbf{Q}$ as well. Since $\mathbf{U}(\alpha) \cap \mathbf{M}_1$ is the unipotent radical of a *maximal* $\mathbf{Q}$-parabolic subgroup, and $\mathbf{V}_2 \subset \mathbf{U}(\alpha) \cap \mathbf{M}_1$, $\mathbf{V}_2 = \mathbf{U}(\alpha) \cap \mathbf{M}_1$ or $\mathbf{V}_2 = \{e\}$; in the former case $\mathbf{V}_1 = \mathbf{U}(\alpha)$. We may thus assume that $\mathbf{V}_2 = \{e\}$. On the other hand, $\mathfrak{V}_1$ is the sum of eigen-spaces for *any* torus normalising $\mathbf{U}$. It follows that $\mathfrak{V}_1 \subset \mathfrak{F}$; but then the considerations of 4.13 show that $\mathbf{V}_1 = \mathfrak{F}$ or $\mathbf{V}_1 = \{e\}$. Since $\mathbf{V}_1 \subset \mathbf{U}(\alpha)$ and $\mathfrak{F} \not\subset \mathbf{U}(\alpha)$, the former possibility is ruled out. This completes the proof of Lemma 4.11.

4.15. Fix the torus $\mathbf{B}$ as in 4.11. Let $W \subset$ Normaliser $\mathbf{B}$ be a set of representatives for the Weyl group. Let $\mathbf{U}^*$ be the maximal unipotent subgroup of $\mathbf{G}$ determined by the simple system $\boldsymbol{\Delta}$ of roots. Then (Bruhat-decomposition) any element $g \in \mathbf{G}$ can be written in the form

$$g = u\, w\, z\, v$$

where $u, v \in \mathbf{U}^*$, $w \in W$ and $z \in \mathbf{B}$. Let $w_0 \in W$ be the unique element of W which takes all the positive roots into negative roots. Then $\mathbf{U}^*\, w_0\, \mathbf{B}\, \mathbf{U}^*$ is Zariski open in G so that $\Gamma \cap (\mathbf{U}^*\, w_0\, \mathbf{B}\, \mathbf{U}^*) = \Gamma^*$ is Zariski-dense in $\mathbf{G}$. Elements of Γ^* are sometimes referred to as *principal elements* of Γ in the sequel.

4.16. We now fix an element $\gamma \in \Gamma^*$ and introduce the following notation :

$$\mathbf{U}^- = \gamma\, \mathbf{U}\, \gamma^{-1}$$
$$\hat{\mathbf{U}}^{-1}(\alpha) = \gamma\, \mathbf{U}\,(\alpha)\, \gamma^{-1}$$

$$\hat{\mathbf{F}} = \gamma \mathbf{F} \gamma^{-1} \text{ and}$$
$$\mathbf{U}_1^- = \mathbf{U}^- \cap \mathbf{N} (= \mathbf{U}^- \cap \mathbf{N}_0).$$

Let $\gamma = u\, w_0\, zv$ be the Bruhat decomposition of γ. We then have $\mathbf{U}^- = u\, w_0\, \mathbf{U}\, w_0^{-1}\, u^{-1}$, $\hat{\mathbf{U}}^-(\alpha) = u\, w_0\, \mathbf{U}(\alpha)\, w_0^{-1}\, u^{-1}$ etc. Since w normalises $\mathbf{B}$, it follows that $\mathbf{B}' = u\, \mathbf{B}\, u^{-1}$ normalises $\mathbf{U}^-$. Since $u \in \mathbf{U}^*$ normalises $\mathbf{U}$, one sees that the Lie algebra $\mathfrak{U}^-$ of $\mathbf{U}^-$ is the linear span of eigen-spaces for $\mathbf{B}$. Moreover, if we set $\mathbf{U}^{*-} = \gamma\, \mathbf{U}^*\, \gamma^{-1}$, the eigen-characters of $\mathbf{B}'$ acting on $\mathfrak{U}^{*-}$ (= Lie algebra of $\mathbf{U}^{*-}$) are precisely the inverses of eigen-characters of $\mathbf{B}'$ acting on $\mathfrak{U}^*$ (= Lie algebra of $\mathbf{U}^*$): this follows from the fact that the automorphism of $\mathbf{B}$ induced by w_0 takes all the positive roots into negative roots. We will now identify the torus $\mathbf{B}$ with $\mathbf{B}'$ by the inner conjugation by the element and, using this identification regard characters on $\mathbf{B}$ as characters on $\mathbf{B}'$ and carry over the order on $X(\mathbf{B})$ to an order $X(\mathbf{B}')$. With these identifications, we may regard the roots of $\mathbf{G}$ w.r.t. $\mathbf{B}$ also as roots of $\mathbf{G}$ w.r.t. $\mathbf{B}'$. For $\chi \in \Sigma$ (regarded as a character on $\mathbf{B}'$) we set

$$\mathfrak{G}'(\chi) = \{v \in \mathfrak{G} \mid \operatorname{ad} t(v) = \chi(t)v \text{ for all } t \in \mathbf{B}'\}.$$

One sees then immediately that

$$\mathfrak{U} = \sum_{\theta \in \Sigma(U)} \mathfrak{G}'(\theta)$$
$$\mathfrak{U}(\alpha) = \sum_{\theta \in \Sigma(\alpha)} \mathfrak{G}'(\theta) \text{ and}$$
$$\mathfrak{F} = \sum_{\theta \in \Sigma(F)} \mathfrak{G}'(\theta).$$

(See 4.11 for notation). The automorphism of $\mathbf{B}$ induced by w_0 takes each element $\varphi \in \mathbf{\Delta}$ into $-\hat{\varphi}$ for some $\varphi \in \Delta$; the map $\varphi \to \hat{\varphi}$ is a bijection of $\mathbf{\Delta}$. For a subset $\mathbf{\Delta}' \subset \mathbf{\Delta}$, let $\hat{\mathbf{\Delta}}'$ or $\mathbf{\Delta}'^{\wedge}$ denote the image of $\mathbf{\Delta}'$ under this bijection. The following direct-sum decompositions are easily seen to hold (notation as in 4.11)

$$\mathfrak{U}^- = \sum_{\theta \in \Sigma(\mathbf{\Delta}(\mathbf{U})^{\wedge})} \mathfrak{G}'(-\theta)$$
$$\hat{\mathfrak{U}}^-(\alpha) = \sum_{\theta \in \Sigma(\mathbf{\Delta}(\alpha)^{\wedge})} \mathfrak{G}'(-\theta)$$
$$\mathfrak{F} = \sum_{\theta \in \Sigma(\mathbf{\Delta}(\mathbf{F})^{\wedge})} \mathfrak{G}'(-\theta).$$

Let $\mathbf{M}' = u\,\mathbf{M}\,u^{-1}$; then $\mathfrak{M}'$ (= Lie algebra of $\mathbf{M}'$) is a sum of eigen-spaces for $\mathbf{B}'$. Moreover $\mathfrak{N} = \mathfrak{M}' \oplus \mathfrak{F}$. It follows from this and the fact that $\mathfrak{F} \cap \mathfrak{U}^- = \{0\}$ that we have $\mathfrak{U}^- \cap \mathfrak{N} = \mathfrak{U}^- \cap \mathfrak{M}'$. Hence $\mathbf{U}^- \cap \mathbf{N} = \mathbf{U}^- \cap \mathbf{M}'$.

4.17. Now the group $\mathbf{U}^-$ being unipotent

$$\mathbf{U}_1^- = \mathbf{U}^- \cap \mathbf{N} = \mathbf{U}^- \cap \mathbf{N}_0.$$

The natural maps

$$U^-/U^- \cap \Gamma \longrightarrow G/\Gamma \text{ and}$$

$$N_0/N_0 \cap \Gamma \longrightarrow G/\Gamma$$

are proper. We claim that if $U_1^- = \mathbf{U}_1^- \cap G$ then

$$U_1^-/U_1^- \cap \Gamma \longrightarrow G/\Gamma$$

is proper as well. This follows in fact from the general Lemma 4.18 below. Assuming this for the moment we conclude from the compactness of $U^-/U^- \cap \Gamma$ that $U_1^-/U_1^- \cap \Gamma$ is compact as well. Since $\mathbf{U}_1^-$ is defined over $\mathbf{R}$, $\mathbf{U}_1^-$ is the Zariski-closure of U_1^- hence also of $U_1^- \cap \Gamma$. Since $\mathbf{U}_1^-$ is contained in $\mathbf{N}$, we find that $\mathbf{U}_1^- \subset \mathbf{D}^0$. It is moreover a $\mathbf{Q}$-subgroup. Before we proceed we will prove the claim made above.

4.18. Lemma. *Let H_1, H_2 be two closed subgroups of G such that the natural inclusions $H_1/H_1 \cap \Gamma \to G/\Gamma$ and $H_2/H_2 \cap \Gamma \to G/\Gamma$ are both proper maps. Then setting $H = H_1 \cap H_2$, the natural map $H/H \cap \Gamma \to G/\Gamma$ is also proper.*

To prove the lemma, it suffices to show the following: let $x_n \in H$ and $\gamma_n \in \Gamma$ be sequences such that $x_n\,\gamma_n$ converges; then there exist $\theta_n \in H \cap \Gamma$ such that $x_n\,\theta_n$ converges. Since the map $H_i/H_i \cap \Gamma \to G/\Gamma$, $i = 1, 2$, are proper, there exist $\alpha_{in} \in H_i \cap \Gamma$, $i = 1, 2$, such that the sequence $x_n \cdot \alpha_{in}$ converges (for $i = 1, 2$). It follows that the sequence $\alpha_{1n}^{-1} \cdot \alpha_{2n} (\in \Gamma)$ converges. Since Γ is discrete, there exists an integer $p > 0$ such that for $n \geqslant p$, $\alpha_{1n}^{-1}\,\alpha_{2n} = \alpha_{1p}^{-1}\,\alpha_{2p}$; equivalently, for $n \geqslant p$,

$$\alpha_{2n}\,\alpha_{2p}^{-1} = \alpha_{1n}\,\alpha_{1p}^{-1} = \theta_n, \text{ say.}$$

Evidently $\theta_n \in H \cap \Gamma$ and $x_n\,\theta_n$ converges. Hence the lemma.

We now go back to the main discussion. We will establish the following:

4.19. LEMMA. *Let* $\pi: \mathbf{N} \to \mathbf{N}/\mathbf{F}$ *be the natural map. Then* $\pi(\mathbf{U}_1^-)$ *is a maximal unipotent* $\mathbf{Q}$*-subgroup of the reductive* $\mathbf{Q}$*-group* $\pi(\mathbf{D}^0)$. *Further let* $\mathbf{E}$ *be the centraliser of* $\mathbf{U}_1^-$ *in* $\mathbf{F}$ *and* $E = \mathbf{E} \cap G$. *Then there exists* $a \in E$ *such that* $a^{-1} u \in \mathbf{D}_\mathbf{Q}$. (*We recall that* $u \in \mathbf{U}^*$ *is defined by the Bruhat-decomposition* $\gamma = u\, w_0\, z\, v$).

4.20. Let $p = \dim \mathbf{F}$ and $q = \dim \mathbf{U}$. Then $\mathbf{N}$ is a subgroup of $\mathbf{G}$ of codimension exactly p. It follows that $\dim(\mathbf{U}^- \cap \mathbf{N}) \geqslant q - p$. Now $\mathbf{U}/\mathbf{F}$ is a maximal unipotent $\mathbf{Q}$-subgroup of $\pi(\mathbf{D}^0)$ and π maps $\mathbf{U}_1^- = \mathbf{U}^- \cap \mathbf{N}$ isomorphically onto a $\mathbf{Q}$-subgroup of $\pi(\mathbf{D}^0)$. We conclude from this that $\pi(\mathbf{U}_1^-)$ is a maximal unipotent $\mathbf{Q}$-subgroup of $\pi(\mathbf{D})$.

For proving the second assertion, we argue as follows. For $\theta \in \mathbf{U}_1^- \cap \Gamma$, we have $\pi(u^{-1}\, \theta\, u) = \pi(\theta)$. Since $\pi(\theta)$ is a $\mathbf{Q}$-rational point of $\pi(\mathbf{D}^0)$, π restricted to $\mathbf{M}_1$ is a $\mathbf{Q}$-isomorphism, and $u^{-1}\theta \in \mathbf{M} \cap D^0 = M_1$, we conclude that $u^{-1}\,\theta\, u$ is a $\mathbf{Q}$-rational point of $\mathbf{D}^0$ for all $\theta \in \mathbf{U}_1^- \cap \Gamma$. But this means that if $\tau: \mathbf{C} \to \mathbf{C}$ is any field automorphism

$$u^{-1}\,\theta u = (u\,\tau)^{-1}\,\theta\, u\,\tau,\ \theta \in \mathbf{U}_1^- \cap \Gamma$$

(note that θ is $\mathbf{Q}$-rational). Equivalently $b_\tau = u(u^\tau)^{-1} \in \mathbf{E}$. Evidently $\{b_\tau\}_{\tau \in \text{Aut}\, \mathbf{C}}$ is a 1-cocycle on Aut $\mathbf{C}$; and since $\mathbf{E}$ is a *unipotent* $\mathbf{Q}$-group, we can find $a \in \mathbf{E}$ such that $a(a^\tau)^{-1} = u(u^\tau)^{-1}$ for all $\tau \in$ Aut $\mathbf{C}$. Clearly $a^{-1}\, u \in \mathbf{D}^0{}_\mathbf{Q}$ and since $u \in G$, $a \in G$ as well. This proves the Lemma 4.19.

4.21. Let $\mathbf{M}_1'' = a^{-1} u \mathbf{M}_1 u^{-1} a$; then $\mathbf{M}_1''$ is a maximal connected reductive $\mathbf{Q}$-subgroup of $\mathbf{D}^0$. Also $\mathbf{U}_1^-$ is contained in $\mathbf{M}_1''$ and is a maximal unipotent $\mathbf{Q}$-subgroup of $\mathbf{M}_1''$. Consider for $\alpha \in \Delta$, the group $\hat{\mathbf{U}}^-(\alpha)$: this is a normal subgroup of $\mathbf{U}^-$; it is the unipotent radical of a parabolic subgroup; this last property is equivalent to the following: the orthogonal complement of $\hat{\mathfrak{U}}^-(\alpha)$ in $\mathfrak{G}$ is a Lie subalgebra ($=$ normaliser $\hat{\mathfrak{N}}^-(\alpha)$ of $\hat{\mathfrak{U}}^-(\alpha)$). It follows that the orthogonal complement of $\hat{\mathfrak{U}}^-(\alpha)$ in $\mathfrak{M}'$ is a Lie subalgebra as well.

Since $\mathfrak{M}$ is reductive and is a sum of eigen-spaces for $\mathbf{B}'$, we conclude that the orthogonal complement of $\hat{\mathfrak{U}}_1^-(\alpha) = \hat{\mathfrak{U}}^-(\alpha) \cap \mathfrak{M}'$ in $\mathfrak{M}'$ is a Lie subalgebra. We conclude therefore that the Lie subgroup $\hat{\mathbf{U}}_1^-(\alpha)$ corresponding to $\hat{\mathbf{U}}_1^-(\alpha)$ is the unipotent radical of a parabolic subgroup of $\mathbf{M}'$. On the other hand since $\mathbf{U}^- \cap \mathbf{M}' = \mathbf{U}^- \cap \mathbf{N}$, we see that $\hat{\mathbf{U}}_1^-(\alpha)$ is precisely $\hat{\mathbf{U}}^-(\alpha) \cap \mathbf{N}$. Since a centralises this last group we find that for $\alpha \in \Delta$, $\hat{\mathbf{U}}_1^-(\alpha)$ is the unipotent radical of a parabolic subgroup of $\mathbf{M}'' = a^{-1}\mathbf{M}''a$. Further $\hat{\mathbf{U}}_1^-(\alpha) \cap \Gamma$ is a uniform lattice in $\hat{\mathbf{U}}_1^-(\alpha) = \hat{\mathbf{U}}_1^-(\alpha) \cap G$, so that $\hat{\mathbf{U}}_1^-(\alpha) \subset \mathbf{M}_1''$ is a subgroup defined over $\mathbf{Q}$ of $\mathbf{M}_1''$. Entirely analogous remarks hold for $\hat{\mathbf{F}}_1^- = \hat{\mathbf{F}}^- \cap \mathbf{N}$. Now consider the normal subgroup $\mathbf{H}$ of $\mathbf{U}^-$ generated by $\hat{\mathbf{U}}_1^-(\alpha)$. This subgroup is evidently defined over $\mathbf{R}$ and contained in $\hat{\mathbf{U}}^-(\alpha)$. Further if $H = \mathbf{H} \cap G$, $H/H \cap \Gamma$ is compact. Finally it is easily seen that $\mathbf{H}$ itself is the unipotent radical of a parabolic subgroup of $\mathbf{G}$ defined over $\mathbf{R}$. According to Lemma 4.12 $\mathbf{H} = \hat{\mathbf{U}}^-(\alpha)$ or $\mathbf{H} = \{e\}$ i.e. the group $\hat{\mathbf{U}}^-(\alpha) \cap \mathbf{M}_1''$ is either trivial or no proper normal subgroup of $\hat{\mathbf{U}}^-(\alpha)$ contains it. Similarly $\mathbf{F}_1^-$ is either trivial or no proper normal subgroup of $\mathbf{F}^-$ contains it. We summarise the remarks made above as follows.

4.22. Lemma. *Let* $\mathbf{V}^-$ *denote one of the groups* $\hat{\mathbf{U}}^-(\alpha)$, $\alpha \in \Delta$, *or* $\mathbf{F}^-$. *Then* $\mathbf{V}_1^- = \mathbf{V}^- \cap \mathbf{M}_1''$ *is the unipotent radical of a parabolic* $\mathbf{Q}$*-subgroup of* $\mathbf{M}_1''$. *Moreover if* $\mathbf{V}_1^-$ *is non-trivial, then* $\mathbf{V}^-$ *is the smallest normal subgroup of* $\mathbf{V}^-$ *containing* V_1^-. *Further if* $\mathbf{V}_1^-$ *is non-trivial, its normaliser in* $\mathbf{M}_1''$ *is a maximal parabolic* $\mathbf{Q}$*-subgroup.*

4.23. The last assertion is proved as follows: If possible let $\mathbf{P}$ be a $\mathbf{Q}$-parabolic of $\mathbf{M}_1''$ properly containing the normaliser of $\mathbf{V}_1^-$ in $\mathbf{M}_1''$. Let $\mathbf{V}_2^-$ be the unipotent radical of $\mathbf{P}$. Then $\mathbf{V}_2^- \subsetneq \mathbf{V}_1^-$ and the normaliser of $\mathbf{V}_2^-$ in $\mathbf{M}_1'$ (and hence $\mathbf{M}'$) is parabolic subgroup of $\mathbf{M}_1'$ (resp. $\mathbf{M}'$). Now from the general structure of parabolic subgroups, one sees easily that the normal subgroup $\mathbf{H}$ of $\mathbf{U}^-$ generated by $\mathbf{V}_2^-$ is properly contained in $\mathbf{V}^-$ and is the unipotent radical of a $\mathbf{R}$-parabolic subgroup of $\mathbf{G}$. In addition $H/H \cap \Gamma$ is compact where $H = \mathbf{H} \cap G$. Lemma 4.12 then leads us to conclude that $\mathbf{H} = \{e\}$ i.e. $\mathbf{V}_2^-$ is trivial. This completes the proof of Lemma 4.22.

4.24. Consider now the group $\hat{\mathbf{F}}^- \cap \mathbf{N}$. There are two possibilities:

CASE A. $\hat{\mathbf{F}}^- \cap \mathbf{N} = \{e\}$, or

CASE B. $\hat{\mathbf{F}}^- \cap \mathbf{N}$ is the unipotent radical of a maximal parabolic $\mathbf{Q}$-subgroup of $\mathbf{M}_1''$.

In the first case we have from dimension considerations

$$\mathfrak{G} = \hat{\mathfrak{F}}^- \oplus \mathfrak{N}.$$

We now discuss Case B in detail. $\hat{\mathbf{F}}^- \cap \mathbf{N}$ is the unipotent radical of a parabolic subgroup of $\mathbf{M}'$ as well; this follows from the fact that $\mathbf{\Delta}$ centralises $\hat{\mathbf{F}}^- \cap \mathbf{N}$. Evidently $\hat{\mathbf{F}}^- \cap \mathbf{N}$ is stable under the torus $\mathbf{B}'$. It follows that we have a subset $\mathbf{\Delta}' \subset \mathbf{\Delta}$ such that for $\theta \in \mathbf{\Delta}'$, $\mathfrak{G}'(\theta) \subset \mathfrak{M}'$ and if $\mathbf{\Sigma}'$ denotes the set $\{ \theta \in \Sigma^+ \mid \mathfrak{G}'(\theta) \subset \mathfrak{M}', \theta = \sum_{\varphi \in \mathbf{\Delta}} m_\varphi . \varphi$ with $m_\varphi > 0$ for some $\varphi \in \mathbf{\Delta}' \}$, then

$$\hat{\mathfrak{F}}^- \cap \mathfrak{N} = \sum_{\theta \in \mathbf{\Sigma}'} \mathfrak{G}'(-\theta) \text{ and } \hat{\mathfrak{F}} = \sum_{\theta \in \mathbf{\Sigma}(\mathbf{\Delta}')} \mathfrak{G}'(-\theta).$$

Further since $\hat{\mathbf{F}}^- \cap \mathbf{N}$ is a $\mathbf{Q}$-subgroup of $\mathbf{M}_1''$, one sees easily that the θ restricted to $\mathbf{T}$ are in fact $\mathbf{Q}$-roots. (We have allowed ourselves to consider elements of Σ as characters on $\mathbf{B}$ as well as $\mathbf{B}'$). Now from the decomposition $\gamma = u\, w_0\, z\, v$, using $\gamma\, \mathbf{F}\, \gamma^{-1} = \mathbf{F}^-$, we conclude that the automorphism of $\mathbf{T}$ defined by w_0 maps $\mathbf{\Delta}(F)$ onto $\{-\theta \mid \theta \in \mathbf{\Delta}'\}$. Further in view of Lemma 4.12, the restrictions $\{r(\theta), \theta \in \mathbf{\Delta}\}$ must all be equal to a single simple $\mathbf{Q}$-root $\alpha_0 \in \Delta$. It follows immediately from this that $\mathbf{U}(\alpha_0)$ has the same dimension as $\mathfrak{F}$. Moreover the automorphism induced by w_0 on $\mathbf{B}$ is of order 2 and maps $\mathbf{\Delta}'$ into $\{-\theta \mid \theta \in \mathbf{\Delta}(F)\}$. It follows easily from this that $\hat{\mathbf{U}}^-(\alpha_0)$ is of the same dimension as $\mathbf{F}$ and $\hat{\mathbf{U}}^-(\alpha_0) \cap \mathbf{N} = \{e\}$. We have thus proved the following:

4.25. LEMMA. *Either* $\mathfrak{G} = \mathfrak{N} \oplus \hat{\mathfrak{F}}^-$ *or there exists a unique* $\alpha_0 \in \Delta$ *such that* $\mathfrak{G} = \mathfrak{N} \oplus \hat{\mathfrak{U}}^-(\alpha_0)$.

4.26. Among the Lie subalgebras $\mathfrak{U}^-(\alpha)$, $\alpha \in \Delta$ and $\hat{\mathfrak{F}}$, there is thus a unique one which is supplementary to $\mathfrak{N}$ (note that all the $\hat{\mathfrak{U}}^-(\alpha)$, $\alpha \in \Delta$ and $\hat{\mathfrak{F}}$ are B'^- stable). We denote this subalgebra $\hat{\mathfrak{F}}^-$ and the corresponding group by $\mathbf{F}^-$. It is now easy to see from

Lemma 4.22 and 4.12 that all the groups $\hat{\mathbf{U}}^-(\alpha)$, $\alpha \in \Delta$, and $\hat{\mathbf{F}}^-$ other than $\mathbf{F}^-$ are generated as normal subgroups of $\mathbf{U}^-$ by their intersections with $\mathbf{M}_1''$ and these latter intersections are unipotent radicals of maximal $\mathbf{Q}$-parabolic subgroups of M_1'' (as well as certain parabolic subgroups of $\mathbf{M}'$). These groups are in addition normalised by $\mathbf{B}'$. These considerations now lead us to the following result:

4.27. Lemma. *For* $\theta \in \mathbf{\Delta}(\mathbf{F}) \cup \bigcup_{\alpha \in \Delta} \mathbf{\Delta}(\alpha)$, $w_0(\theta) = -\hat{\theta}$ *with* $\hat{\theta} \in \mathbf{\Delta}(F) \cup \bigcup_{\alpha \in \Delta} \mathbf{\Delta}(\alpha)$. *Moreover the map* $\theta \to \hat{\theta}$ *defines a bijection of each* $\mathbf{\Delta}(\alpha)$, $\alpha \in \Delta$, *onto some* $\mathbf{\Delta}(\alpha')$, $\alpha' \in \Delta$, *or onto* $\mathbf{\Delta}(F)$. *Similarly* $\mathbf{\Delta}(\mathbf{F})$ *is mapped bijectively some* $\mathbf{\Delta}(\alpha)$, $\alpha \in \Delta$, *or* $\mathbf{\Delta}(\mathbf{F})$.

4.28. It follows easily from Lemma 4.27 that for each $\alpha_0 \in \Delta$, there is a unique Lie subalgebra among the $\hat{\mathfrak{U}}^-(\alpha)$, $\alpha \in \Delta$, and $\hat{\mathfrak{F}}^-$ which is supplementary to $\mathfrak{N}(\alpha_0)$ in $\mathfrak{G}$. We denote this subalgebra in the sequel by $\mathfrak{U}^-(\alpha_0)$ and the corresponding Lie subgroup by $\mathbf{U}^-(\alpha_0)$. Now $\mathbf{U}^-(\alpha)$, $\alpha \in \Delta$, is the unipotent radical of a $\mathbf{R}$-parabolic subgroup of $\mathbf{G}$. Its centraliser in $\mathbf{G}$ is hence equal to the centre of $\mathbf{U}^-(\alpha)$. Further $U^-(\alpha)/U^-(\alpha) \cap \Gamma\,(U^-(\alpha) = \mathbf{U}^-(\alpha) \cap G)$ is compact. Let $\mathbf{N}^-(\alpha)$ denote the normaliser of $\mathbf{U}^-(\alpha)$ in $\mathbf{G}$ and $\sigma^-(\alpha)$ the adjoint representation of $\mathbf{N}^-(\alpha)$ on $\mathfrak{U}^-(\alpha)$. Similarly, let $\sigma(\alpha)$ denote the adjoint representation of $\mathbf{N}(\alpha)$ on $\mathfrak{U}(\alpha)$. Let $\mathbf{N}_0^-(\alpha) = \{x \in \mathbf{N}^-(\alpha) \mid \det \sigma^-(\alpha)(x) = \pm 1\}$ and analogously $\mathbf{N}_0(\alpha) = \{x \in \mathbf{N}(\alpha) \mid \det \sigma(\alpha)(x) = \pm 1\}$. Let $\mathscr{L}(\alpha)$ (resp. $\mathscr{L}^-(\alpha)$) denote the $\mathbf{Z}$-span of $\exp^{-1}(U(\alpha) \cap \Gamma)$ (resp. $\exp^{-1}(U^-(\alpha) \cap \Gamma)$) in $\mathfrak{U}(\alpha)$ (resp. $\mathfrak{U}^-(\alpha)$). $\mathscr{L}(\alpha)$ (resp. $\mathscr{L}^-(\alpha)$) spans $\mathfrak{U}(\alpha)$ (resp. $\mathfrak{U}^-(\alpha)$) as a $\mathbf{C}$-vector space and hence defines a natural $\mathbf{Q}$-structure on $GL(\mathfrak{U}(\alpha))$ (resp. $GL(\mathfrak{U}^-(\alpha))$. Analogous remarks apply to $\mathfrak{F}^-$: we denote by $\mathbf{N}^-$ the normaliser of $\mathbf{F}^-$, by σ^- the adjoint representation of $\mathbf{N}^-$ on $\mathfrak{F}^-$ and by $\mathbf{N}_0^-$ the group $\{x \in \mathbf{N}^- \mid \det \sigma(x) = \pm 1\}$. With this notation, the following maps induced by natural inclusions and $\sigma(\alpha)$, σ, $\sigma^-(\alpha)$ σ^- etc. are proper:

$$N_0 / N_0 \cap \Gamma \longrightarrow G/\Gamma$$

$$N_0 / N_0 \cap \Gamma \longrightarrow GL(\mathfrak{f}) / GL\,\mathscr{L}$$

$$N_0^- / N_0^- \cap \Gamma \longrightarrow G/\Gamma$$

$$N_0^- / N_0^- \cap \Gamma \longrightarrow GL(\mathfrak{f}^-)/GL(\mathscr{L}^-)$$

$$N_0(\alpha) / N_0(\alpha) \cap \Gamma \longrightarrow G/\Gamma$$

$$N_0(\alpha) / N_0(\alpha) \cap \Gamma \longrightarrow GL(\mathfrak{U}(\alpha))/GL(\mathscr{L}(\alpha))$$

$$N_0^-(\alpha) / N_0^-(\alpha) \cap \Gamma \longrightarrow G/\Gamma$$

$$N_0^-(\alpha) / N_0^-(\alpha) \cap \Gamma \longrightarrow GL(\mathfrak{U}^-(\alpha))/GL(\mathscr{L}^-(\alpha)).$$

(Here $N_0(\alpha) = \mathbf{N}_0(\alpha) \cap G$, $N_0^-(\alpha) = \mathbf{N}_0^-(\alpha) \cap G$ and $N_0^- = \mathbf{N}_0^- \cap G$). Using Lemma 4.18 we now conclude the following

4.29. Lemma. *Let* $\mathbf{H}^* = \mathbf{N}_0 \cap \mathbf{N}_0^-$ *and for* $\alpha \in \Delta$, $\mathbf{H}^*(\alpha) = \mathbf{N}_0(\alpha) \cap \mathbf{N}_0^-(\alpha)$. *Let* $H^* = \mathbf{H}^* \cap G$ *and* $H^*(\alpha) = \mathbf{H}^*(\alpha) \cap G$. *Then the maps* $H^*/H^* \cap \Gamma \to G/\Gamma$, $H^*(\alpha)/H^*(\alpha) \cap \Gamma \to G/\Gamma$, $H^*/H^* \cap \Gamma \to GL(\mathfrak{f})/GL(\mathscr{L})$ *and* $H^*(\alpha)/H^*(\alpha) \cap \Gamma \to GL(\mathfrak{U}(\alpha))/GL(\mathscr{L}(\alpha))$ *are proper.*

4.30. Let $\mathbf{H}$ (resp. $\mathbf{H}(\alpha)$) denote the Zariski-closure of $H^* \cap \Gamma$ (resp. $H^*(\alpha) \cap \Gamma$) in $\mathbf{G}$ and set $H = \mathbf{H} \cap G$, $H(\alpha) = \mathbf{H}(\alpha) \cap G$. Then $\sigma(\mathbf{H})$ (resp. $\sigma(\alpha)(\mathbf{H}(\alpha))$) is a $\mathbf{Q}$-subgroup of $GL(\mathfrak{F})$ (resp. $GL(\mathfrak{U}(\alpha))$). Since $\sigma(H \cap \Gamma)$ (resp. $\sigma(\alpha)(H(\alpha) \cap \Gamma)$) is contained in $GL(\mathscr{L})$ (resp. $GL(\mathbf{L}(\alpha))$) and the map $H/H \cap \Gamma \to GL(\mathfrak{f})/GL(\mathscr{L})$ (resp. $H(\alpha)/H(\alpha) \cap \Gamma \to GL(\mathfrak{U}(\alpha))/GL(\mathscr{L}(\alpha))$) is proper, $\sigma(H \cap \Gamma)$ (resp. $\sigma(H(\alpha) \cap \Gamma)$) is an arithmetic subgroup of $\sigma(\mathbf{H})$ (resp. $\sigma(\alpha)(\mathbf{H}(\alpha))$). Since $\sigma_{\mathbf{H}}$ (resp. $\sigma(\alpha)|_{\mathbf{H}(\alpha)}$) is an isomorphism onto the image, we conclude that $\mathbf{L} = \mathbf{H}.\mathbf{F}$ (resp. $\mathbf{L}(\alpha) = \mathbf{H}(\alpha).\mathbf{U}(\alpha)$) carries a natural $\mathbf{Q}$-structure such that $\mathbf{H}$ (resp. $\mathbf{H}(\alpha)$) is a $\mathbf{Q}$-subgroup and $\mathbf{L} \cap \Gamma$ (resp. $\mathbf{L}(\alpha) \cap \Gamma$) is an arithmetic subgroup of $\mathbf{L}$ (resp. $\mathbf{L}(\alpha)$). Similarly $\mathbf{L}^- = \mathbf{H}.\mathbf{F}^-$ and $\mathbf{L}^-(\alpha) = \mathbf{H}(\alpha).\mathbf{U}^-(\alpha)$ have natural $\mathbf{Q}$-group structures in which $\mathbf{H}$ and $\mathbf{H}(\alpha)$ are $\mathbf{Q}$-subgroups. The $\mathbf{Q}$-structures on $\mathbf{H}$ (resp. $\mathbf{H}(\alpha)$) induced from $\mathbf{L}$ and $\mathbf{L}^-$ (resp. $\mathbf{L}(\alpha)$ and $\mathbf{L}^-(\alpha)$) coincide.

4.31. Lemma. *The group* $\mathbf{H}$ (*resp.* $\mathbf{H}(\alpha)$, $\alpha \in \Delta$) *is reductive. Let* $\mathbf{H}'$ (*resp.* $\mathbf{H}'(\alpha)$) *be the smallest algebraic subgroup of* $\mathbf{H}$ (*resp.* $\mathbf{H}'(\alpha)$) *generated by all the* $\mathbf{Q}$-*rational unipotents in* $\mathbf{H}$ (*resp.* $\mathbf{H}(\alpha)$). *Then the Lie algebra* $\mathfrak{H}^*$ *of* $\mathbf{H}^*$ (*resp.* $\mathfrak{H}^*(\alpha)$ *of* $\mathbf{H}^*(\alpha)$) *decomposes into a direct sum of the form* $\mathfrak{H}^* = \mathfrak{h} \oplus \mathfrak{K}$ (*resp.* $\mathfrak{H}^*(\alpha) = \mathfrak{H}(\alpha) + \mathfrak{K}(\alpha)$) *where* $\mathfrak{K}$ (*resp.* $\mathfrak{K}(\alpha)$) *consists of* $\mathbf{H}'$ (*resp.* $\mathbf{H}'(\alpha)$) *invariants.*

4.32. We prove the assertions in the lemma for $\mathbf{H}$ and $\mathbf{H}^*$. The proofs for $\mathbf{H}(\alpha)$ and $\mathbf{H}^*(\alpha)$ are identical except for notational differences. The unipotent radical $\mathbf{V}$ of $\mathbf{H}$ is a $\mathbf{Q}$-group. Since $H \cap \Gamma$ is arithmetic in H, $V/V \cap \Gamma$ is compact where $V = \mathbf{V} \cap G$. Now we have $\mathbf{U} = (\mathbf{U} \cap \mathbf{H}).\, \mathbf{F}$ and $\mathbf{U}^- = (\mathbf{U}^- \cap \mathbf{H}).\, \mathbf{F}^-$. Since $\mathbf{V} \subset \mathbf{H}$, $\mathbf{V}$ normalises $\mathbf{F}$ and $\mathbf{F}^-$. It follows that $(\mathbf{U} \cap \mathbf{H}).\, \mathbf{V}.\, \mathbf{F}$ and $(\mathbf{U}^- \cap \mathbf{H}).\, \mathbf{V}.\mathbf{F}$ are $\mathbf{R}$-unipotent subgroups of $\mathbf{G}$ such that their intersections with Γ are Zariski-dense in them. In view of the maximality of $U \cap \Gamma$ and $U^- \cap \Gamma$ among unipotent subgroups of Γ, we conclude that $\mathbf{V} \subset \mathbf{U}^- \cap \mathbf{U}$ i.e. $\mathbf{V} = \{e\}$. This proves that $\mathbf{H}$ is reductive. Consequently, $\mathbf{H}'$ is semisimple. Moreover $\mathbf{H}'$ is non-trivial since $\mathbf{U} \cap \mathbf{H}$ is non-trivial and $(U \cap H)/(U \cap H \cap \Gamma)$ is compact. We can then decompose the Lie algebra $\mathfrak{h}^*$ of H^* $(= \mathbf{H}^* \cap G)$ into a direct sum $\mathfrak{h}^* = \mathfrak{h} \oplus \mathfrak{k}$ where $\mathfrak{k}$ is stable under H' $(= \mathbf{H}' \cap G)$. It suffices to show that $\mathfrak{k}$ is trivial as an H'-module. Let $\mathbf{S} \subset \mathbf{H}'$ be a maximal $\mathbf{Q}$-split torus in $\mathbf{H}'$ and $S = \mathbf{S} \cap G$. Then $\mathfrak{K}$ $(= \mathbf{C}$-span of $\mathfrak{k})$ decomposes into a direct sum of eigen-spaces for $\mathbf{S}: \mathfrak{K} = \sum_i \mathfrak{K}_i$, and, since $\mathbf{S}$ is $\mathbf{R}$-split, $\mathfrak{k}$ decomposes correspondingly into eigen-spaces for $S: \mathfrak{k} = \sum_i \mathfrak{k}_i$. Now, $\mathfrak{F}$ decomposes into a direct sum of eigen-spaces for $\mathbf{S}$; since $\mathbf{S}$ is $\mathbf{Q}$-split, this decomposition is defined over $\mathbf{Q}: \mathfrak{F} = \sum_{j \in J} \mathfrak{F}_i$. It follows that $\mathscr{L}' = \sum_{j \in J} \mathfrak{k}_j \cap \mathscr{L}$ has finite index in $(\mathscr{L})$.

4.33. Suppose now that $X \in \mathfrak{k}_i$ and that the eigen-character χ_i defined by $\operatorname{ad} t\, (X) = \chi_i(t)\, X$, for $t \in S$, is non-trivial. Consider the one-parameter group $\{\exp t\, X \in H^* \mid t \in \mathbf{R}\}$. The map $\mathbf{R} \times H \to H^*$ defined by $(t, h) \to \exp tX.\, h$, is a homeomorphism onto a closed subset of H^* : in fact $\mathbf{H}$ being reductive, we can find a representation ρ of $\mathbf{G}$ defined over $\mathbf{R}$ and a $\mathbf{R}$-rational vector v_0 in the representation space such that

$$\mathbf{H} = \{g \in \mathbf{G} \mid \rho(g)v_0 = v_0\};$$

then H^*/H is homeomorphic under the orbit map to $H^* v_0$ and the orbit $\{\exp t\, X v_0 \mid t \in \mathbf{R}\}$ under the one-parameter group is closed in the representation space hence *à fortiori* in $H^* v_0$ (note that $\exp tX$ is unipotent). Since the map $H^*/H^* \cap \Gamma \to GL(\mathfrak{f})/GL\ (\mathscr{L})$ is proper we conclude that $g_n = \sigma(\exp nX)$ is discrete modulo $GL(\mathscr{L})$; further $\sigma(\exp nX) \subset SL(\mathfrak{f})$. It follows from Mahler's criterion that we can find vectors $v_n \in \mathscr{L}'$ such that $g_n(v_n)$ tends to zero. Now we can write v_n in the form

$$v_n = \sum_{j \in J} v_n(j)$$

with $v_n(j) \in \mathscr{L} \cap \mathfrak{f}_j$. Let λ_j be the character defined as follows: for $v \in L \cap \mathfrak{f}_j$, ad $(t)(v) = \lambda_j(t) . v$ for all $t \in \mathbf{S}$. Introduce an ordering on $X(\mathbf{S})$ = group of rational characters on $\mathbf{S}$ such that w.r.t. this ordering χ_i is greater than 0. Since J is a finite set, we can find a subsequence $h_n = g_{r_n}$ such that the following holds: let $w_n = v_{r_n}$; then $w_n(j_0) \neq 0$ for all n and if $w_n(j) \neq 0$ for any j and n, then $\lambda_j > \lambda_{j_0}$. One sees now from the assumption that $\chi_i > 0$. Then

$$h_n w_n = w_n(j_0) + w'_n$$

where $w'_n \in \Sigma_{j \neq j_0} \mathfrak{f}_j$. Thus, since $w_n(j_0)$ are lattice points, $h_n w_n$ cannot tend to zero, a contradiction. We conclude therefore that χ_i cannot be non-trivial for any i. This proves Lemma 4.31.

4.34. COROLLARY. *Let ρ (resp. $\rho(\alpha)$) denote the adjoint representation of $\mathfrak{H}'$ (resp. $\mathfrak{H}'(\alpha)$) on $\mathfrak{H}$ (resp. $\mathfrak{H}(\alpha)$) and ρ^* (resp. $\rho^*(\alpha)$) that on $\mathfrak{H}^*$ (resp. $\mathfrak{H}^*(\alpha)$). Let $X, Y \in \mathfrak{H}'$ (resp. $\mathfrak{H}'(\alpha)$). Then trace $\rho(X)\rho(Y)$ = trace $\rho^*(X)\rho^*(Y)$ (resp. trace $\rho^*(\alpha)(X)\rho^*(\alpha)(Y)$). If $X, Y \in \mathfrak{H}'$ (resp. $\mathfrak{H}'(\alpha)$) are $\mathbf{Q}$-rational for the $\mathbf{Q}$-structure on $\mathfrak{H}$ (resp. $\mathfrak{H}(\alpha)$), then $\langle X, Y \rangle \in \mathbf{Q}$ where $\langle\, ,\rangle$ denotes the Killing form on $\mathfrak{G}$. In particular if $X, Y \in \mathfrak{H}$ are $\mathbf{Q}$-rational and* ad X, ad Y *are nilpotent, then $\langle X, Y \rangle \in \mathbf{Q}$.*

The first assertion is immediate from Lemma 4.31. The second assertion follows from the fact

$$\text{trace ad } X \text{ ad } Y = \text{trace } \rho^*(X)\ \rho^*(Y) + 2 \text{ trace } \sigma(X)\ \sigma(Y)$$

for $X, Y \in \mathfrak{H}'$. The last assertion follows from the definition of $\mathfrak{R}'$.

4.35. PROPOSITION. *Let $\mathfrak{U}_{\mathbf{Q}}$ (resp. $\mathfrak{U}_{\mathbf{Q}}^-$) be the $\mathbf{Q}$-linear span of* $\exp^{-1}(U \cap \Gamma)$ *(resp.* $\exp^{-1}(U^- \cap \Gamma)$ *in $\mathfrak{U}$ (resp. $\mathfrak{U}^-$). Then $\langle X, Y \rangle \in \mathbf{Q}$ for all $X \in \mathfrak{U}_{\mathbf{Q}}^-$, $Y \in \mathfrak{U}_{\mathbf{Q}}$.*

4.36. Let $\mathfrak{U}'_{\mathbf{Q}} = \{X \in \mathfrak{U}^-_{\mathbf{Q}} \mid \langle X, Y \rangle \in \mathbf{Q}$ for all $Y \in \mathfrak{U}_{\mathbf{Q}}\}$. We have to show that $\mathfrak{U}'_{\mathbf{Q}} = \mathfrak{U}^-_{\mathbf{Q}}$. Enumerate the elements of Δ in some order as $\alpha_1,\dots,\alpha_l$ and set $E_0 = \mathfrak{H}' \cap \mathfrak{U}^-_{\mathbf{Q}}$ and for $1 \leqslant i \leqslant l$, $E_i = \mathfrak{H}'(\alpha) \cap \mathfrak{U}^-_{\mathbf{Q}}$. Also let $\mathbf{L}_0 = \mathbf{L}$ and for $1 \leqslant i \leqslant l$, $\mathbf{L}_i = \mathbf{L}(\alpha_i)$; similarly, let $\mathbf{F}_0 = \mathbf{F}$, $\mathbf{H}_0 = \mathbf{H}$ and for $1 \leqslant i \leqslant l$, $\mathbf{F}_i = \mathbf{U}(\alpha_i)$, $\mathbf{H}_i = \mathbf{H}(\alpha_i)$. Analogous notation will be used for the Lie algebras as well. We first show that $E_i \subset \mathfrak{U}'_{\mathbf{Q}}$ for $1 \leqslant i \leqslant 0$. Suppose then that $X \in E_i$ and $Y \in \mathfrak{U}_{\mathbf{Q}}$ since $\mathfrak{U} = \mathfrak{U} \cap \mathfrak{H}_i \oplus \mathfrak{F}_i$ for $0 \leqslant i \leqslant l$ and the decomposition is defined over $\mathbf{Q}$ (for the natural $\mathbf{Q}$-structure on $\mathbf{L}_i$), we can write $Y = Y_1 + Y_2$ with $Y_1 \in \mathfrak{U}_{\mathbf{Q}} \cap \mathfrak{H}_i$, $Y_2 \in \mathfrak{F}_i$ since $\mathfrak{F}_i$ is orthogonal to all of $\mathfrak{N}_i$ and $E_i \subset \mathfrak{N}_i$, we have

$$\langle X, Y \rangle = \langle X_1, Y_1 \rangle \in \mathbf{Q}$$

in view of Corollary 4.34. Thus $E_i \subset \mathfrak{U}'_{\mathbf{Q}}$ for $0 \leqslant i \leqslant 1$.

4.37. We next show that if $Z \in \mathfrak{U}'_{\mathbf{Q}}$ and $X \in E_i$ for some i with $1 \leqslant i \leqslant l$, then $[X, Z] \in \mathfrak{U}'_{\mathbf{Q}}$. As before, let $Y = Y_1 + Y_2$ be any element of $\mathfrak{U}_{\mathbf{Q}}$ with $Y_1 \in \mathfrak{h}_i \cap \mathfrak{U}_{\mathbf{Q}}$ and $Y_2 \in \mathfrak{f}_i \cap \mathfrak{U}_{\mathbf{Q}}$. We then have

$$\langle [X, Z], Y_1 + Y_2 \rangle = -\langle Z, [X, Y_1] \rangle - \langle Z, [X, Y_2] \rangle.$$

The second term on the right is in $\mathbf{Q}$ since $[X, Y_2] \in \mathfrak{U}_{\mathbf{Q}}$ and $Z \in \mathfrak{U}'_{\mathbf{Q}}$. We have thus to show that $\langle Z, [X, Y_1] \rangle \in Q$. Now $[X, Y_1]$ is a $\mathbf{Q}$-rational element in $\mathfrak{l}_i = \mathfrak{h}_i \oplus \mathfrak{f}_i$ contained in the $\mathbf{Q}$-subspace $\mathfrak{h}'_i \oplus \mathfrak{f}_i$. We can therefore write $[X, Y_1] = Y_3 + Y_4$ with $Y_3 \in \mathfrak{H}'_i$, $Y_4 \in \mathfrak{F}_i$ with Y_2, Y_3 $\mathbf{Q}$-rational. Clearly $\langle Z, Y_4 \rangle \in \mathbf{Q}$. On the other hand, we can write $Z = X_1 + X_2$ with $X_1 \in E_i \subset \mathfrak{H}'_i$ and $X_2 \in \mathfrak{U}_{\mathbf{Q}} \cap \mathfrak{F}_i$ (*cf.* 4.30) $\mathfrak{F}_i$ is orthogonal to its normaliser $\mathfrak{N}_i$ hence in particular to $\mathfrak{H}'_i$. Hence $\langle Z, Y_3 \rangle = \langle X_1, Y_3 \rangle$; and $\langle X_1, Y_3 \rangle \in \mathbf{Q}$ in view of Corollary 4.34.

4.38. In view of 4.36 and 4.37, to prove Lemma 4.35, it suffices to show that the $\{E_i \mid 0 \leqslant i \leqslant 1\}$ generate $\mathfrak{U}_{\mathbf{Q}}$ as a Lie algebra. This follows from the fact that if $\theta \in \mathbf{\Delta}$ is a (simple) root w.r.t. $\mathbf{B}$ which is non-trivial on $\mathbf{T}^*$, then $\theta \in \mathbf{\Delta}(\mathbf{F})$ or $\theta \in \mathbf{\Delta}(\alpha)$ for some $\alpha \in \Delta$. This proves Proposition 4.37. We have therefore proved in fact the following:

4.39. **Proposition.** *For* $\gamma \in \Gamma^*$ *and* $X, X' \in \mathfrak{U}$ *such that* $\exp X$, $\exp X' \in U \cap \Gamma$ $\langle \operatorname{ad} \gamma\, X, X' \rangle \in \mathbf{Q}$.

4.40. Γ being irreducible and Zariski-dense in $\mathbf{G}$ and Γ^* being the intersection of Γ with a Zariski open subset of $\mathbf{G}$, we can find a finite number $\gamma_1, \ldots, \gamma_r$ of elements in Γ^* such that $\gamma_i^{-1}\, \gamma_j \in \Gamma^*$ for $1 \leqslant i, j \leqslant r$, $i \neq j$ and $\operatorname{ad} \gamma_i$ $(\exp^{-1}(U \cap \Gamma))$ span all of $\mathfrak{G}$ as a complex vector space. Now let $\Gamma^{**} = \{\gamma \in \Gamma^* \mid \gamma_j^{-1}\, \gamma^{-1} \gamma_i \in \Gamma^*$ for $1 \leqslant i, i \leqslant r\}$. Then Γ^{**} is Zariski-dense in $\mathbf{G}$. Further if $X_1, \ldots, X_n$ is a basis of $\mathfrak{G}$ chosen from among the elements

$$\{\operatorname{ad} \gamma_i\,(X) \mid X \in \mathfrak{U},\ \exp X \in \Gamma,\ 1 \leqslant i \leqslant r,\ \gamma_i \in \Gamma^*\},$$

then $\langle X_i, X_j \rangle \in \mathbf{Q}$. We see now that for $\gamma \in \Gamma^{**}$,

$$\langle \operatorname{ad} \gamma\,(X_i), X_j \rangle \in \mathbf{Q}$$

for all i, j with $1 \leqslant i, j \leqslant n$. It follows that w.r.t. the $\mathbf{Q}$-structure on the vector space $\mathfrak{G}$ defined by $X_1, \ldots, X_n$, $\operatorname{ad} \gamma$ is a $\mathbf{Q}$-rational element of $GL(\mathfrak{G})$ for all $\gamma \in \Gamma^{**}$. Since Γ^{**} is open in the topology on Γ induced from the Zariski topology of $\mathbf{G}$ and this induced topology is quasi-compact, a finite number of translates of Γ^{**} cover all of Γ. It follows that the group Γ_1 generated by Γ^{**} has finite index in Γ. Clearly $\operatorname{ad} \Gamma_1 \subset GL(\mathfrak{G}_{\mathbf{Q}})$ and $\operatorname{ad} \mathbf{G}$ $(\simeq \mathbf{G})$ is the Zariski-closure of $\operatorname{ad} \Gamma_1$. If we set $\mathbf{G}^* = \operatorname{ad} \mathbf{G}$ and u to be the inverse of the isomorphism $\operatorname{Ad} : \mathbf{G} \to \mathbf{G}^*$, we find that G^* carries a natural $\mathbf{Q}$-structure such that $u^{-1}(\Gamma_1) \subset \mathbf{G}^*_{\mathbf{Q}}$. Further Γ normalises some subgroup Γ_2 of finite index in Γ_1, and Γ_2 is Zariski-dense in $\mathbf{G}$. It follows from this, since $\mathbf{G}^*$ has trivial centre, that $u^{-1}(\Gamma) \subset \mathbf{G}^*_{\mathbf{Q}}$. This proves the first part of Theorem 4.4.

4.41. From the definition of the $\mathbf{Q}$-structure one sees immediately that the $\mathbf{F}_i$, $\mathbf{F}_i^-$, $0 \leqslant i \leqslant 1$ and $\mathbf{U}$, $\mathbf{U}^-$ (F_i^-, $\mathbf{U}^-$ etc. are defined now simply as conjugates $\gamma F_i\, \gamma^{-1}$, $\gamma\, U\, \gamma^{-1}$ for some fixed $\gamma \in \Gamma^*$) are $\mathbf{Q}$-subgroups and that $F_i \cap \Gamma$, $F_i^- \cap \Gamma$, $0 \leqslant i \leqslant 1$, and $U \cap \Gamma$ are arithmetic subgroups of $\mathbf{F}_i$ and $\mathbf{U}$ respectively. We claim now that $\mathbf{U}$ is a maximal unipotent $\mathbf{Q}$-subgroup of $\mathbf{G}$ for the $\mathbf{Q}$-structure obtained above on G. To see this we consider the adjoint representation σ of $\mathbf{N}$ on $\mathfrak{F}(=\mathfrak{F}_0)$. The groups $\mathbf{N}$, $\mathbf{N}^-$ as well as $\mathbf{N}_0$ and $\mathbf{N}_0^-$ are defined on $\mathbf{Q}$ and $\mathbf{H}^* = \mathbf{N}_0 \cap \mathbf{N}_0^-$ is also defined over $\mathbf{Q}$. Now the natural map $H^*/H^* \cap \Gamma \to GL\ (\mathfrak{f})/GL(\mathscr{L})$ is defined over $\mathbf{Q}$ and $\sigma(H^*)$ is a $\mathbf{Q}$-subgroup of $GL(\mathfrak{F})$. It follows that $H^* \cap \Gamma$ is arithmetic in $\mathbf{H}^*$ ($\sigma|_{H^*}$ is injective). We conclude that since $\mathbf{N}_0 = \mathbf{H}^*. \mathbf{F}$, $\mathbf{N}_0 \cap \Gamma$ is arithmetic. Consequently $\mathbf{N} \cap \Gamma$ is arithmetic as well. Since $\mathbf{N} \cap \Gamma$ is arithmetic in $\mathbf{N}$, the Zariski-closure of a maximal unipotent subgroup of Γ is necessarily a maximal unipotent $\mathbf{Q}$-subgroup of $\mathbf{N}$. Thus $\mathbf{U}$ is a maximal unipotent $\mathbf{Q}$-subgroup. Further the preceding arguments can be carried over verbatim to conclude that for $0 \leqslant i \leqslant l$, $\mathbf{N}_i \cap \Gamma$ is arithmetic in $\mathbf{N}_i$. Further, from Lemma 4.12, it is easy to deduce that the $\mathbf{N}_i$, $0 \leqslant i \leqslant l$ are precisely the *maximal* parabolic subgroups of $\mathbf{G}$ defined over $\mathbf{Q}$ and containing $\mathbf{U}$.

4.42. Suppose now that $\mathbf{P}$ is a maximal parabolic subgroup of $\mathbf{G}$ defined over $\mathbf{Q}$. Then there exists a unique i, $0 \leqslant i \leqslant l$ such that $\mathbf{P} \cap \mathbf{N}_i$ is a maximal reductive subgroup of $\mathbf{P}$ (we may have to replace N_i by $\gamma\, N_i\, \gamma^{-1}$ for some $\gamma \in \Gamma$ for this; but since $\gamma \in \mathbf{G}_{\mathbf{Q}}$, this is permissible). Further if $j \neq i$ (such a j exists: $l \geqslant 1$ since Γ has rank atleast 2), it is easy to see that the unipotent radical $\mathbf{V}$ of $\mathbf{P}$ is the smallest $\mathbf{Q}$-subgroup of $\mathbf{P}$ containing $\mathbf{V} \cap \mathbf{N}_j$ and stable under an arithmetic subgroup of $\mathbf{P} \cap \mathbf{N}_i$. It follows that the smallest subgroup of $\mathbf{V}$ containing $\mathbf{V} \cap \Gamma$ and normalised by $\mathbf{P} \cap N_i \cap \Gamma$ is Zariski-dense in $\mathbf{V}$. Hence $\mathbf{V} \cap \Gamma$ is Zariski-dense in $\mathbf{V}$. Thus $V/V \cap \Gamma$ is compact and $V \cap \Gamma$ is arithmetic in $\mathbf{V}$. It follows that $\mathbf{P} \cap \Gamma$ is arithmetic in P. Since any $\mathbf{Q}$-parabolic subgroup of $\mathbf{G}$ is contained in a maximal $\mathbf{Q}$-parabolic subgroup of $\mathbf{G}$, this completes the proof of Theorem 4.4.

We will now establish the following result.

4.43. THEOREM. *Let* $\mathbf{G}$ *be an algebraic group defined and simple over* $\mathbf{Q}$ *and* $\Gamma \subset G_{\mathbf{Q}}$ *a subgroup such that for every parabolic subgroup* $\mathbf{P} \subset \mathbf{G}$ *defined over* $\mathbf{Q}$, $\mathbf{P} \cap \Gamma$ *is arithmetic in* $\mathbf{P}$. *Assume that* $\mathbf{Q}$*-rank* $G > 1$. *Then* Γ *is arithmetic.*

PROOF. Let $\mathbf{U}$ be a maximal unipotent $\mathbf{Q}$-subgroup of $\mathbf{G}$ and $\mathbf{C}_0$ its centre. Let $\gamma \in \Gamma$ be any element and $\mathbf{T}$ a maximal $\mathbf{Q}$-split torus normalising $\mathbf{U}$ and $\gamma\, U\, \gamma^{-1}$. Then the Lie algebras $\mathfrak{C}_0$ of $\mathbf{C}_0$ and

$\mathfrak{C}_0^- = \text{ad}\, \gamma(\mathfrak{C}_0)$ of $\gamma\, C_0\, \gamma^{-1}$ are eigen-spaces for **T**. Let α and α^- denote the corresponding **Q**-roots. Then the two roots are conjugate under the Weyl group and hence of the same length. Now since α and α^- are of the same length, they are proportionate if and only if $\alpha^- = \pm\, \alpha$. Thus if $\alpha^- \neq -\alpha$, we conclude that trace ad X^k. ad $Y^l = 0$ for all integers $k, l > 0$, $X \in \mathfrak{C}_0$ and $Y \in \mathfrak{C}_0^-$. Hence

$$\text{trace ad } (\gamma\, x\, \gamma^{-1}) \text{ ad } y = \dim \mathbf{G} \in \mathbf{Z} \qquad (*)$$

for all $x, y \in C_0 \cap \Gamma$. Suppose next that $\alpha^- = -\alpha$; in this case since **Q**-*rank* $G > 1$, one can find a **Q**-parabolic subgroup **P** of **G** containing $\mathbf{C}_0$ and $\mathbf{C}_0^-$. Since $\mathbf{P} \cap \Gamma$ is arithmetic, we have trace $\theta \in \mathbf{Z}$ for all $\theta \in \mathbf{P} \cap \Gamma$. In particular, we get

$$\text{trace ad } \gamma\, x\, \gamma^{-1}.\ \text{ad } y \in Z \qquad (**)$$

for all $x, y \in \mathbf{C} \cap \Gamma$.

Suppose now $X, Y \in \exp^{-1} (\mathbf{C}_0 \cap \Gamma) = E$ in $\mathfrak{C}_0$. Then setting $\exp X = x$, $\exp Y = y$, we have for $\gamma \in \Gamma$,

$$\text{trace ad } (\text{ad}\, \gamma(X))\ \text{ad } Y$$

$$= \text{trace} \left\{ \sum_{n=1}^{p} (-1)^p (\text{ad}\gamma x \gamma^{-1} - 1)^p/p \right\} . \left\{ \sum_{n=1}^{p} (-1)^p (\text{ad } y - 1)^p/p \right\};$$

where $\dim \mathfrak{G} = p$. The right hand side belongs to $(p!)^{-1}\, \mathbf{Z}$ in view of (*) and (**) above. We conclude from this that on the **Z**-linear span of $\mathscr{L}$ of $\{ \text{ad}\, \gamma\, (E) \mid \gamma \in \Gamma \}$ the Killing bilinear form takes values which are rational numbers with denominators dividing $p!$. This shows immediately that $\mathscr{L}$ is a lattice in $\mathfrak{g}$. Since Γ leaves $\mathscr{L}$ stable, Γ is arithmetic (one uses Mahler's criterion).

We have thus shown the following:

4.44. THEOREM. *An L-subgroup of rank $\geqslant 2$ is arithmetic.*

5. Existence of Γ-rational parabolics. Our aim in this section is to establish the following result.

MAIN THEOREM. *Let* **G** *be a connected semisimple algebraic group defined over* **R** *and with trivial centre and no compact factors. Let* $\Gamma \subset G$ *(= identity component of* $\mathbf{G}_\mathbf{R}$*) be an irreducible non-uniform L-subgroup of G. Then any* Γ*-adapted parabolic subgroup of* **G** *is* Γ*-rational.*

("Γ-adapted" and "Γ-rational" were defined in 4.3.)

5.1. We introduce the following notation for the entire section:

$\mathbf{N}$, a Γ-adapted $\mathbf{R}$-parabolic subgroup of $\mathbf{G}$, $N = \mathbf{N} \cap G$,

$\mathbf{F}$, the unipotent radical of $\mathbf{N}$, $F = \mathbf{F} \cap G$.

$\mathfrak{N}$ (resp. $\mathfrak{F}$), the Lie algebra of $\mathbf{N}$(resp. F); $\mathfrak{n} = \mathfrak{N} \cap \mathfrak{g}$, $\mathfrak{f} = \mathfrak{F} \cap \mathfrak{g}$,

σ, the adjoint representation of $\mathbf{N}$ on $\mathfrak{F}$,

$\mathbf{N}_0 = \{x | \det \sigma(x) = \pm 1\}$, $N_0 = \mathbf{N}_0 \cap G$,

$\mathbf{D}$ = Zariski-closure of $N \cap \Gamma$ $(= N_0 \cap \Gamma)$, $D = \mathbf{D} \cap G$,

$\mathscr{L}$ = $\mathbf{Z}$-span of $\exp^{-1}(F \cap \Gamma)$ in $\mathfrak{f}$.

$\mathscr{L}$ is a lattice in $\mathfrak{f}$ and defines a natural $\mathbf{Q}$-structure on $\mathfrak{F}$ and hence on $GL(\mathfrak{f})$. For this $\mathbf{Q}$-structure $\sigma(\mathbf{D})$ is a $\mathbf{Q}$-subgroup. With this notation we will now establish a criterion for the Γ-rationality of $\mathbf{N}$.

5.2. Definition. *A parabolic subgroup* $\mathbf{P}$ *of* $\mathbf{G}$ *is 'symmetric' if the following holds: there exists* $g \in \mathbf{G}$ *such that* $g \mathbf{V} g^{-1} \cap \mathbf{P} = \{e\}$ *where* V *is the unipotent radical of* $\mathbf{P}$. $\mathbf{G}$ *is 'symmetric' if every parabolic group is symmetric.*

5.3. Proposition. *If* $\mathbf{N}$ *is symmetric,* $\mathbf{N}$ *is* Γ*-rational.*

5.4. Since $\mathbf{N}$ is symmetric and Γ is Zariski-dense in $\mathbf{G}$, we can find $\gamma \in \Gamma$ such that $\gamma \mathbf{N} \gamma^{-1} \cap \mathbf{F} = \{e\}$. In this case, it is well known—and easy to see—that $\mathbf{N} \cap \gamma \mathbf{N} \gamma^{-1} = \mathbf{M}$ is a maximal reductive subgroup of $\mathbf{N}$. Let $\mathbf{M}_0 = \mathbf{M} \cap \mathbf{N}_0$ and $M_0 = \mathbf{M}_0 \cap G$. It is easy to see that $\mathbf{M}_0 = \mathbf{N}_0 \cap \gamma \mathbf{N}_0 \gamma^{-1}$. Since the map $N_0/N_0 \cap \Gamma \to G/\Gamma$ is proper we conclude from Lemma 4.18 that $M_0/M_0 \cap \Gamma \to G/\Gamma$ and hence $M_0/M_0 \cap \Gamma \to GL(\mathfrak{f})/GL(\mathscr{L})$ are proper maps. Evidently $M_0 \cap \Gamma \subset D$. Since $D/D \cap \Gamma \to G/\Gamma$ is proper, the map $D \cap M_0/D \cap M_0 \cap \Gamma \to G/\Gamma$ is proper as well. Composing with the proper map $D/D \cap \Gamma \to D/F(D \cap \Gamma)$, the map $D \cap M_0/D \cap M_0 \cap \Gamma \to D/(D \cap \Gamma)F$ is proper. Since $N_0 = M_0.F$ and $D \supset F$, $D = (D \cap M_0).F$. Thus the natural map $\pi : \mathbf{D} \to \mathbf{D}/\mathbf{F}$ maps $D \cap M_0$ onto D/F. Further $(D \cap \Gamma)$ is a Zariski-dense arithmetic subgroup of $\mathbf{D}/\mathbf{F}$ for the natural $\mathbf{Q}$-structure on $\mathbf{D}/\mathbf{F}$ $(= \sigma(\mathbf{D})/\sigma(\mathbf{F}))$. It follows that $\pi(M_0 \cap D \cap \Gamma)$ is arithmetic in $\mathbf{D}/\mathbf{F}$. Since σ restricted $\mathbf{M}_0$ is an injection, $M_0 \cap D \cap \Gamma$ is Zariski-dense in $\mathbf{M}_0 \cap \mathbf{D}$. It is now evident that $\sigma(\mathbf{M}_0 \cap \mathbf{D})$ is a maximal reductive $\mathbf{Q}$-subgroup of $\sigma(\mathbf{D})$ and that $\sigma(M_0 \cap \Gamma)$ is arithmetic in $\sigma(\mathbf{M}_0 \cap \mathbf{D})$. This proves Proposition 5.3.

5.5. Remarks. Proposition 5.3 is applicable in a large number of cases. Let w_0 be the unique element in the Weyl group of **G** which takes a positive system of roots (w.r.t. a torus etc.) into negative roots. If $w_0(\alpha) = -\alpha$ for all roots α, then every parabolic subgroup of **G** is symmetric. (In fact since **N** itself is defined over **R** the same criterion replaced by **R**-Weyl group **R**-roots etc. holds). This shows that if **G** has no absolutely simple factors isomorphic to A_n, D_n(n odd) and E_6, then **N** is Γ-rational. (i.e. A_n, D_n(n odd) and E_6 are the only non-symmetric groups). Notice that these groups admit a unique non-trivial automorphism of the Dynkin diagram of order 2.

5.6. As remarked earlier $\mathscr{L}$ is a lattice in $\mathfrak{f}$ and defines a **Q**-structure on $\mathfrak{F}$ and hence also on $GL(\mathfrak{F})$. For this **Q**-structure, $\sigma(\mathbf{D})$ is a **Q**-subgroup. We introduce now the following additional notation.

$\mathbf{M}_1$, a maximal connected reductive **Q**-subgroup of $\sigma(D)$,

$\Gamma_1 = \mathbf{M}_1 \cap \sigma(N \cap \Gamma)$,

$\tilde{\mathbf{M}}_1 = \sigma^{-1}(\mathbf{M}_1)$, $\tilde{\mathbf{M}}_1 = \tilde{\mathbf{M}}_1 \cap G$,

$\Gamma_1 = \mathbf{M}_1 \cap \Gamma$,

$\mathbf{K}$ = Centre of $\mathbf{F}$, $K = \mathbf{K} \cap G$.

The group Γ_1 is an arithmetic subgroup of the **Q**-group $\mathbf{M}_1$. Also $M_1/M_1 \cap \Gamma \to GL(\mathfrak{f})/GL(\mathscr{L})$ and $M_1/M_1 \cap \Gamma \to G/\Gamma$ are both proper maps. We have the following exact sequences

$$e \longrightarrow K \cap \Gamma \longrightarrow \tilde{\Gamma}_1 \xrightarrow{\sigma} \Gamma_1 \longrightarrow e \tag{A}$$

$$e \longrightarrow \mathbf{K} \longrightarrow \tilde{\Gamma}_1 . \mathbf{K} \longrightarrow \Gamma_1 \longrightarrow e \tag{B}$$

$$e \longrightarrow \mathbf{K} \longrightarrow \tilde{\mathbf{M}}_1 \longrightarrow \mathbf{M}_1 \longrightarrow e \tag{C}$$

The sequence (C) is split (as a sequence of algebraic groups). It follows that the sequence (B) is also split. We will now give a second condition under which the sequence (A) "almost" splits. We first observe that **K** being abelian, it can be identified with its Lie algebra $\mathfrak{K}$. Under this identification $K \cap \Gamma$ is identified with $\mathfrak{k} \cap \mathscr{L}$. Let $\mathfrak{K}_\mathbf{Q}$ denote the **Q**-linear span of $\mathfrak{k} \cap \mathscr{L}$ in $\mathfrak{k}$ and $\mathbf{K}_\mathbf{Q}$ the corresponding subgroup of **K**. We will prove a second criterion for Γ-rationality of N.

5.7. PROPOSITION. *If* $H^1(\Gamma_1, \mathfrak{K}) = 0$, *then* **N** *is* Γ-*rational.*

5.8. Let $r: \Gamma_1 \to \Gamma_1$ be a set theoretic section. Then the map $(\theta, \theta') \to r(\theta, \theta')\, r(\theta')^{-1} r(\theta)^{-1}$ defines a 2-cocycle u on Γ_1 with coefficients in $K \cap \Gamma$. This cocycle considered as a cocycle with values in **K** is cohomologous to zero since (B) splits. It follows that the cocycle is cohomologous to zero in $K_{\mathbf{Q}}$ as well (the map $H^2(\Gamma_1, K_{\mathbf{Q}}) \to H^2(\Gamma_1, \mathbf{K})$ is injective). Now $K_{\mathbf{Q}}$ (written additively) is the inductive limit of $\{n^{-1}(K \cap \Gamma) \mid n \text{ a positive integer}\}$. According to Raghunathan ([3], Corollary 4), $H^2(\Gamma_1, K_{\mathbf{Q}})$ is the inductive limit of $H^2(\Gamma_1, n^{-1}(K \cap \Gamma))$: notice that Γ_1 is an arithmetic subgroup of a reductive group and is hence "upto finite index" the direct product of a free abelian group and an arithmetic subgroup of a semisimple group. It follows that we can find an integer $n > 0$ such that u considered as a cocycle with values in $n^{-1}(K \cap \Gamma)$ is a coboundary. This means that the sequence

$$e \to n^{-1}(K \cap \Gamma) \to \tilde{\Gamma}_1(n^{-1}(K \cap \Gamma)) \to \Gamma_1 \to e$$

is split. Let $r_0: \Gamma_1 \to \tilde{\Gamma}_1(n^{-1}(K \cap \Gamma))$ be a splitting. Since $H^1(\Gamma_1, \mathfrak{K}) = 0$, any two splittings of the sequence (B) in 5.6 are conjugate. Let τ' be a splitting of (C) as algebraic groups and τ the induced splitting of (B). Then τ and r_0 (considered as a splitting of (B)) are conjugates. It follows that the Zariski-closure **H** of r_0 (Γ_1) in $\mathbf{M}_1$ is a maximal reductive subgroup of $\mathbf{M}_1$. Further $r_0(\Gamma_1) \subset \mathbf{H}$. Now $\tilde{\Gamma}_1$ has finite index in $\Gamma_1(n^{-1}(K \cap \Gamma))$ so that $\Gamma_1 \cap \mathbf{H}$ is commensurable with $r_0(\Gamma_1)$. It is now immediate that $\sigma(\mathbf{H}) = \mathbf{M}_1$ and $\sigma(\mathbf{H} \cap \Gamma (= \mathbf{H} \cap \tilde{\Gamma}_1))$ is an arithmetic subgroup of $\mathbf{M}_1$. Hence the proposition.

The following result is proved in Appendix II.

5.9. LEMMA. *Suppose* $\theta \in \Gamma_1$ *is a central element such that* $(\theta - \text{Identity})$ *is an automorphism of* $\mathfrak{K}$, *then* $H^p(\Gamma_1, \mathfrak{K}) = 0$ *for all* $p \geqslant 0$.

Lemma 5.9 can be used to establish Γ-rationality of N in a number of cases.

5.10. PROPOSITION. *Assume that* **N** *is not 'symmetric' and not a maximal* **R**-*parabolic subgroup of* **G**. *Then* **N** *is* Γ-*rational.*

5.11. In view of Lemma 5.9 and Proposition 5.7 we need only show that Γ_1 contains a central element θ which acts on $\mathfrak{K}$ such that 1 is not an eigen-value. Let $\mathbf{M}_1^* \subset \mathbf{M}_1$ be a maximal $\mathbf{R}$-reductive subgroup of $\mathbf{M}_1$. Let $\mathbf{M}$ be a maximal reductive subgroup of $\mathbf{N}$ containing $\mathbf{M}_1^*$. Let $\mathbf{S}$ be the maximal central $\mathbf{R}$-split torus in $\mathbf{M}$. Let $\mathbf{M}_0 = \mathbf{M} \cap \mathbf{N}_0$ and $\mathbf{S}_0$ the identity component of $\mathbf{S} \cap \mathbf{M}_0$. Then (since M_0/M_1^* is compact $M_1^* = M_1^* \cap M_0$) $\mathbf{S}_0 \subset \mathbf{M}_1^*$. Clearly $\mathbf{S}_0$ is central in $\mathbf{M}_1^*$. Hence $\sigma(\mathbf{S}_0)$ is central in $\mathbf{M}_1$. Let $\mathbf{S}_0^*$ be the smallest (central) $\mathbf{Q}$-subgroup of $\mathbf{M}_1$ containing $\sigma(\mathbf{S}_0)$. Then, since Γ_1 is Zariski-dense in $\mathbf{M}_1$, $\Gamma_1 \cap \mathbf{S}_0^*$ is Zariski-dense in S_0^*. Thus to find an element θ as above, it suffices to show the following

5.12. CLAIM. $\mathbf{S}_0$ *acts non-trivially on every irreducible* $\mathbf{M}$-*stable* $\mathbf{R}$-*subspace of* $\mathfrak{K}$ (*note that* $\sigma(S_0)$ *is diagonalisable over* $\mathbf{R}$ *so that it acts as scalars on every* $\mathbf{M}$-*stable simple component of* $\mathfrak{K}$ *defined over* $\mathbf{R}$).

5.13. When G is simple this is precisely Lemma B of Appendix I. We now establish the claim and hence Proposition 5.10 under the following hypothesis: $\mathbf{G}$ is *not* simple over $\mathbf{R}$. Since Γ is irreducible F is not contained in any proper connected normal subgroup of G. The Central torus $\mathbf{S}$ decomposes into a product $\prod_{i \in I} \mathbf{S}_i$ following the decomposition $\mathbf{G} = \prod_{i \in I} \mathbf{G}_i$ of $\mathbf{G}$ into $\mathbf{R}$-simple factors and for all $i \in I$, S_i is non-trivial and acts nontrivially on $\mathbf{K}_i$, the projection $\mathbf{K}$ on $\mathbf{G}_i$ ($\mathbf{K}_i$ are in fact the irreducible $\mathbf{M}$-stable subgroups of $\mathbf{K}$ defined over $\mathbf{R}$). Let λ_i be the character on $\mathbf{S}_i$ defined by $\lambda_i(x) = 1$ for $x \in \mathbf{S}_j$, $j \neq i$, $\lambda_i(x) = \lambda(x) = \det \sigma(x)$. Then $\lambda = \prod_{i \in I} \lambda_i$; the λ_i are all non-trivial and (multiplicatively) linearly independent so that kernel λ cannot contain any $\mathbf{S}_i$. This proves the assertion.

5.14. Proposition 5.10 now reduces the proof of our main theorem of this section to the following situation which will be the hypothesis in the sequel.

The group $\mathbf{G}$ is *simple* over $\mathbf{R}$. The group $\mathbf{N}$ is a *non-symmetric maximal* $\mathbf{R}$-parabolic subgroup of $\mathbf{G}$.

The remarks in 5.5 show that the absolutely simple components of **G** (there are at the most two of them and they are mutually isomorphic) are one of the following types:

$$A_n (n \geqslant 2),\ D_n (n \text{ odd},\ n > 3) \text{ and } E_6.$$

5.15. We introduce in addition to the notation introduced in 5.1 and 5.6, the following further notation. As in 5.11 **M** will denote a maximal reductive subgroup of **N** such that $\mathbf{M}_1^* = \mathbf{M} \cap \mathbf{M}_1$ maps onto $\mathbf{M}_1$ isomorphically. Let ${}_{\mathbf{R}}\mathbf{T}$ (resp. **T**) be a maximal **R**-split(resp. maximal torus defined over **R**) in **M** so chosen that ${}_{\mathbf{R}}\mathbf{T} \subset \mathbf{T}$. Again (as in 5.11) **S** will denote the maximal **R**-split *central* torus in **M**. We fix compatible orders on $X(\mathbf{T})$ (= group of characters on **T**) and $X({}_{\mathbf{R}}\mathbf{T})$ (= group of characters on ${}_{\mathbf{R}}\mathbf{T}$) and denote by Σ, Σ^+ and Δ respectively the system roots, positive roots and simple roots respectively. We assume, as we may, that the following holds: for $\alpha \in \Sigma$, let $\mathfrak{G}(\alpha)$ denote the root space corresponding to α; then if $\mathfrak{G}(\alpha) \subset \mathfrak{F}$, α is positive. With such a choice, it is well known that we can find a subset $\Delta' \subset \Delta$ with the following property: let $\Sigma(\Delta') = \{\alpha \in \Sigma \mid \alpha = \sum_{\varphi \in \Delta} m(\varphi)\, \varphi$ with $m(\varphi) > 0$ for some $\varphi \in \Delta'\}$; then $\mathfrak{F}$ is the linear span of

$$\{\mathfrak{g}(\varphi) \mid \varphi \in \Sigma(\Delta)\}.$$

5.16. We denote by θ the involution on $X(T)$ induced by complex conjugation. Then if $\alpha \in \Delta$ is non-trivial on ${}_{\mathbf{R}}\mathbf{T}$, then $\theta(\alpha) > 0$ and we have

$$\theta(\alpha) = \bar{\alpha} + \sum_{\varphi \in \Delta} m(\varphi).\ \varphi$$

with $m(\varphi) > 0$ only if φ is trivial on ${}_{\mathbf{R}}\mathbf{T}$. Now since **N** is defined over **R**, it is easy to see that every $\alpha \in \Delta'$ is non-trivial on ${}_{\mathbf{R}}\mathbf{T}$; moreover if $\alpha \in \Delta'$, $\alpha \in \Delta'$. On the other hand, let τ denote the unique non-trivial automorphism of the Dynkin diagram of Δ which leaves irreducible component of Δ stable. Then since **N** is *not* symmetric, Δ' is *not* stable under τ. The automorphism $\alpha \longmapsto \bar{\alpha}$ of the subset of Δ consisting of roots non-trivial on ${}_{\mathbf{R}}\mathbf{T}$, it is well known, extends to an automorphism of the Dynkin diagram of Δ as well; we denote by $\bar{\alpha}$ the image of $\alpha \in \Delta$ under this automorphism. This automorphism cannot be equal to τ, since Δ' is stable under it. Since the types under consideration admit only one non-trivial symmetry of the irreducible components of their Dynkin-diagrams, $\bar{\alpha} = \alpha$ for all $\alpha \in \Delta$ or $\alpha \to \bar{\alpha}$ switches the two components — the second alternative holds if and only if **G** is not absolutely simple.

5.17. The table on p. 305 classifies the Δ' which will satisfy the conditions we have obtained on it in the preceding discussion. β is (one of) the highest root(s) of **G** w.r.t. **T**.

5.18. From Table I one reads off the following (the diagram of $\mathbf{M}_0$ is obtained by deletion of Δ'). $\mathbf{M}_0$ has at the most two **R**-simple factors at least one of which is isotropic over **R** since $\mathbf{M}_1 \subset \sigma(\mathbf{M}_0)$ is isotropic even over **Q**. Further Γ_1, an arithmetic subgroup of $\mathbf{M}_1$ is Zariski-dense in $\mathbf{M}_1$. Since $\mathbf{M}_1$ contains all the **R**-isotropic factors of $\sigma(\mathbf{M}_0)$, we conclude that either (i) $\mathbf{M}_0$ has no **R**-isotropic factors and $\sigma(\mathbf{M}_0) = \mathbf{M}_1$ or $\mathbf{M}_0$ is the unique **R**-isotropic factor of $\sigma(\mathbf{M}_0)$. In either case, $\mathbf{M}_1$ is a normal subgroup of $\sigma(\mathbf{M}_0)$. Now $\mathfrak{K}$ being **R**-irreducible as a module over $\mathbf{M}_0$, it is **R**-isotypical as a module over $\mathbf{M}_1$. When **G** is absolutely simple, $\mathfrak{K}$ is absolutely simple over $\mathbf{M}_0$ and absolutely isotypical over $\mathbf{M}_1$; when **G** is not absolutely simple, $\mathbf{M}_0$ has both factors isotropic over **R** so that $\mathbf{M}_0 = \mathbf{M}_1$; moreover, $\mathfrak{K}$ breaks into two simple components $\mathfrak{K}' + \theta(\mathfrak{K}')$ and $\mathbf{M}_0 = \mathbf{M}_1$ breaks into a corresponding product $\mathbf{M}'_0$ and $\theta(\mathbf{M}'_0)$ being conjugate under θ and the representations of $\mathbf{M}'_0$ and $\theta(\mathbf{M}'_0)$ on $\mathfrak{K}'$ and $\theta(\mathfrak{K}')$ respectively are irreducible and equivalent under θ. Also $\mathbf{M}'_0$ (resp. $\theta(\mathbf{M}'_0)$) acts trivially on $\theta(\mathfrak{K}')$ (resp. $\mathfrak{K}'$). Table II shows the simple roots of $\mathbf{M}_0$ and $\mathbf{M}'_0$ to which the dominant weights of the irreducible representations of $\mathbf{M}_0$ and $\mathbf{M}_1$ on $\mathfrak{K}$ are connected. (The diagrams of $\mathbf{M}_0$ and $\mathbf{M}_1$ are realised as sub-diagrams of Δ in a natural fashion). Notice that the dominant weight of the representation in question are simply the restrictions of the highest root β to the corresponding maximal tori and this enables one to read off the information contained in Tables II and III from Table I.

5.19 [Notation. μ_0 (resp. μ) in the table below denotes one of the dominant weights in $\mathfrak{K}$; if there are two, the other is necessarily $\theta(\mu_0)$ (resp. $\theta(\mu)$). The diagrams of Type A_n, D_n (n odd >3) and E_6 etc. are taken as represented in Table I. (The table gives one representative from each pair of symmetric situations)].

TABLE I

Type of G	Extended diagram of Δ	Δ'
A_n	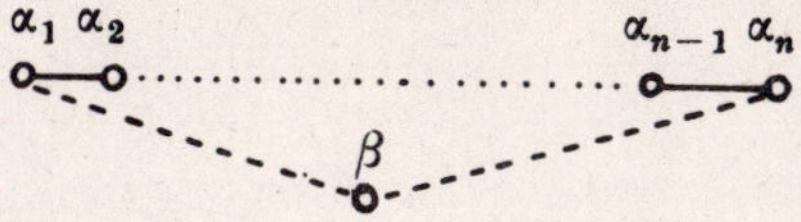	$\{\alpha_i\}$ $2i \neq n+1$
$A_n \times A_n$	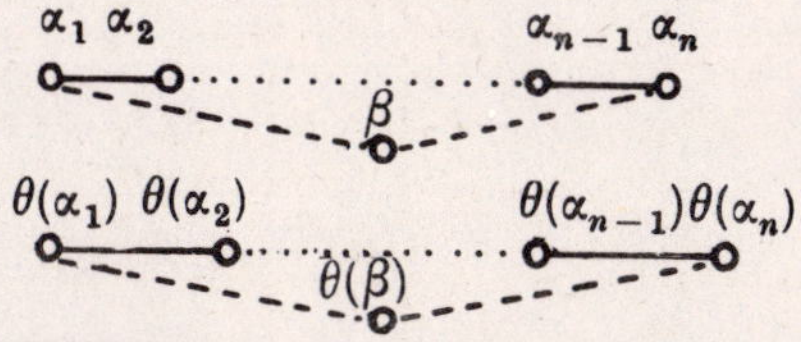	$\{\alpha_i,\ \theta(\alpha_i)\}$ $2i \neq n+1$
D_n, n odd $n > 3$	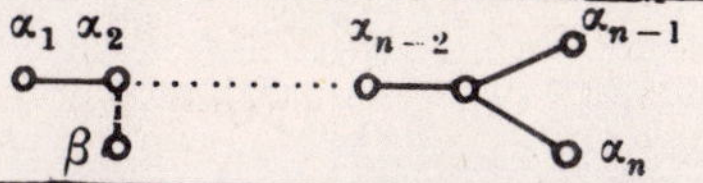	$\{\alpha_{n-1}\}$ or $\{\alpha_n\}$
$D_n \times D_n$ n odd $n > 3$	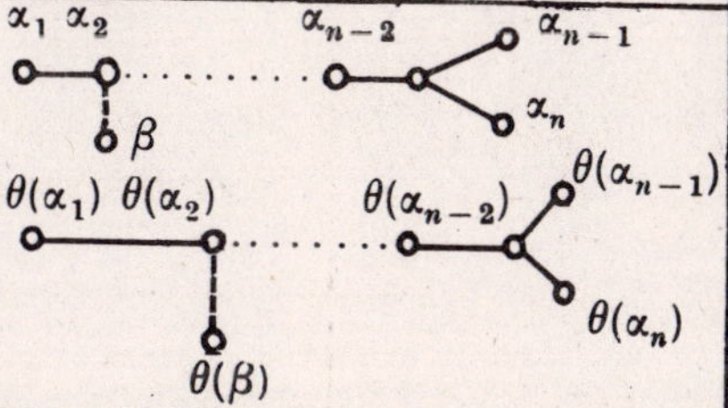	$\{\alpha_{n-1},\ \theta(\alpha_{n-1})\}$ or $\{\alpha_n,\ \theta(\alpha_n)\}$
E_6	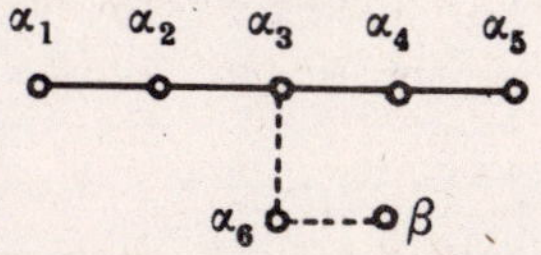	α_1 or α_2 or α_4 or α_5
$E_6 \underline{\times} E_6$	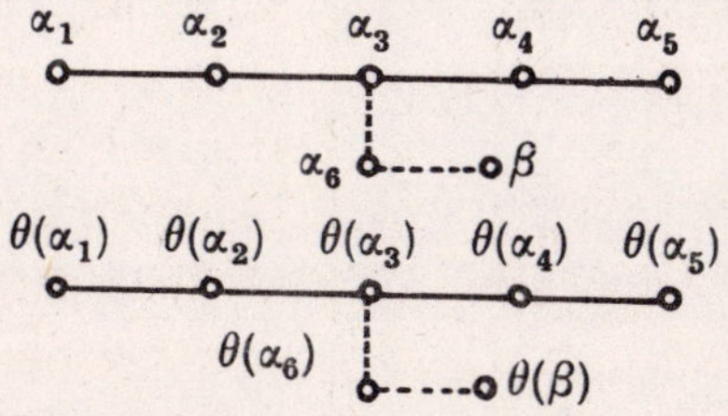	$\{\alpha_1,\ \theta(\alpha_1)\}$ or $\{\alpha_2,\ \theta(\alpha_2)\}$ or $\{\alpha_4,\ \theta(\alpha_4)\}$ or $\{\alpha_5,\ \theta(\alpha_5)\}$

5.20 TABLE II (G absolutely simple)

<table>
<tr><th>Type</th><th>Possible diagrams of $\mathbf{M}_0$ as subdiagram of Δ and position of μ_0, $\theta(\mu_0)$</th><th colspan="2">Possible diagrams of $\mathbf{M}_1$ as subdiagram of Δ and position of μ, $\theta(\mu)$.</th></tr>
<tr><td rowspan="2">A_n—I</td><td rowspan="2">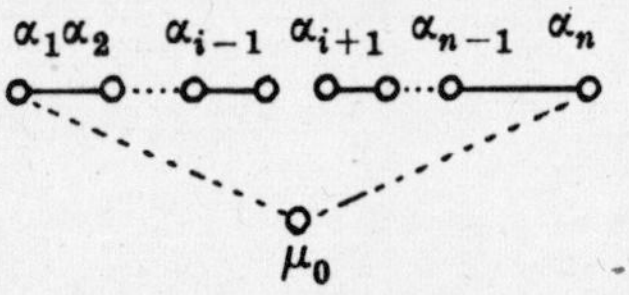

$1 < i < n/2$ and $2i \neq n+1$.</td><td colspan="2">(a) $\mathbf{M}_1 = \mathbf{M}_0$; diagram as in second column</td></tr>
<tr><td>(b)†
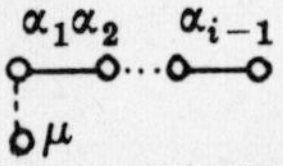
</td><td>(c)
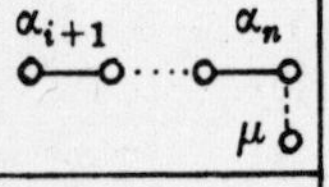
</td></tr>
<tr><td>A_n—II</td><td>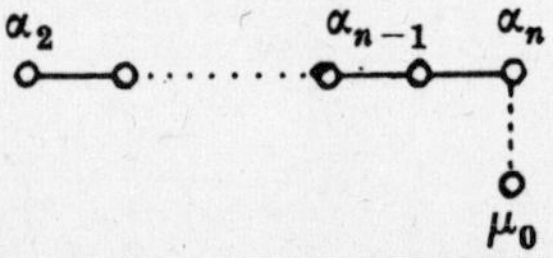
</td><td colspan="2">$\mathbf{M}_1 = \mathbf{M}_0$; diagram as in second column</td></tr>
<tr><td>D_n
$n > 3$</td><td>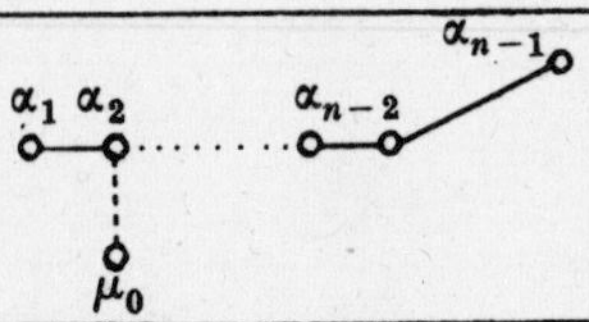
</td><td colspan="2">$\mathbf{M}_1 = \mathbf{M}_0$; diagram as in second column</td></tr>
<tr><td>E_6—I</td><td>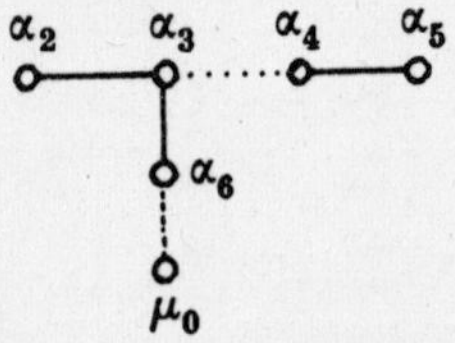
</td><td colspan="2">$\mathbf{M}_1 = \mathbf{M}_0$; diagram as in second column</td></tr>
<tr><td>E_6—II</td><td>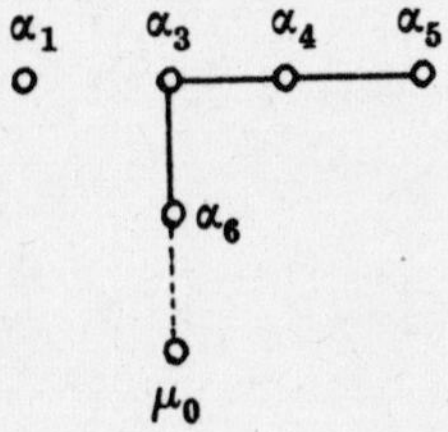
</td><td colspan="2">$\mathbf{M}_1 = \mathbf{M}_0$; diagram is same as in second column. Here we use the classification of real forms of E_6: if the parabolic group corresponding to α_2 is defined over $\mathbf{R}$, real form is split‡.</td></tr>
</table>

(†) It is shown later that (b) cannot in fact occur.

(‡) See for example Tits [1].

TABLE III (G non-absolutely simple)

In this case $\mathbf{M}_0 = \mathbf{M}_1$ always.

<table>
<tr><th>Type</th><th>Diagram of $\mathbf{M}_0 = \mathbf{M}_1$ and position of μ.</th></tr>
<tr><td>$A_n \times A_n$—I</td><td>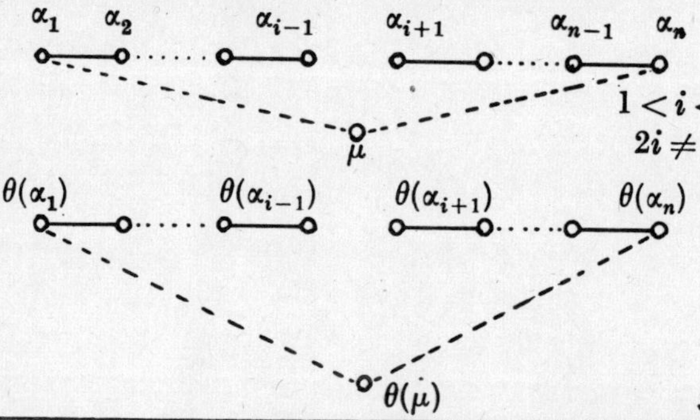

$1 < i < n-2$
$2i \neq n+1$</td></tr>
<tr><td>A_n—II</td><td>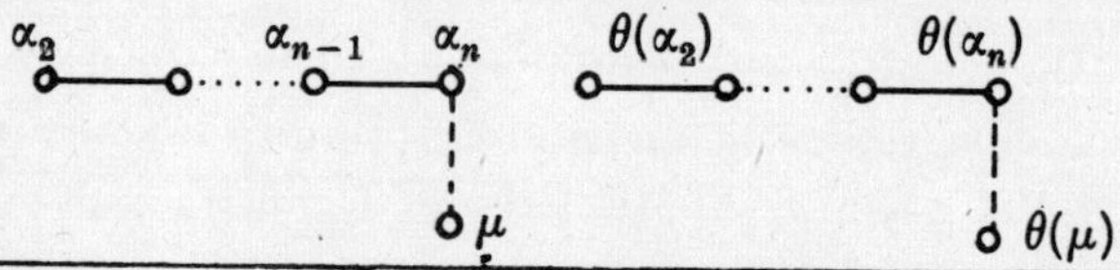
</td></tr>
<tr><td>D_n</td><td>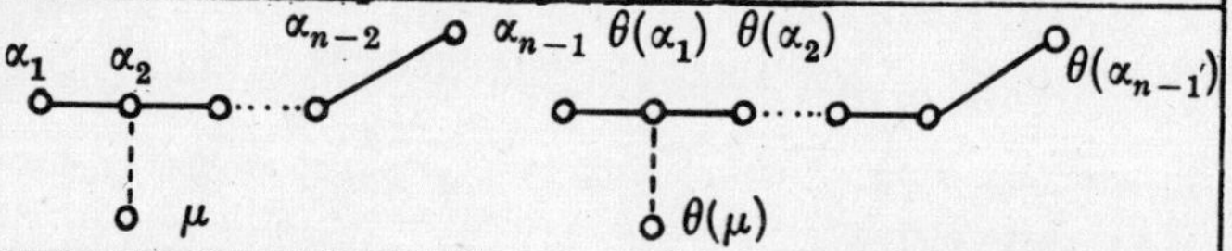
</td></tr>
<tr><td>E_6—I</td><td>
</td></tr>
<tr><td>E_6—II</td><td>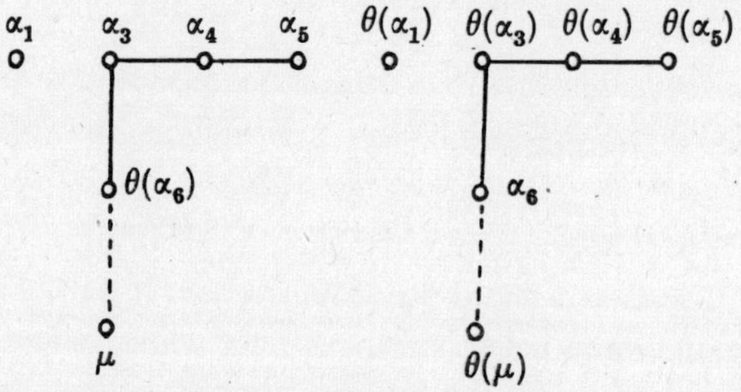
</td></tr>
</table>

5.21. From Tables II and III, it is clear that if $\mathbf{M}_0$ has two $\mathbf{R}$-simple factors, then their absolute types are not the same. It follows that if $\mathbf{M}_1 = \mathbf{M}_0$ is not $\mathbf{R}$-simple, then it breaks up also over $\mathbf{Q}$ into exactly two factors. We will now treat two cases separately. Case A: $\mathbf{Q}$-rank $\mathbf{M}_1 > 1$; Case B: $\mathbf{Q}$-rank $M_1 = 1$.

5.22. Case A. $\mathbf{Q}$-rank $M_1 > 1$. If $\mathbf{M}_1$ is $\mathbf{R}$-simple, it is $\mathbf{Q}$-simple. Now the following result is known (Raghunathan [2], §3, Theorem 2).

Let Φ be an arithmetic subgroup of a $\mathbf{Q}$-simple algebraic group $\mathbf{H}$ of $\mathbf{Q}$-rank > 1. Then for any non-trivial irreducible representation ρ of $\mathbf{H}$, $H^1(\Phi, \rho) = 0$.

It is seen from Tables II and III (5.20) that M_1 acts non-trivially on every simple factor of $\mathfrak{K}$. Hence in this case $H^1(\Gamma_1, \mathfrak{K}) = 0$. Consider next the case when M_1 is not simple. In Tables II and III this corresponds to A_n—I (a) and E_6—II. In the case A_n—I, one sees that $\mathfrak{K}$ is (over $\mathbf{R}$) a tensor product $\rho_1 \otimes \rho_2$ where ρ_1 and ρ_2 are faithful $\mathbf{R}$-irreducible representations of the simple factors of $\mathbf{M}_1$. Let $\mathbf{M}_1 = \mathbf{M}_1' \times \mathbf{M}_1''$. Then a subgroup Γ_2 of Γ_1 of finite index decomposes into a direct product of the form $\Gamma_1' \times \Gamma_2''$, $\Gamma_2' \subset \mathbf{M}_1'$ and $\Gamma_2'' \subset \mathbf{M}_2''$. According to Kunneth formula for cohomology we then find

$$H^1(\Gamma_2, \rho) \simeq H^1(\Gamma_2', \rho_1) \otimes H^0(\Gamma_2'', \rho_2) \otimes H^0(\Gamma_2', \rho_1) \oplus H^1(\Gamma_2'', \rho_2)$$

and $H^0(\Gamma_2', \rho_1) = H^0(\Gamma_2'', \rho_2) = 0$ since ρ_1 and ρ_2 are nontrivial. Finally, $H^1(\Gamma_1, \rho) \to H^1(\Gamma_2, \rho)$ is injective since Γ_2 has finite index in Γ_1. We are left with the case E_6—II now. In the discussion below in 5.23 — 5.24 we will prove the following. In the case of E_6—II, the (unique) simple component of type A_4 has $\mathbf{Q}$-rank *at least* 2. Once this claim is admitted, we observe that this factor acts non-trivially on $\mathfrak{K}$ while the other factor of type A_1 acts trivially on $\mathfrak{K}$. We can appeal to the theorem quoted above (applied to the factor of type A_4) and the Kunneth formula to conclude that $H^1(\Gamma_1, \mathfrak{K}) = 0$. Thus in case $\mathbf{Q}$-rank $\mathbf{M}_1 \geqslant 2$, $\mathbf{N}$ is Γ-rational.

5.23. We introduce some additional notation now; this notation will be used in the entire sequel. Let $\mathbf{U}_1 \subset \mathbf{M}_1$ be a maximal unipotent $\mathbf{Q}$-subgroup. Let $\mathbf{U} = \sigma^{-1}(\mathbf{U}_1)$. Evidently, then $U/U \cap \Gamma$ is compact. Moreover $U \cap \Gamma$ is a maximal unipotent subgroup of Γ: in fact if θ

generates with $U \cap \Gamma$ a unipotent group and $X \in \mathfrak{g}$ is the nilpotent element such that $\exp X = \theta$, then X is orthogonal to $\mathfrak{f}$ w.r.t. $\langle , \rangle$ and hence belongs to $\mathfrak{n}$ so that $\exp X = \theta \in N$; it follows that $\sigma(\theta) \in \mathbf{U}_1$ so that $\theta \in U \cap \Gamma$. Let $\mathbf{U}^*$ be a maximal unipotent subgroup of $\mathbf{G}$ containing $\mathbf{U}$. Let $\gamma \in \Gamma$ be an element such that $\gamma U^* \gamma^{-1} \cap U^* = \{e\}$. For a subgroup H of $\mathbf{G}$, let $H^- = \gamma H \gamma^{-1}$. Then clearly $U^-/U^- \cap \Gamma$ is compact. Since $N_0/N_0 \cap \Gamma \to G/\Gamma$ is proper, $N_0 \cap U^- \cap \Gamma$ is a lattice in $N_0 \cap U^-$. Let $U_1^- = \sigma(N_0 \cap U^-)$. Then $\mathbf{U}_1^-$ is defined over $\mathbf{Q}$. Now from the structure of parabolic subgroups and our choice of γ one deduces easily that $\dim \mathbf{U}_1^- = \dim \mathbf{U}_1$. Thus $\mathbf{U}_1^-$ is also a maximal unipotent subgroup of $\mathbf{M}_1$. Let $\mathbf{P}_1^-$ (resp. $\mathbf{P}_1$) be the normaliser of $\mathbf{U}_1$ (resp. $\mathbf{U}_1^-$) in σ $(\mathbf{M}_0)$. Then $\mathbf{P}_1$ and $\mathbf{P}_1^-$ are easily seen to be opposed to each other. Choosing $\mathbf{M}$ so that it normalises $\mathbf{F}$ and $\mathbf{F}^-$, we can also choose $\mathbf{T}$ (cf 5.15) to normalise $\mathbf{U}$ and $\mathbf{U}^-$ and to have all roots in $\mathfrak{U}$ positive. We assume such a choice made; then the parabolic subgroup P_1 can be described as the maximal parabolic subgroup of $\mathbf{M}_0$ associated to certain roots in the diagrams of $\mathbf{M}_0$ and as indicated below:

5.24. We use the notation of Tables II and III.

TABLE IV

(A). *G absolutely simple*

Type of $(\mathbf{G}, \mathbf{M}_0)$	Roots defining the parabolic P_1
A_n—I	α_{n+1-i}
A_n—II	α_n
D_n	α_{n-1}
E_6—I	α_5
E_6—II	α_4

(B) *G non-absolutely simple*

Type of G	Roots associated to P_1
$A_2 \times A_n$—I	$(\alpha_{n+1-i}, \theta(\alpha_{n+1-i}))$
$A_n \times A_n$ II	$(\alpha_n, \theta(\alpha_n))$
$D_n \times D_n$	$(\alpha_{n-1}, \theta(\alpha_{n-1}))$
$E_6 \times E_6$—I	$(\alpha_5, \theta(\alpha_5))$
$E_6 \times E_6$—II	$(\alpha_4, \theta(\alpha_4))$

Table IV shows immediately—since $\mathbf{U}_1 \subset \mathbf{M}_1$—that the possibility (b) in A_n—I, Table II, is ruled out. Also, in the case of E_6 — II, this shows that the **R**-simple component of type A_4 in $\mathbf{M}_0$ does occur in $\mathbf{M}_1$ as an isotropic factor. Moreover, the parabolic group $\mathbf{P}_1$ in this case is non-symmetric. It follows that this component of $\mathbf{M}_1$ (and hence $\mathbf{M}_1$ itself) cannot have **Q**-rank 1 in this case. (This was what was claimed in 4.22) Symmetry arguments show that $\mathbf{M}_1$ can have **Q**-rank 1 only in the following cases:

(1) $A_{3r-1} - \mathrm{I}, i=r.$ (1′) $A_{3r-1} \times A_{3r-1} - \mathrm{I}, i = r.$

(2) $A_2 - \mathrm{II}$ (2′) $A_2 \times A_2 - \mathrm{II}$

(3) $E_6 - \mathrm{II}$ (3′) $E_6 \times E_6 - \mathrm{II}.$

5.25. We shall consider now the cases $(A_{3r-1} - \mathrm{I},\ i = r)$ and $(A_{3r-1} \times A_{3r-1},\ i = r)$ with $r \geqslant 2$. In the case $(A_{3r-1} - \mathrm{I},\ i = r)$, since the maximal parabolic subgroups corresponding to the roots α_r and α_{2r} are defined over **R**, the classification of **R**-simple groups of Type A_n show that over **R** both factors of $\mathbf{M}_0$ are isotropic except when $r = 2$ and **G** is not quasi-split over **R** (see for instance Tits [1]).

As already shown, in the non-absolutely simple case $A_n \times A_n - \mathrm{I}$, $n \geqslant 2$, $\mathbf{M}_0$ has two factors both isotropic over **R**. Moreover, as remarked earlier again, these factors do not have isomorphic absolutely simple components. Thus $\mathbf{M}_0 = \mathbf{M}_1$ decomposes over **Q** into two factors $\mathbf{M}'_0 \times \mathbf{M}''_0$ both of which are isotropic over **R**. The group Γ_1 decomposes—upto finite index—into a product and since the representation on $\mathfrak{R}$ is a tensor product of non-trivial **R**-irreducible representations of the factors $\mathbf{M}'_0$ and $\mathbf{M}''_0$, the Kunneth formula can again be applied to establish that $H^1(\Gamma_1, \mathfrak{R}) = 0$.

5.26. The preceding discussion shows that except possibly in the case of $E_6 - \mathrm{II}$ and $E_6 \times E_6 - \mathrm{II}$, if $\mathbf{M}_1$ has **Q**-rank 1, **G** has **R**-rank exactly 2. In all these cases we argue as follows: let α (resp. $(\alpha, \theta(\alpha))$) be the unique element (resp. pair) in Δ determining the parabolic subgroup **N** in **G**. Let α (resp. $(\alpha, \theta(\alpha))$) denote the unique element (resp. pair) equivalent to α (resp. $(\alpha, \theta(\alpha))$) under the non-trivial symmetry of Δ. Let λ (resp. $\lambda^\wedge$) denote the fundamental weight associated to α (resp. $\hat{\alpha}$). Let w_1 be the element in the Weyl group of $\mathbf{M}_0$ which changes the sign of all the roots of $\mathbf{M}_0$. A simple computation then shows the following:

$$w_1(\lambda-\mu) = \mu \text{ and } w_1\,\theta(\lambda-\mu) = \theta(\mu). \qquad (*)$$

(In fact one has, $w_1(\lambda) = \lambda$ while $w_1(\mu) = \lambda - \mu$). Now let $\mathfrak{T}$ be the Lie algebra of $\mathbf{T}$ and H_λ (resp. H_μ) the unique element in $\mathfrak{T}$ such that $\langle H, H_\lambda \rangle = \lambda(H)$ (resp. $\langle H, H_\mu \rangle = \mu(H)$). Let $H^*_\lambda = (H_\lambda + \theta(H_\lambda))/2$ and $H^*_\mu = (H_\mu + \theta(H_\mu)/2$; then ad H^*_λ and ad H^*_μ are semisimple and their eigen-values are ± 1 and 0. Further $\mathfrak{F}$(resp. $\mathfrak{F}^-$) is the eigen-space of ad H^*_λ (resp. ad H^*_μ) corresponding to the eigen-value $+1$ (resp. -1). It is also easy to verify the following: $\mathfrak{U}$ (resp. $\mathfrak{U}^-$) is the span of eigen-spaces of ad$(H^*_\lambda + H^*_\mu)$ corresponding to the eigen-values 1 and 2 (resp. -1 and -2); moreover, the centre $\mathfrak{C}$(resp. $\mathfrak{C}^-$) of $\mathfrak{U}$(resp. $\mathfrak{U}^-$) is the eigen-space corresponding to the eigen-value 2 (resp. -2). Let $H = H^*_\lambda - H^*_\mu$. Let $\mathfrak{V}$ be the eigen-space of H corresponding to the eigen-value 1. According to (*), this eigen-space is conjugate to the parabolic group *opposed* to $\mathfrak{F}^-$ (w.r.t. $\mathbf{T}$) (note that $\mathfrak{F}$ and $\mathfrak{F}^-$ are not opposed to each other). The corresponding subgroup $\mathbf{V}$ of $\mathbf{G}$ is the unipotent radical of a maximal $\mathbf{R}$-parabolic subgroup $\mathbf{P}$ of $\mathbf{G}$. Moreover, it is easy to see that $\mathfrak{v} = \mathfrak{n}_0 \cap \mathfrak{f}^- + \mathfrak{n}^-_0 \cap \mathfrak{f}$. Since $(N^-_0 \cap F)/(N_0 \cap F \cap \Gamma)$ and $(N_0 \cap F^-)/(N_0 \cap F^- \cap \Gamma)$ are compact, $V/V \cap \Gamma$ is compact. It follows that if ρ denotes the adjoint representation of $\mathbf{P}$ on $\mathfrak{V}$ and $\mathbf{P}_0 = \{x \in \mathbf{P} \mid \det \rho(x) = \pm 1\}$ and $P_0 = \mathbf{P}_0 \cap G$, then $P_0/P_0 \cap \Gamma \to G/\Gamma$ is proper. Also $\mathfrak{V}$ carries a $\mathbf{Q}$-structure determined by a lattice $\mathscr{M}$ in $\mathfrak{V}$ such that the map $P_0/P_0 \cap \Gamma \to GL(\mathfrak{v})/GL(\mathscr{M})$ is proper. Now, by general position arguments, we can find $\varphi \in \Gamma$ such that $\varphi\, \mathbf{P}\, \varphi^{-1}$ is opposed to $\mathbf{F}$ (as also to $\mathbf{F}^-$). The arguments in 5.4 with minor modifications now show that $\mathbf{N}$ is Γ-rational.

This completes the proof of the main theorem.

APPENDIX I

A.1. Let **G** be a connected algebraic group defined, simple and of rank $\geqslant 2$ over **R**. Let **P** be an **R**-parabolic subgroup of **G** and **U** the unipotent radical of **P**. Let $\mathfrak{U}$ be the Lie algebra of **U** and $\sigma : \mathbf{P} \to GL(\mathfrak{U})$ the adjoint representation of **P** on $\mathfrak{U}$. Let **M** be a maximal reductive **R**-subgroup of **P** and **M*** the product of the maximal **R**-split central torus **S** in **M** and all the **R**-isotropic **R**-simple components of **M**. Let $\mathbf{M}_0^*$ be the identity component of

$$\mathbf{M}_0^{*\prime} = \{x \in \mathbf{M}^* \mid \det \sigma(x) = \pm 1\},$$

and $\mathbf{S}_0$ the identity component of

$$\mathbf{S}_0' = \{x \in \mathbf{S} \mid \det \sigma(x) = \pm 1\} = \mathbf{S} \cap \mathbf{M}_0^{*\prime}.$$

With this notation we will prove the following two results.

A. 2. Lemma A. *Assume that* **G** *has no absolutely simple component of type* $\mathbf{G}_2$. *Then the centraliser of* $\mathbf{M}_0^*$ *in* $\mathfrak{U}$ *is contained in the centre of* $\mathfrak{U}$.

A. 3. Lemma B. *Assume that the absolutely simple components of* **G** *are all of the following types* : A_n, D_n, *n odd* > 3, *or* E_6. *Let* **P** *be non-maximal and 'non-symmetric' — i. e.* $\mathbf{P} \cap g\,\mathbf{U}\,g^{-1} \neq \{e\}$ *for any* $g \in G$. *Then* $\mathbf{S}_0$ *acts non-trivially and as scalar automorphisms on the centre* $\mathfrak{C}$ *of* $\mathfrak{U}$.

A. 4. Notation. Choose tori ${}_{\mathbf{R}}\mathbf{T}$ and **T** such that $\mathbf{S} \subset {}_{\mathbf{R}}\mathbf{T} \subset \mathbf{T} \subset \mathbf{M}$ and ${}_{\mathbf{R}}\mathbf{T}$ (resp. **T**) is maximal **R**-split (resp. maximal and defined over **R**) in **G**. Introduce compatible orderings on $X(\mathbf{T})$ (= group of characters on **T**) and $X({}_{\mathbf{R}}\mathbf{T})$ (= group of characters on ${}_{\mathbf{R}}\mathbf{T}$). Let Φ denote the roots of G w.r.t. **T**. For $\alpha \in \Phi$, let $\mathfrak{G}(\alpha)$ denote the root space of α in the Lie algebra $\mathfrak{G}$ of **G**. Let $\mathfrak{M}$ denote the Lie subalgebra corresponding to **M** and let $\Phi(\mathbf{M}) = \{\alpha \in \Phi \mid \mathfrak{G}(\alpha) \subset \mathbf{M}\}$. We assume the order on $X(\mathbf{T})$ so chosen that for $\alpha \in \Phi$, if $\mathfrak{G}(\alpha) \subset \mathfrak{U}$ (= Lie algebra of **U**), $\alpha > 0$. Let Δ denote the system of simple roots and $\Delta(M) = \Delta \cap \Phi(\mathbf{M})$. Let $\Delta' = \Delta - \Delta(\mathbf{M})$; then

(i) $\{\mathfrak{G}(\alpha) \mid \alpha \in \Delta(\mathbf{M})\}$ generate $[\mathfrak{M}, \mathfrak{M}]$ as a Lie algebra ($\Delta(\mathbf{M})$ is a simple root system for $\mathfrak{M}$) and

(ii) $\mathfrak{U}$ is the linear span of

$$\left\{\mathfrak{G}(\alpha) \mid \alpha = \sum_{\varphi\in\Delta} m(\varphi)\,.\,\varphi \text{ with } m(\varphi) > 0 \text{ for some } \varphi \in \Delta'\right\}.$$

Let 2ρ(resp. $2\rho(\mathbf{M})$) denote the sum of the positive roots of $\mathbf{G}$ (resp. $\mathbf{M}$). Then one sees from the definition of σ that for $x \in \mathbf{T}$,

$$\det \sigma(x) = (2\rho - 2\rho(\mathbf{M}))(x). \qquad (*)$$

Let θ be the involution of $X(\mathbf{T})$ defined by complex conjugation; then $\theta(\Phi) = \Phi$. Further if $\alpha \in \Delta$ is non-trivial on ${}_{\mathbf{R}}\mathbf{T}$, there exists a unique element $\hat{\alpha} \in \Delta$ such that $\hat{\alpha}$ is non-trivial on ${}_{\mathbf{R}}\mathbf{T}$ and

$$\theta(\alpha) = \hat{\alpha} + \sum_{\varphi\in\Delta} m(\varphi)\,\varphi \qquad (**)$$

with $m(\varphi) \neq 0$ only if φ is trivial on ${}_{\mathbf{R}}\mathbf{T}$; if $\mathbf{G}$ is *not* absolutely simple $\theta(\alpha) = \hat{\alpha}$. Finally all the elements in Δ' are non-trivial on ${}_{\mathbf{R}}\mathbf{T}$ and for $\alpha \in \Delta'$, $\hat{\alpha} \in \Delta'$ (this is true since $\mathbf{P}$ is defined over $\mathbf{R}$). Finally we also note that since $\mathbf{M}$ is defined over $\mathbf{R}$,

$$\theta(\alpha) \in \Phi(\mathbf{M}) \text{ for all } \alpha \in \Phi(M); \qquad (***)$$

in particular $\Phi(\mathbf{M})$ and $\theta(\Phi(\mathbf{M}))$ are linear combinations of elements in $\Delta(\mathbf{M})$. After these preliminary common remarks, we will now first take up the proof of Lemma A.

A. 5. Proof of Lemma A. Let $\mathfrak{C}$ be the centre of $\mathfrak{U}$. Suppose now that $E \not\subset \mathfrak{C}$. Since $\mathbf{M}_0^*$ is normal in $\mathbf{M}$, E is $\mathbf{M}$-stable. On the other hand $\mathfrak{C}$ is evidently $\mathbf{M}$-stable. Let E' be an $\mathbf{M}$-stable supplement to $E \cap \mathfrak{C}$ in E (E' is in fact unique since $\mathbf{T} \subset \mathbf{M}$ and $\mathfrak{U}$ is a sum of mutually inequivalent simple $\mathbf{T}$-modules). Let β' be the highest root in Φ such that $\mathfrak{G}(\beta') \subset E'$. We then have (for the natural scalar product on $X(T)$),

$$\left.\begin{aligned} &\langle \varphi, \beta' \rangle \geqslant 0 \text{ for all } \varphi \in \Delta(\mathbf{M}) \text{ and} \\ &\langle \varphi, \beta' \rangle = 0 \text{ for } \varphi \in \Delta(\mathbf{M}^*) = \{\alpha \in \Delta \mid \mathfrak{G}(\alpha) \subset \mathfrak{M}_0^*\} \end{aligned}\right\} \qquad (\mathrm{I})$$

(here $\mathfrak{M}_0^*$ = Lie algebra of M_0^*). Further, since $\beta'(x) = 1$ for all $x \in {}_{\mathbf{R}}\mathbf{T}^0 = \{y \in \mathbf{T} \mid \det \sigma(y) = 1\}$ $(\subset \mathbf{M}_0^*)$ we have necessarily

$$\beta \mid_{\mathbf{R}}\mathbf{T} = p(2\rho - 2\rho(\mathbf{M}))\mid_{\mathbf{R}}\mathbf{T} \text{ with } p > 0.$$

If λ is any character on $\mathbf{T}$, $\lambda + \theta(\lambda)$ is defined over $\mathbf{R}$ so that if μ and ν are characters on $\mathbf{T}$ which coincide on ${}_{\mathbf{R}}\mathbf{T}$, we have $\nu + \theta(\nu) = \mu + \theta(\mu)$. Also $\theta(\lambda) = \lambda$ if and only if λ is defined over $\mathbf{R}$. These considerations show that

$$\beta' + \theta(\beta') = q(2\rho - 2\rho(M)),\ q > 0.$$

Now as is well known we have

$$\left.\begin{aligned} 2\langle \rho, \alpha \rangle &= \langle \alpha, \alpha \rangle \quad \text{for } \alpha \in \Delta \text{ and} \\ 2\langle \rho(\mathbf{M}), \alpha \rangle &= \langle \alpha, \alpha \rangle \quad \text{for } \alpha \in \Delta(\mathbf{M}). \end{aligned}\right\} \qquad \text{(II)}$$

It follows that

$$\begin{aligned} \langle \beta' + \theta(\beta'), \alpha \rangle &= 0 \quad \text{for } \alpha \in \Delta(\mathbf{M}) \\ &= r \langle \alpha, \alpha \rangle \text{ for } \alpha \in \Delta' \end{aligned}$$

for a suitable *positive* number r. Suppose now that $\langle \beta, \alpha \rangle < 0$ for some $\alpha \in \Delta'$; then we have

$$2\langle \beta', \alpha \rangle = -k \langle \alpha, \alpha \rangle$$

where $k = \langle \beta', \beta' \rangle / \langle \alpha, \alpha \rangle$; since $\langle \theta(\beta'), \alpha \rangle / \langle \alpha, \alpha \rangle \leqslant k \langle \alpha, \alpha \rangle$, in this case, we are led to a contradiction. Thus we necessarily have (combining with (*))

$$2\langle \beta', \alpha \rangle \geqslant 0 \text{ for all } \alpha \in \Delta.$$

Suppose $\alpha \in \Delta'$ is a root such that $\beta' + \alpha$ is again a root, then we have

$$\langle \beta' + \alpha, \beta' + \alpha \rangle = \langle \beta', \beta' \rangle + \langle \alpha, \alpha \rangle + 2\langle \beta', \alpha \rangle.$$

Since β' is not the highest root, we see that such a root α exists; it follows that there are *necessarily two root lengths*. Moreover, if $\langle \beta', \alpha \rangle \neq 0$, $\beta' + \alpha$ is atleast thrice as long as the shorter of the roots (α, β'). Thus since $\mathbf{G}$ has no components of type G_2 we are led to the following conclusion: (absolutely) simple components of $\mathbf{G}$ are of type B_n, C_n or F_4 and

$$\left.\begin{aligned} \langle \beta', \alpha \rangle &\geqslant 0 \quad \text{for all } \alpha \in \Delta \\ &= 0 \quad \text{for } \alpha \in \Delta(\mathbf{M}^*) \cup \Delta'. \end{aligned}\right\} \qquad \text{(III)}$$

Now if β denotes the highest root, $\beta = \beta' + \sum_{\varphi \in \Delta} m(\varphi).\varphi$ with $m(\varphi) \geqslant 0$ so that from (III) we deduce that $\langle \beta, \beta \rangle > \langle \beta', \beta' \rangle$. Thus β' is the unique short root which is a dominant weight. Now since $\mathbf{M}_0^*$ contains all the isotropic simple factors of $\mathbf{M}$, one sees that the roots $\alpha \in \Delta$ which are non-trivial on ${}_{\mathbf{R}}\mathbf{T}$ are necessarily contained in $(\Delta(\mathbf{M}^*) \cup \Delta')$. *In particular if* $\langle \beta', \alpha \rangle > 0$ *for* $\alpha \in \Delta$, α *is trivial on* ${}_{\mathbf{R}}\mathbf{T}$. We now list below the possible Tits diagrams for $\mathbf{G}$ showing also to which simple roots β and β' are linked. Roots in Δ which are non-trivial on ${}_{\mathbf{R}}\mathbf{T}$ are indicated by drawing a circle around them. The arrows point towards the short root.

CASE I. *G absolutely simple* (note that *R*-rank $G \geqslant 2$). $r = \mathbf{R}$*-rank* G.

Type	Diagram
$B_{n,r}$	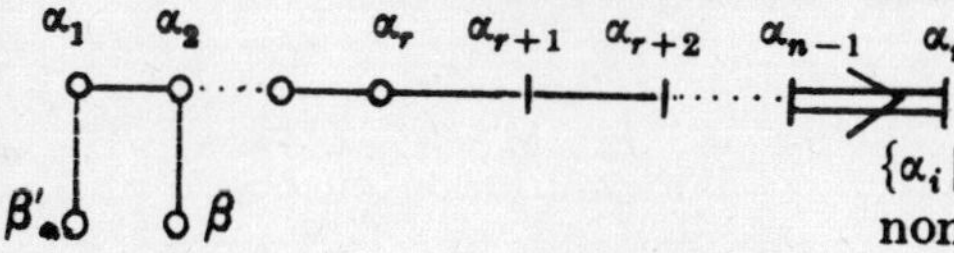 $\{\alpha_i \mid 1 \leqslant i \leqslant r\}$ are non-trivial on ${}_{\mathbf{R}}\mathbf{T}$
$C^{(1)}_{n,r}$ $(n = r)$	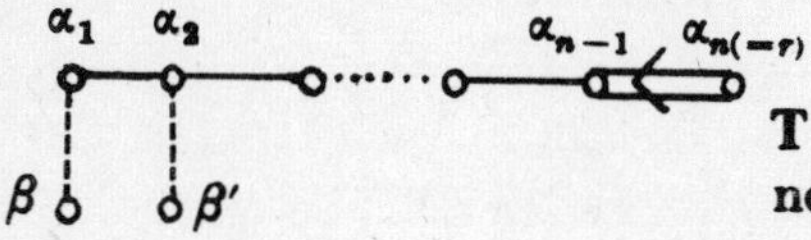 $\mathbf{T} = {}_{\mathbf{R}}\mathbf{T}$; all roots are non-trivial on ${}_{\mathbf{R}}\mathbf{T}$
$C^{(2)}_{n,r}$ $(n > 2r)$	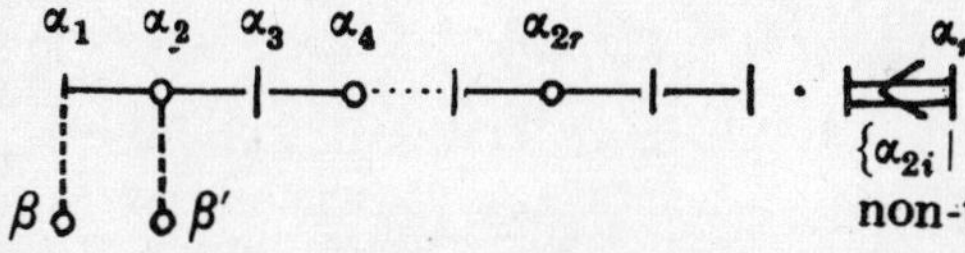 $\{\alpha_{2i} \mid 1 \leqslant i \leqslant r\}$ are non-trivial on ${}_{\mathbf{R}}\mathbf{T}$
$C^{2}_{n,r}$ $(n = 2r)$	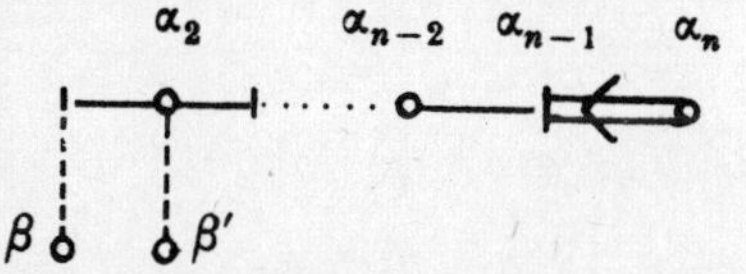 $\{\alpha_{2i} \mid 1 \leqslant i \leqslant r\}$ are non-trivial on ${}_{\mathbf{R}}\mathbf{T}$.
F^{0}_{44} $(n = r = 4)$	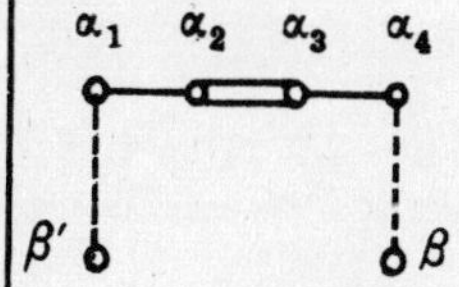 ${}_{\mathbf{R}}\mathbf{T} = \mathbf{T}$; all roots are non-trivial on ${}_{\mathbf{R}}\mathbf{T}$.

CASE II. **G** *non-absolutely simple*. In this situation every root of **G** is necessarily non-trivial on ${}_{\mathbf{R}}\mathbf{T}$.

The table on p. 315 shows that the short dominant root has always a non-trivial scalar product with a simple root which is non-trivial on ${}_{\mathbf{R}}\mathbf{T}$, leading to a contradiction.

A.6. PROOF OF LEMMA B. The fact that **P** is not symmetric leads us in the first place immediately to the conclusion that the (absolutely) simple components of **G** are of one of the following types: A_n, D_n(n odd) or E_6: these are the groups in which the Weyl group element that takes all simple roots into negative roots, is different from $x \longmapsto -x$ (on the character group of the maximal torus); this automorphism is in fact of the form $\alpha \longmapsto -\tau(\alpha)$, where τ is the unique *non-trivial symmetry* of the Dynkin diagram of each simple factor. The subset Δ' corresponding to the parabolic group, one now observes, must necessarily be *not* stable under τ. Since **P** is defined over **R**, α is non-trivial on ${}_{\mathbf{R}}\mathbf{T}$ for all $\alpha \in \Delta'$ and for $\alpha \in \Delta'$, $\hat{\alpha} \in \Delta'$. Finally since **P** is non-maximal, there exist a pair $\alpha_1, \alpha_2 \in \Delta'$ such that $\alpha_1 \neq \hat{\alpha}_2$. The torus **S** is precisely the identity component of the intersection ${}_{\mathbf{R}}\mathbf{T} \bigcap_{\alpha \in \Delta - \Delta'} \ker \alpha$. Now it is easy from the description of $\mathfrak{U}$ as a sum of root spaces of $\mathfrak{G}$ that the centre $\mathfrak{C}$ of $\mathfrak{U}$ is in fact an **R**-irreducible **M**-module. Since **S** is **R**-split and *central* in **M**, **S** acts as scalars on $\mathfrak{C}$ (Schur's lemma). Finally if β is a highest root of $\mathfrak{G}$, $\mathfrak{G}(\beta) \subset \mathfrak{U}$. In view of our remarks in A.4 (in particular (*) and (***)) Lemma B reduces to the following assertion:

The set of roots $(\Delta - \Delta') \cup \{\beta, 2\rho - 2\rho(\mathbf{M})\}$ restricted to ${}_{\mathbf{R}}\mathbf{T}$ are linearly independent. Equivalently,

$$\Delta - \Delta' \cup \{\beta + \theta(\beta), 2\rho\}$$

are linearly independent in $X(\mathbf{T})$ (note that for $\alpha \in \Delta - \Delta' = \Delta(\mathbf{M})$, $\theta(\alpha)$ is again a linear combination of elements in $\Delta(M)$). From (**) of A.4, we see immediately that if

$$\beta = \sum_{\varphi \in \Delta} p(\varphi) \cdot \varphi$$

and

$$\theta(\beta) = \sum_{\varphi \in \Delta} \hat{p}(\varphi)\, \varphi,$$

then $\hat{p}(\varphi) = p(\varphi)$ for all φ which are non-trivial on ${}_{\mathbf{R}}\mathbf{T}$, in particular for $\varphi \in \Delta'$. Now, let

$$\beta + \theta(\beta) = \sum_{\varphi \in \Delta} r(\varphi) \cdot \varphi$$

and

$$2\rho = \sum_{\varphi \in \Delta} q(\varphi) \cdot \varphi.$$

We have to show then that there exist $\varphi_1, \varphi_2 \in \Delta'$ such that $r(\varphi_1)/r(\varphi_2) \neq q(\varphi_1)/q(\varphi_2)$. Let Δ_0 be the set of elements of Δ which are trivial on ${}_{\mathbf{R}}\mathbf{T}$. Let κ be the Weyl group element of the corresponding root system Φ_0 (associated to Δ_0) which takes all of Δ_0 into negative roots. For $\alpha \in \Delta_0$, let $\hat{\alpha} = -\kappa(\alpha)$. Then $\alpha \longmapsto \hat{\alpha}$ is an automorphism of the diagram of Δ; this automorphism of Δ cannot be the same as τ since Δ' is stable under it. It follows that in the absolutely simple case $\alpha = \hat{\alpha}$ for $\alpha \in \Delta$; in the non-absolutely simple case Δ breaks into two components and θ exchanges the two. One sees immediately from this that for all $\varphi \in \Delta'$, we have $p(\varphi) = \hat{p}(\varphi)$. The problem reduces once again to show that there exist $\varphi_1, \varphi_2 \in \Delta'$ such that $p(\varphi_1)/p(\varphi_2)$ is not equal to $q(\varphi_1)/q(\varphi_2)$. This can be seen immediately from the following table. (The first three are absolutely simple and the last three are not).

Type	Diagram of G	$p_i = p(\alpha_i), p'_i = p(\beta_i)$	$q_i = q(\alpha_i), q'_i = q(\beta_i)$
A_n	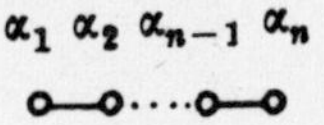	$p_i = 1$	$q_i = i(n-i+1)$
D_n n odd $n > 3$	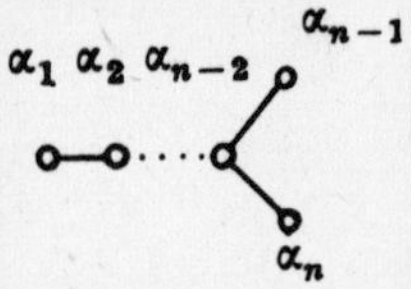	$p_i = 2$ for $1 < i < n-1$ $= 1$ for $i = 1, n-1$ or n	$q_i = i(2n-i-1)$ $1 \leqslant i < n-1$ $q_{n-1} = q_n = n(n-1)/2$
E_6	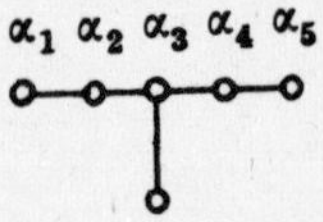	$p_1 = p_5 = 1, p_2 = p_4 =$ $p_6 = 2, p_3 = 3$	$q_1 = q_5 = 16, q_2 = q_4 = 15$ $q_3 = 21, q_6 = 11$
A_n	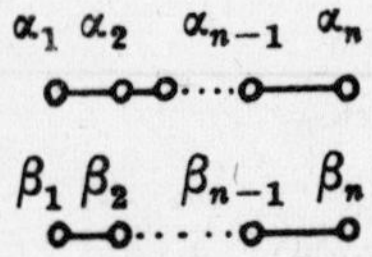 $\theta(\alpha_i) = \beta_i$	$p_i = p'_i = 1$	$q_i = q'_i = i(n-i+1)$
D_n n odd $n > 3$	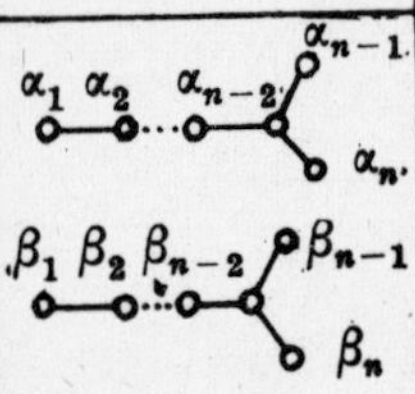 $\theta(\alpha_i) = \beta_i$	$p_i = p'_i = 2$ for $1 < i < n-1$ $= 1$ for $i = 1, n-1$ or n	$q_i = q'_i = i(2n-i-1)$ $q_{n-1} = q'_{n-1} = q_n =$ $q'_n = n(n-1)/2$
E_6	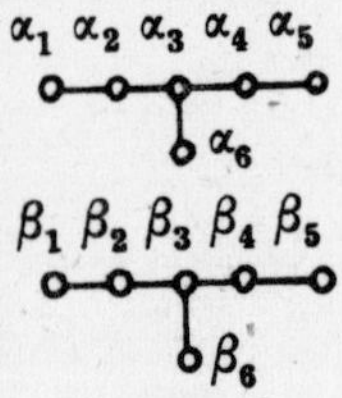 $\theta(\beta_i) = \alpha_i$	$p_1 = p'_1 = p_5 = p'_5 = 1,$ $p_2 = p'_2 = p_4 = p'_4 = 2$ $p_6 = p'_6 = 2, p_3 = p'_3 = 3$	$q_1 = q'_1 = q_5 = q'_5 = 16,$ $q_2 = q'_2 = q_4 = q'_4 = 15$ $q_3 = q'_3 = 21, q_6 = q'_6 = 11$

APPENDIX II

A.7. Let A be a ring and B the centre of A. Let M and N be A-modules. Then a homothesy by an element $b \in B$ induces a A-linear map of M in M. Consequently it induces a map

$$\bar{b}\colon \mathrm{Ext}^p_A(M, N) \longrightarrow \mathrm{Ext}^p_A(M, N).$$

It is a simple consequence of the definition of the groups Ext that this defines a structure of a B-module on $\mathrm{Ext}^p_A(M, N)$ and that this structure is the same as that obtained by treating an element $b \in B$ as the A-linear homothesy of N. An immediate result of this observation is the following:

A.8. Proposition. *Suppose $b \in B$ is such that the corresponding homothesy of M is trivial while that of N is an automorphism of N; then* $\mathrm{Ext}^p_A(M, N) = 0$ *for* $p \geqslant 0$.

One has only to remark that $\bar{b}$ is both the zero map and an automorphism.

A.9. Corollary. *Let k be a field and π (resp. $\mathfrak{g}$) a group (resp. a k-Lie algebra). Let ρ be an 'irreducible' representation of π (resp. $\mathfrak{g}$) on a k-vector space M. Suppose there is a central element x (resp. X) in π (resp. $\mathfrak{g}$) such that $\rho(x) \neq$ identity (resp. $\rho(X) \neq 0$). Then $H^p(\pi, \rho) = 0$ (resp. $H^p(\mathfrak{g}, \rho) = 0$) for all $p \geqslant 0$.*

By definition $H^p(\Gamma, \rho) \simeq \mathrm{Ext}^p_{k(\pi)}(k, M)$ (resp. $H^p(\mathfrak{g}, \rho) \simeq \mathrm{Ext}^p_{k(\mathfrak{g})}(k, M)$) where $k(\pi)$ (resp. $k(\mathfrak{g})$) is the group-algebra (resp. the enveloping algebra) of π (resp. $\mathfrak{g}$), and M is considered a $k(\pi)$- (resp. $k(\mathfrak{g})$-) module through the trivial action of π (resp. $\mathfrak{g}$) on k. The element $b = x - 1$ (resp. $= X$) is then a central element of $A = k(\pi)$ (resp. $= k(\mathfrak{g})$) and the corollary follows directly from the proposition and Schur's lemma (note that ρ is *irreducible*).

A.10. Corollary. *Let π (resp. $\mathfrak{g}$) be an 'abelian' group (resp. Lie algebra) and ρ 'any nontrivial' irreducible representation of π (resp. $\mathfrak{g}$); then $H^p(\pi, \rho) = 0$ (resp. $H^p(\mathfrak{g}, \rho) = 0$) for all $p \geqslant 0$.*

This follows from Corollary 1 choosing for x (resp. X) any element of π (resp. $\mathfrak{g}$) such that $\rho(x) \neq 1$ (resp. $\rho(X) \neq 0$).

REFERENCES

L. Auslander : [1] Bieberbach's theorem on space groups and discrete uniform subgroups, *Ann. of Math.* 71 (1960), 579-590 II, *Amer. J. Math.*, 83 (1961), 276-280.

A. Borel : [1] Density properties of certain subgroups of semisimple groups, *Ann. of Math.*, 72 (1960), 179-188.

[2] Sousgroupes discrets de groupes semisimple, *Seminaire Bourbaki* (1968-69), *Expose* 358.

[3] *Linear algebraic groups*, Benjamin, New York, (1969).

A. Borel, and J. Tits : [1] Groupes reductifs, *Publ. Mathematiques, I.H.E.S.*, 27 (1965), 55-150.

[2] Éléments unipotents et sous-groupes paraboliques de groupes réductifs. I., *Inv. Math.*, 12 (1971), 95-104.

H. Garland and M. S. Raghunathan : [1] Fundamental domains for lattices in (R-) rank 1 semisimple Lie groups, *Ann. of Math.*, 92 (1970), 279-326.

D. A. Kazdan and G. A. Margolis : [1] A proof of Selberg's, hypothesis, *Math. Sbornik* (*N. S.*), 75 (117) (1968), 162-168 (Russian).

A. I. Mal'cev : [1] On a class of homogeneous spaces, *Amer. Math. Soc. Transl. No.* 39 (1951).

G. A. Margolis : [1] On arithmeticity of discrete groups, *Soviet. Math. Dokl.*, 10 (1969), 900-902 (Russian).

[2] On the action of unipotent groups in the space of lattices, *Math. Sb*, 86 (1969), 552 (Russian).

G. D. Mostow : [1] On maximal subgroups of real Lie groups. *Ann. of Math.*, 74 (1961), 503-517.

I. I. Pjatetskĭ-Shapiro : [1] Discrete subgroups of Lie groups. *Trudy Moskov. Mat. Obsc.*, 18 (1968), 3-18.

M. S. Raghunathan : [1] *Discrete subgroups of Lie groups*, Springer-Verlag, New York (1972).

[2] Cohomology of arithmetic subgroups of algebraic groups, I., *Ann. of Math.*, 86 (1967), 409-424.

[3] A note on quotients of real algebraic groups by arithmetic subgroups, *Inv. Math.*

[4] Lattices in semisimple Lie groups, *Actes du Congrès International des mathématiciens* (Nice 1970), 2, 337-341, Gautier-Villars, Paris (1971).

M. ROSENLICHT : [1] Some rationality questions on algebraic groups, *Ann. di. Math.*, 43 (1957), 25-50.

[2] On quotient varieties and affine embedding of certain homogeneous spaces, *Trans. Amer. Math. Soc.*, 101 (1961), 211-233.

J. TITS : [1] Classification of semisimple algebraic groups, algebraic groups and discontinuous groups, *A. M. S., Proceedings of Symposia in Pure Mathematics*, IX.

H. C. WANG : [1] On the deformation of lattices in a Lie group, *Amer. J. Math.*, 91 (1967), 921-937.

H. ZASSENHAUS : [1] Beweis eines Satzes uber diskrete Gruppen, *Abh. Math. Sem. Hansisch. Univ.*, 12 (1938), 289-312.

SOME ARITHMETICAL DISCRETE GROUPS IN LOBAĈEVSKIĬ SPACES

By È. B. VINBERG

SOME TERMINOLOGY.

A reflection (in a vector space or in a simply connected Riemannian space of constant curvature) — a reflection with respect to a hyperplane (the mirror).

A reflection group—group generated by reflections.

An integral quadratic form—a form

$$f(x) = \Sigma\, a_{ij}x_ix_j \text{ where } a_{ij} = a_{ji} \in \mathbf{Z}.$$

An integral automorphism, or a unit, of the form f — an integral linear transformation, which preserves this form.

INTRODUCTION. The subject of this report is an application of the theory of discrete reflection groups to the study of the groups of units of some indefinite integral quadratic forms.

The basic propositions of Coxeter's theory of discrete reflection groups in Euclidean spaces [1] may be transferred without difficulty to discrete reflection groups in Lobaĉevskiĭ spaces. This enables us to find a fundamental polyhedron, generators and defining relations of any such group.

On the other hand, let f be an integral quadratic form of signature $(n, 1)$, i.e. equivalent over $\mathbf{R}$ to the form

$$f_n(x) = -x_0^2 + x_1^2 + \dots + x_n^2.$$

Then the group $\mathcal{O}(f, \mathbf{Z})$ of units of the form f or, more precisely, its subgroup of index 2, may be regarded as a discrete group of motions of n-dimensional Lobaĉevskiĭ space. It is known that it has a fundamental domain of finite volume [2]. If the group $\bar{\mathcal{O}}(f, \mathbf{Z})$ contains a reflection subgroup of finite index, we have the means for defining its fundamental domain, generators and relations.

An integral quadratic form $f(x) = \Sigma a_{ij} x_i x_j$ is called unimodular if $\det(a_{ij}) = \pm 1$. For unimodular forms of signature $(n,1)$ the following two cases are possible [6]:

(1) f is odd; then f is equivalent over $\mathbf{Z}$ to f_n;

(2) f is even; then $n = 8k + 1$ and when n is fixed, all such forms are equivalent over $\mathbf{Z}$.

For the group $\mathcal{O}(f, \mathbf{Z})$ of units of an unimodular integral quadratic form f of signature $(n, 1)$ we shall prove the two following theorems.

THEOREM A. *If $n \leqslant 17$, the group $\mathcal{O}(f, \mathbf{Z})$ contains a reflection subgroup of finite index.*

The fundamental polyhedron of the maximal reflection subgroup of $\mathcal{O}(f, \mathbf{Z})$ will be described explicitly in all these cases.

THEOREM B. *If $n \geqslant 25$, the group $\mathcal{O}(f, \mathbf{Z})$ contains no reflection subgroup of finite index.*

For $18 \leqslant n \leqslant 24$ the question is open.†

A more detailed exposition of these and some other results will appear in [13, 14].

1. Discrete reflection groups.

1. Let X^n be an n-dimensional simply connected Riemannian space of constant curvature, i.e. a sphere S^n, Euclidean space E^n or Lobačevskiĭ space Λ^n.

Let Γ be any discrete reflection group (d.r.g.) in the space X^n. The mirrors of all reflections belonging to Γ decompose X^n into Γ-equivalent convex polyhedra, called Γ-cells. Each of these cells is a fundamental domain for Γ.

† J. M. Kaplinskaya has proved that the groups $\mathcal{O}(f_{18}, \mathbf{Z})$ and $\mathcal{O}(f_{19}, \mathbf{Z})$ contain a reflection subgroup of finite index. On the other hand, it follows from a consideration communicated to me by M. Kneser, and Theorem 3.5 below that the group $\mathcal{O}(f_{20}, \mathbf{Z})$ contains no such subgroup. Thus the group $\mathcal{O}(f, \mathbf{Z})$ (where f is such as in the text) contains a reflection subgroup of finite index if and only if $n < 19$.

For some Γ-cell P, let us denote

$P_i (i \in I)$—all $(n-1)$-dimensional faces of P,

$H_i (i \in I)$—the corresponding hyperplanes,

$R_i (i \in I)$—the corresponding reflections.

It is known that

(1) for any pair $\{P_i, P_j\}$ of adjacent faces the angle between P_i and P_j is of the form $\frac{\pi}{n_{ij}}$, where $n_{ij} \in Z$;

(2) the R_i generate Γ;

(3) the relations

$$R_i^2 = 1, \quad (R_i R_j)^{n_{ij}} = 1$$

are defining relations for the R_i.

It is also known [9,10] that if P_i and P_j are not adjacent, then H_i and H_j are either parallel, or diverging (in the case $X^n = \Lambda^n$).

Conversely, any convex polyhedron, all the dihedral angles of which are submultiples of π, is a cell of some d.r.g.

The Coxeter's diagram $\Sigma(\Gamma)$ *of a* d.r.g. Γ. In the above notation, $\Sigma(\Gamma)$ is a graph with vertices $v_i (i \in I)$ which are joined as follows:

if	the vertices v_i and v_j are joined
P_i and P_j are adjacent and the angle between them is equal to π/n_{ij}	by an $(n_{ij}-2)$-tuple line or by a simple line with index n_{ij}
H_i and H_j are parallel	by a thick line or by a simple line with index ∞
H_i and H_j are diverging	by a dotted line

If P_i and P_j are orthogonal ($n_{ij} = 2$), v_i and v_j are not joined.

The cosines matrix $\cos \Gamma$ *of a* d.r.g. Γ is defined as follows:

$$\cos \Gamma = \left(-\cos \frac{\pi}{n_{ij}} \right)$$

where we put $n_{ij} = 1$ and $\cos \dfrac{\pi}{n_{ij}} = 1$ if n_{ij} is not defined. Evidently the cosines matrix may be reconstructed after the Coxeter's diagram.

2. We shall consider now three remarkable classes of d.r.g.

Finite reflection groups. Such are all the d.r.g. in S^n and those d.r.g. in E^n and Λ^n, which have a fixed point. The number of generators of a finite reflection group is called its rank.

It is known [1] that a d.r.g. is finite if and only if its cosines matrix is positive definite, so the finiteness property of a d.r.g. depends only on its diagram.

A diagram of a finite reflection group is called an elliptic diagram. Its rank is by definition the rank of the corresponding group. It is equal to the number of vertices.

A Coxeter's diagram is elliptic if and only if all its connected components are such. All the connected elliptic diagrams are given in Table 1.

Parabolic Reflection Groups. A diagram of a d.r.g. in E^n with a bounded cell is called a parabolic diagram of rank n. A d.r.g. is said to be parabolic reflection group of rank n if its diagram is parabolic of rank n. For example, such is a d.r.g. in Λ^n with a fixed improper point if it has a bounded cell on an orysphere with center at this point.

A Coxeter's diagram is parabolic if and only if all its connected components are such. Its rank is equal to the number of vertices minus the number of connected components. All connected parabolic diagrams are given in Table 2.

It is known [1] that the connected Coxeter's diagram is parabolic if and only if the corresponding cosines matrix is degenerate non-negative definite.

LANNER'S GROUPS. In the work [3] by Lanner were firstly enumerated all d.r.g. in Λ^n with simplicial bounded cells. We shall call the diagrams of these groups the Lanner's diagrams. They are given in Table 3. Any d.r.g., whose Coxeter's diagram is a Lanner's diagram, will be called a Lanner's group.

3. Let Γ be a d.r.g. in X^n. It is known that the stable subgroup $\Gamma_x \subset \Gamma$ of any point $x \in X^n$ is generated by reflections. More precisely, let P be a Γ-cell containing x; P_i, H_i and $R_i (i \in I)$ are the same as in 1.1. Let us denote by H_i^- the halfspace bounded by H_i and containing P. So $P = \bigcap_{i \in I} H_i^-$. If $J = \{i \in I \colon H_i \ni x\}$, then Γ_x is generated by the reflections R_i, $i \in J$, and $P_x = \bigcap_{i \in J} H_i^-$ is a Γ_x-cell.

In the case $X^n = \Lambda^n$ the above assertions hold true for some improper points namely those for which Γ_x has a bounded fundamental domain on an orysphere with center at x. Such improper points will be said to possess the compactness property with respect to Γ.

4. Let Θ be an arbitrary discrete group in X^n. We denote by Γ the group generated by all reflections belonging to Θ. Let P be a Γ-cell and Sym P be the symmetry group of P.

It is trivial that Γ is a normal subgroup of Θ and that Θ is a semi-direct product

$$\Theta = \Gamma . H, \tag{1.1}$$

where H is some subgroup in Sym P.

How to find P? Let us fix a point $x_0 \in X^n$ and suppose that for any $r > 0$ we can enumerate all reflections in Θ, whose mirrors H satisfy the condition $\rho(x_0, H) \leqslant r$, where ρ denotes the distance. Then we can determine P as described below.

SOME CONVENTIONS. For any hyperplane H, we shall denote by H^+ and H^- the halfspaces bounded by H. We shall say that halfspaces H_1^- and H_2^- are opposite if one of the following cases take place:

1) H_1 and H_2 are crossing and the dihedral angle $H_1^- \cap H_2^-$ does not exceed $\frac{\pi}{2}$;

2) $H_1^- \supset H_2$ and $H_2^- \supset H_1$;

3) $H_1^- \cap H_2^- = \varnothing$.

AN ALGORITHM FOR CONSTRUCTING A Γ-CELL.

"First we consider the stable subgroup Γ_0 of x_0 in Γ, i.e. the group generated by all reflections in Θ, whose mirrors contain x_0. Let

$$P_0 = \bigcap_{i=1}^{k} H_i^-$$

be some Γ_0-cell, each of the H_i^- being essential. There exists a unique Γ-cell containing x_0 and contained in P_0. This Γ-cell we denote by P.

Now we shall construct one by one hyperplanes $H_{k+1}, H_{k+2}, \ldots$ and halfspaces $H_{k+1}^-, H_{k+2}^-, \ldots$ such that

$$P = \bigcap_{\text{all } i} H_i^-$$

each of the H_i^- being essential. The H_i will be ordered by increase of $\rho(x_0, H_i)$.

The rules for constructing H_m and H_m^- for $m \geqslant k+1$ are the following.

1⁰. H_m^- is that of two halfspaces bounded by H_m, which contains x_0.

2⁰. If the H_i^- with $i < m$ have been constructed, then H_m is chosen as the nearest to x_0 mirror of a reflection belonging to Θ for which the halfspaces H_m^- and H_i^- are opposite for all $i < m$.

The procedure may be finite or infinite."

In the rule 2⁰ one may consider only those H_i, for which $\rho(x_0, H_i) < \rho(x_0, H_m)$.

In the case $X^n = \Lambda^n$ one may take for x_0 an improper point possessing the compactness property with respect to Γ. The Algorithm remains in force if we replace $\rho(x_0, H) \overset{\text{def}}{=} \min_{x \in H} \rho(x_0, x)$ by

$b(x_0, H) \overset{\text{def}}{=} \min_{x \in H} b(x_0, x)$, where $b(x_0, x)$ is a positive function satisfying the following conditions:

(a) for each motion φ of space Λ^n leaving x_0 fixed, there exists such a number $c > 0$ that

$$b(x_0, \varphi x) = c\, b(x_0, x)$$

for all $x \in \Lambda^n$;

(b) $x_p \to x_0$, $x_p \in \Lambda^n$, implies $b(x_0, x_p) \to 0$.

(Explicit formulas for $b(x_0, x)$ and $b(x_0, H)$ see in 2.1).

Finding the Algorithm is not much longer than its formulation. Obviously $P \subset \bigcap H_i^-$. It is sufficient to prove that each of the H_i bounds P. Let m be the smallest index, for which H_m does not bound P, and let x be the point of H_m nearest to x_0. It is easy to prove

LEMMA 1.4. *Let $\{H_i^-\}$ be a set of halfspaces, each two of them being opposite. If $x_0 \in \bigcap H_i^-$ and x is the point of H_m nearest to x_0, then $x \in \bigcap H_i^-$. Moreover, $x \in H_i$, $i \neq m$, implies $x_0 \in H_i$.*

Thus $x \in \bigcap H_i^-$. From the rule 2^0 of choosing H_m we can deduce that neither hyperplane bounding P separates x and x_0. Furthermore if a hyperplane bounding P contains x, then it is one of the hyperplanes $H_1, \ldots, H_k$. Hence P contains some neighbourhood of x in P_0. This is evidently impossible.

If the procedure by the Algorithm is finite, how to recognize its end? This problem is of peculiar interest for the case $X^n = \Lambda^n$.

We shall discuss it in 2.4.

2. The Gram matrix of a convex polyhedron in Lobaĉevskiĭ space.

1. A MODEL OF LOBAĈEVSKIĬ SPACE. Let $E^{n,1}$ be an $(n+1)$-dimensional vector space with scalar multiplication of signature $(n, 1)$. We have

$$\{v \in E^{n,1} : (v, v) < 0\} = \mathfrak{C} \cup (-\mathfrak{C}),$$

where $\mathfrak{C}$ is an open convex cone. Let us denote by $\mathbf{P}$ the group of positive real numbers. Then we identify n-dimensional Lobaĉevskiĭ space Λ^n with $\mathfrak{C}/\mathbf{P}$ in such a way that motions of Λ^n are induced by orthogonal transformations of $E^{n,1}$ preserving $\mathfrak{C}$.

If we consider "projective sphere" $PSE^{n,1} = (E^{n,1}\backslash\{0\})/\mathbf{P}$, then by definition $\Lambda^n \subset PSE^{n,1}$. The closure $\bar{\Lambda}^n$ of Λ^n in $PSE^{n,1}$ is the natural compactification of Λ^n. Points of $\bar{\Lambda}^n \backslash \Lambda^n$ are called improper points of Λ^n.

We shall denote by π the natural mapping

$$\pi: E^{n,1} \to PSE^{n,1}.$$

The distance $\rho(x_0, x)$ between two points $x_0 = \pi(v_0)$ and $x = \pi(v)$ of space Λ^n is defined from the formula

$$\text{ch } \rho(x_0, x) = -(v_0, v), \tag{2.1}$$

v_0 and v being normed in such a way that $(v_0, v_0) = (v, v) = -1$. For an improper point $x_0 = \pi(v_0)$, $(v_0, v_0) = 0$, the function $\delta(x_0, x)$ mentioned in 1.4 may be also defined from the formula (2.1), with replacing ρ by b, v being normed as above and v_0 being chosen arbitrarily (so b depends on the choice of v_0).

Any hyperplane of space Λ^n is of the form

$$H_e = \{ \pi(x) \colon x \in \mathfrak{C}, \ (x, e) = 0 \}.$$

where $e \in E^{n,1}$, $(e, e) > 0$. For $x_0 = \pi(v_0) \in \Lambda^n$, $(v_0, v_0) = -1$, the distance $\rho(x_0, H_0)$ is defined from the formula

$$\text{sh } \rho(x_0, H_e) = |(v_0, e)|, \tag{2.2}$$

e being normed in such a way that $(e, e) = 1$.

For an improper point $x_0 = \pi(v_0)$, $(v_0, v_0) = 0$, this formula holds good, if we replace ρ by b.

Let us suppose $(e, e) = (f, f) = 1$, $f \neq \pm e$. Then H_e and H_f are crossing (resp. parallel, diverging) if and only if $|(e, f)| < 1$ (resp. $= 1$, > 1). If $|(e, f)| < 1$, then the angle α (H_e, H_f) between H_e and H_f is determined from the formula

$$\cos \alpha (H_e, H_f) = |(e, f)|.$$

If $|(e, f)| > 1$, then the distance $\rho(H_e, H_f)$ between H_e and H_f is determined from the formula

$$\text{ch } \rho(H_e, H_f) = |(e, f)|.$$

For any $e \in E^{n,1}$ such that $(e, e) > 0$ we put

$$H_e^- = \{ \pi(x) : x \in \mathfrak{C}, (x, e) \leqslant 0 \}. \tag{2.3}$$

This is a halfspace bounded by H_e. The halfspaces H_e^- and H_f^- are opposite (see 1.4) if and only if $(e, f) \leqslant 0$.

The reflection with respect to H_e, in space Λ^n is induced by orthogonal reflection R_e in $E^{n,1}$, which is written by the formula

$$R_e v = v - \frac{2(v, e)}{(e, e)} e. \tag{2.4}$$

2. Let P be a convex polyhedron in Λ^n. Suppose that

$$P = \bigcap_{i \in I} H_i^-, \tag{2.5}$$

each of the H_i^- being essential. Take vectors $e_i \in E^{n,1}$ $(i \in I)$ such that

$$H_i^- = H_{e_i}^-, \ (e_i, e_i) = 1.$$

Then the Gram matrix of the set $\{e_i : i \in I\}$ will be called the Gram matrix of the polyhedron P and will be denoted by $G(P)$.

It is known ([9], [12]) that if all the dihedral angles of P do not exceed $\pi/2$, then H_i^- and H_j^- are opposite for all i, j $(i \neq j)$, so all the non-diagonal elements of $G(P)$ are non-positive.

The polyhedron P will be called non-degenerate, if

(1) the H_i have no common point, proper or improper ;

(2) there exists no hyperplane, orthogonal to each H_i.

It is easy to see, that these conditions are equivalent to the strict convexity of the cone

$$K = \{ v \in E^{n,1} : (v, e_i) \leqslant 0 \text{ for all } i \in I \}.$$

If P is non-degenerate, then $G(P)$ is a symmetric matrix of signature $(n, 1)$.

Conversely, one can prove

THEOREM 2.2. ([10], [12]). *Let $G = (g_{ij})$ be a symmetric matrix of signature $(n, 1)$ with*

$$g_{ii} = 1, \ g_{ij} \leqslant 0 \ (i \neq j).$$

Then G is the Gram matrix of a convex polyhedron in Λ^n determined uniquely up to a congruence.

THE CONNECTION BETWEEN THE GRAM MATRIX AND THE COXETER'S DIAGRAM. Let P be a cell of a d.r.g. Γ in Λ^n, and $G(P) = (g_{ij})$ be its Gram matrix. It is clear from the definition of the Coxeter's diagram $\Sigma\,(\Gamma)$ that

if the vertices v_i and v_j of $\Sigma\,(\Gamma)$ are joined	then
by a m-tuple line	$g_{ij} = -\cos \dfrac{\pi}{m+2}$
by a thick line	$g_{ij} = -1$
by a dotted line	$g_{ij} < -1$

(In particular, if v_i and v_j are not joined, then $g_{ij} = 0$.)

Thus the Gram matrix can be reconstructed after the Coxeter's diagram modulo elements corresponding to dotted lines.

If the Coxeter's diagram contains no dotted lines, then the Gram matrix coincides with the cosines matrix and is completely determined by the Coxeter's diagram.

Note that, if P is bounded, $G(P)$ is always completely determined by $\Sigma(\Gamma)$ [7].

3. From now on, we shall deal only with finite polyhedra, i.e. polyhedra with a finite number of faces.

Let P be a convex polyhedron in Λ^n. We shall assume that P is defined by the formula (2.5) with a finite set I.

An improper point q of space Λ^n will be called an improper vertex of P, if $q \in \overline{P}$ and the intersection of P with any orysphere with center in q is compact.

For an improper point q and a subset M of Λ^n we agree to write $q \subset M$ if $q \in \overline{M}$.

Let F be a proper face or improper vertex of P. We define the subset $b(F)$ of I as follows :

$$b(F) = \{i \in \mathrm{I} : F \subset H_i\}.$$

Obviously F is determined by $b(F)$. More generally,

$$F_1 \subset F_2 \Leftrightarrow b(F_1) \supset b(F_2).$$

The family of the subsets $b(F)$ describe the combinational structure of P.

We denote by $G = (g_{ij})$ the Gram matrix of P and by G_S for any subset S of I, its principal submatrix formed by elements g_{ij} with $i, j \in S$.

It is remarkable that, if all the dihedral angles of P do not exceed $\pi/2$, the property of a subset S of I being one of the $b(F)$ depends only on the matrix G_S. For the formulation of this result we need

SOME DEFINITIONS. A square matrix A is said to be a direct sum of matrices $A_1, \dots, A_k$ (write : $A = A_1 \oplus \dots \oplus A_k$), if A may be reduced to the form $\begin{pmatrix} A_1 & & 0 \\ & & \\ 0 & & A_k \end{pmatrix}$ by a suitable transposition of the rows and the same transposition of the columns. A matrix A is called indecomposable if it is not a direct sum of two matrices.

Any matrix may be uniquely represented as a direct sum of indecomposable matrices, called its indecomposable components.

Let A be a symmetric matrix, all non-diagonal elements of which are non-positive. We shall say the matrix A is parabolic if all its indecomposable components are degenerate non-negative definite.

Now we can formulate

THEOREM 3.3. [12]. *Let all the dihedral angles of P not exceed $\pi/2$. Then for a subset S of* I *being the $b(F)$, where F is a k-dimensional proper face (resp. an improper vertex) of P, it is necessary and sufficient that G_S be a positive definite matrix of rank $n-k$ (resp. a parabolic matrix of rank $n-1$).*

Note that the rank of a parabolic matrix is equal to its order minus the number of its indecomposable components [1].

In the case when P is a cell of a d.r.g. Γ, the matrix G_S is positive definite (resp. parabolic) if and only if the corresponding subdiagram of the Coxeter's diagram $\Sigma(\Gamma)$ is elliptic (resp. parabolic) in the sense of 1.2. The rank of G_S coincides with the rank of the corresponding subdiagram. If $S = b(F)$, the subgroup Γ_S of Γ, generated by reflections R_i with $i \in S$, is the stable subgroup of any interior point of F.

4. A convex polyhedron in Λ^n is of finite volume if and only if it is the convex hull of a finite number of points, proper or improper.

LEMMA 2.4. *Let P be a convex polyhedron of finite volume in Λ^n. Further let P' be a convex polyhedron with the following properties :*

(1) $P' \subset P$;

(2) *each hyperplane bounding P bounds P';*

(3) *all the dihedral angles of P' do not exceed $\pi/2$.*

Then $P' = P$.

It is sufficient to prove that any vertex of P is a vertex of P'. This follows from Theorem 2.3, because $G(P)$ is by the condition a principal submatrix of $G(P')$.

Now we can complement the Algorithm described in 1.4. by

A SUFFICIENT CONDITION FOR THE TERMINATION OF THE PROCEDURE BY THE ALGORITHM: "If for some m the polyhedron

$$P^{(m)} = \bigcap_{i=1}^{m} H_i^-$$

is of finite volume, then $P^{(m)} = P$."

Indeed, suppose that there is some H_{m+1}. Then we may apply the Lemma above to $P^{(m)}$ and $P^{(m+1)}$, which leads to a contradiction.

5. We need to consider some special types of matrices.

A symmetric matrix with non-positive elements beyond the diagonal will be called critical if it is not positive definite but its every proper principal submatrix is such. It is clear that any critical matrix is indecomposable.

There are two types of critical matrices.

FIRST TYPE: *Non-negative definite critical matrices.* These are the same as indecomposable parabolic matrices. The diagonal elements being equal to 1, a symmetric matrix is a critical matrix of the first type if and only if it is the Gram matrix of a simplex in E^n.

SECOND TYPE: *Non-definite critical matrices.* The diagonal elements being equal to 1 and the order being $\geqslant 3$, a symmetric matrix is a critical matrix of the second type if and only if it is Gram matrix of a simplex in Λ^n. This follows from Theorems 2.2 and 2.3.

Let Γ be some d.r.g., and G be the Gram matrix of a Γ-cell. The matrix G is critical in the following three cases :

(1) $\Sigma(\Gamma)$ is a connected parabolic diagram, i.e. is found in Table 2;

(2) $\Sigma(\Gamma)$ is a Lanner's diagram, i.e. is found in Table 3;

(3) $\Sigma(\Gamma)$ is the diagram ○---○.

In the first case P is non-negative definite, in two other cases — non-definite.

6. Let P be a non-degenerate convex polyhedron in Λ^n, defined by means of its Gram matrix G. Under some additional assumptions, we shall give below a simple criterion for P to have finite volume.

In the notations of 2.2, we put

$$P^c = \pi(K) \subset PSE^{n,1}.$$

This is a convex polyhedron in projective sphere $PSE^{n,1}$. Any k-dimensional face of P is a part of a unique k-dimensional face of P^c. From the assumption of non-degeneracy of P it follows that P^c is the convex hull of its vertices. We shall call P^c the completion of P.

Obviously,

P is bounded $\Leftrightarrow P^c \subset \Lambda^n$,

P has finite volume $\Leftrightarrow P^c \subset \overline{\Lambda}^n$.

THEOREM 2.6. *Let G contain no non-definite critical principal submatrices. Then P is of finite volume if and only if every indecomposable parabolic principal submatrix of G is an indecomposable component of some parabolic principal submatrix of rank $n-1$.*

PROOF OF THE "IF" PART. It is sufficient to show that any vertex of P^c lies in $\overline{\Lambda}^n$. Let $q = \pi(e_0)$, $e_0 \in E^{n,1}$, be a vertex of P^c, which is not a proper vertex of P, i.e. does not belong to Λ^n. Put

$$S = \{i \in \mathrm{I} : (e_0, e_i) = 0\}.$$

Then the matrix G_S is not positive definite and contains a critical principal submatrix, say G_T $(T \subset S)$. By our assumptions G_T is parabolic and is a component of a parabolic principal submatrix of rank $n-1$, say $G_{\tilde{S}}(\tilde{S} \supset T)$. By Theorem 2.3, $\tilde{S} = b(\tilde{q})$ where $\tilde{q}$ is an improper vertex of P.

Let us consider the face

$$F = \{\pi(v) : v \in K, (v, e_i) = 0 \ \forall \ i \in T\}$$

of P^c. Evidently $F \ni q, \tilde{q}$. If $F \neq \tilde{q}$, then $F \cap \Lambda^n$ is a proper face of P, and G_T is positive definite, which is not true. Hence

$$F = \tilde{q} = q \in \overline{\Lambda}^n.$$

PROOF OF THE "ONLY IF" PART. Let P be of finite volume, and let S be a subset of I such that G_S is an indecomposable parabolic matrix. It is known [1] that the rows of G obey a linear dependence with positive coefficients a_i, $i \in S$. Put

$$e_0 = \sum_{i \in S} a_i e_i.$$

Then $(e_0, e_0) = 0$ and $e_0 \in K$. Further $e_0 \neq 0$, or else $(v, e_i) = 0$ for all $v \in K$, $i \in S$, which is impossible. Hence $q = \pi(e_0)$ is an improper vertex of P.

By Theorem 2. 3, $G_{b(q)}$ is a parabolic matrix of rank $n-1$. On the other hand, it is clear from the definition of e_0 that $S \subset b(q)$. Hence G_S is a component of $G_{b(q)}$.

THE SPECIALIZATION TO THE CASE, WHERE P IS A Γ-CELL. Let Γ be a d.r.g. in Λ^n and P be its cell. Suppose that P is non-degenerate. Then the theorem above may be formulated in terms of the Coxeter's diagram $\Sigma(\Gamma)$ as follows.

THEOREM 2.6 BIS. *Let $\Sigma(\Gamma)$ contain no dotted lines and Lanner's subdiagrams. Then P is of finite volume if and only if every connected* *parabolic subdiagram of $\Sigma(\Gamma)$ is a connected component of some parabolic subdiagram of rank $n-1$.*

3. The groups of units of unimodular integral quadratic forms of signature $(n, 1)$.

1. Let $f(x) = \Sigma\, a_{ij}\, x_i\, x_j$ be a non-degenerate integral quadratic form of signature (p, q). Then in the pseudo-euclidean space $E^{p,q}$ there exists such a basis $\{u_i\}$ that $(u_i, u_j) = a_{ij}$. The group L generated by the u_i is a lattice in $E^{p,q}$. We shall call L the lattice corresponding to the form f.

The group $\mathcal{O}(f, \mathbf{Z})$ of integral automorphisms of the form f is naturally identified with the group $\mathcal{O}(L)$ of orthogonal transformations of $E^{p,q}$ preserving L.

Let e be a primitive non-isotropic vector of L. The reflection R_e defined by the formula (2.4) preserves L if and only if $f(e) \mid 2(v, e)$ for all $v \in L$. If f is unimodular, this is equivalent to the condition $|f(e)| = 1$ or 2.

2. From now on,

$$f(x) = \sum_{i,j=0}^{n} a_{ij}\, x_i\, x_j$$

is a unimodular integral quadratic form of signature $(n, 1)$ and L is the corresponding lattice in the space $E^{n,1}$. By definition, L has a basis $\{u_0, u_1, \dots, u_n\}$ for which

$$(u_i, u_j) = a_{ij}.$$

We have

$$\mathcal{O}(L) = \{1, -1\} \times \Theta,$$

where Θ is the subgroup consisting of transformations preserving $\mathfrak{C}$ (see 2.1). The group Θ may be considered as a discrete group of motions of Lobačevskiĭ space Λ^n. It is known [2] that its fundamental domain is of finite volume and may be chosen as a finite polyhedron.

Reflections (in the sense of Lobačevskiĭ geometry) belonging to Θ are exactly the reflections R_e, where e runs over all the primitive vectors of L satisfying the condition

$$f(e) = 1 \text{ or } 2. \tag{3.1}$$

We have a decomposition (1.1) for Θ. To determine a Γ-cell P one may use the Algorithm described in 1.4. Let $e_1, e_2, \ldots$ be such primitive vectors of L that

$$H_i^- = H_{e_i}^-$$

(see (2.3)). Then the e_i are solutions of (3.1) such that $(v_0, e_i) \leqslant 0$.

Thus by the Algorithm we have to look for solutions of the Diophantine equation (3.1) such that

$$(v_0, e) \leqslant 0.$$

These solutions must be ordered by increase of the value

$$\nu(e) = \frac{(v_0, e)^2}{f(e)}$$

(see (2.2)). A solution e is included in the sequence $\{e_1, e_2, \ldots\}$ if and only if it satisfies the condition

$$(e, e_i) \leqslant 0$$

for all terms e_i of this sequence which have been already constructed.

The vector v_0 may be chosen arbitrarily in $\mathfrak{C}\backslash\{0\}$. The first vectors $e_1, e_2, \ldots, e_k$ of the sequence $\{e_1, e_2, \ldots\}$ are chosen in such a way that the cone

$$K_0 = \{v \in E^{n,1} : (v, e_i) \leqslant 0 \text{ for } i = 1, \ldots, k\}$$

be a cell for the group Γ_0 generated by all reflections $R_e \in \Theta$ with $(v_0, e) = 0$.

To recognize the end of the procedure we may use the sufficient condition given in 2.4, combining it with Theorem 2.6 bis.

It is easy to show that the following three conditions are equivalent :

(1) $[\Theta : \Gamma] < \infty$

(2) the sequence $\{e_1, e_2, \ldots\}$ is finite;

(3) for some m the volume of $P^{(m)}$ is finite.

3. First we shall consider the case, when f is odd. One may assume that

$$f(x) = -x_0^2 + x_1^2 + \ldots + x_n^2$$

(see the introduction).

Take $v_0 = u_0$. Then Γ_0 consists of all transpositions of $u_1, \dots, u_n$ combined with multiplications by ± 1. One may put

$$K_0 = \{v = \sum_{i=0}^{n} x_i u_i : u_1 \geqslant u_2 \geqslant \dots \geqslant u_n \geqslant 0\}.$$

Hence

$$e_i = -u_i + u_{i+1} (i = 1, \dots, n-1), \; e_n = -u_n.$$

The following terms of the sequence $\{e_1, e_2, \dots\}$ must be of the form $e = \sum_{i=0}^{n} x_i u_i$ where

$$x_0 > 0, \; x_1 \geqslant x_2 \geqslant \dots \geqslant x_n \geqslant 0,$$
$$x_1^2 + \dots + x_n^2 = x_0^2 + \epsilon, \; \epsilon = 1 \text{ or } 2.$$

They must be ordered by increase of $\frac{x_0^2}{\epsilon}$.

For $n \leqslant 17$ the procedure is finite. The vectors e_i with $i > n$ are given in the following table.

i	e_i	ϵ	for which n	$\frac{x_0^2}{\epsilon}$
$n+1$	$u_0 + u_1 + u_2$	1	$n = 2$	1
	$u_0 + u_1 + u_2 + u_3$	2	$n \geqslant 3$	0,5
$n+2$	$3u_0 + u_1 + \dots + u_{10}$	1	$n = 10$	9
	$3u_0 + u_1 + \dots + u_{11}$	2	$n \geqslant 11$	4,5
$n+3$	$4u_0 + 2u_1 + u_2 + \dots + u_{14}$	1	$n = 14$	16
	$4u_0 + 2u_1 + u_2 + \dots + u_{15}$	2	$n \geqslant 15$	8
$n+4$	$4u_0 + u_1 + \dots + u_{17}$	1	$n = 17$	16
	$6u_0 + 2(u_1 + \dots + u_7) + u_8 + \dots + u_{16}$	1	$n = 16$	36
$n+5$	$6u_0 + 2(u_1 + \dots + u_7) + u_8 + \dots + u_{17}$	2	$n = 17$	18

The corresponding Coxeter's diagrams are given in Table 4A. They all satisfy the conditions of Theorem 2.6 bis.

In each of these cases, the group H in the decomposition (1.1) coincides with the group Sym $P \simeq$ Sym $\Sigma(\Gamma)$. This follows from the following facts, which are seen immediately:

(1) any symmetry of $\Sigma(\Gamma)$ induces a permutation of $e_1, \dots, e_m$ preserving the lengths;

(2) $e_1, \dots, e_m$ generate L.

4. In the case, where f is even (and hence $n = 8k + 1$) we shall deal otherwise.

For $n = 9$ and 17 we consider two diagrams of Table 4B. Denote them by Σ' and Σ''. The corresponding cosines matrix we denote by G' and G'' respectively.

We see immediately that

(1) Σ' and Σ'' are not elliptic or parabolic diagrams;

(2) they contain elliptic subdiagrams of rank 8 and 16 respectively;

(3) $\det G'' = 0$.

Hence G' and G'' are of signature (8,1) and (16,1) respectively. By Theorem 2.2 they are the Gram matrices of convex polyhedra $P' \subset \Lambda^9$ and $P'' \subset \Lambda^{17}$ respectively.

The diagrams Σ' and Σ'' satisfy the conditions of Theorem 2.6 bis. Hence P' and P'' are of finite volume. By means of Theorem 2.3 it is easy to show that

(1) P' is a simplex with a single improper vertex q which is opposite to the face P'_{10};

(2) P'' is a pyramid over the direct product of two 8-dimensional simplexes, with an improper apex q; its base is P''_{19}.

In each of the two considered cases, we define vectors $e_1, \dots, e_m \in E^{n,1}$ as in 2.2, but normed in such a way that $(e_i, e_i) = 2$. Then $(e_i, e_j) \in \mathbf{Z}$ for all i, j. Hence the group L generated by $e_1, \dots, e_m$ is an integral lattice in $E^{n,1}$.

The Gram matrix of the set $\{e_1, \dots, e_m\}$ is $2G'$ or $2G''$ respectively. A computation shows that

$$\det 2G' = \det 2G''_{19} = -1,$$

where G''_{19} is G'' without the last row and the last column. Therefore the lattice L is the lattice corresponding to an even unimodular integral quadratic form f of signature $(n, 1)$.

The group $\mathcal{O}(L)$ is naturally isomorphic to the group of units of the form f.

Obviously, $R_{e_i} \in \mathcal{O}(L)$ for $i = 1, \dots, m$. The group Γ generated by $R_{e_1}, \dots, R_{e_m}$ having a fundamental domain of finite volume in Λ^n, is a subgroup of finite index in $\mathcal{O}(L)$.

Really Γ is the maximal reflection subgroup in $\mathcal{O}(L)$. To prove this we may apply the Algorithm taking for x_0 the improper point q defined above.

As well as in 3.3, one can show that $H = \operatorname{Sym} \Sigma(\Gamma)$, i.e. H is trivial for $n = 9$ and $H \simeq \mathbf{Z}/2\mathbf{Z}$ for $n = 17$.

Thus we have proved Theorem A formulated in the introduction.

5. In the rest of this paragraph we shall prove Theorem B formulated in the introduction.

Let f be an odd unimodular integral quadratic form of signature $(n, 1)$ and L be the corresponding lattice in $E^{n,1}$.

Further let f_0 be an arbitrary unimodular positive definite integral quadratic form on $n-1$ variables and L_0 be the corresponding lattice in E^{n-1}.

We agree to denote by the symbol $\perp$ an orthogonal direct sum of metric vector spaces or lattices.

LEMMA 3.5. *There exists an isometric imbedding $\tau: E^{n-1} \to E^{n,1}$ such that*

$$L = \tau(L_0) \perp M,$$

where M is some 2-dimensional lattice.

To prove this, we consider the lattice M' in $E^{1,1}$ corresponding to the quadratic form

$$2\,y_0 y_1 + y_1^2. \tag{3.2}$$

Obviously the lattice $L_0 \perp M'$ in $E^{n-1} \perp E^{1,1}$ corresponds to an odd unimodular integral quadratic form of signature $(n, 1)$. In view of the integral equivalence of all such forms, there exists an isometry $\tau\colon E^{n-1} \perp E^{1,1} \to E^{n,1}$ such that $\tau(L_0 \perp M') = L$. We put $\tau(M') = M$. Then $\Gamma = \tau(L_0) \perp M$. q. e. d.

Now let us assume that $\mathcal{O}(L)$ contains a reflection subgroup Γ of finite index. Then every improper point of Λ^n possessing the compactness property with respect to $\mathcal{O}(L)$ (see the definition in 1.3) possesses this property with respect to Γ

On the other hand, it is well known (and it is easy to show) that an improper point x of Λ^n possesses the compactness property with respect to $\mathcal{O}(L)$ if and only if it is of the form $x = \pi(v)$ where $v \in L$.

In the notations of the lemma above, take for v any isotropic vector of M. It follows from the compactness property of the point $x = \pi(v)$ that there exists such reflections $R_{e_1}, \dots, R_{e_k} \in \mathcal{O}(L)$ leaving v fixed, that the rank of the Gram matrix of the set $\{e_1, \dots, e_k\}$ is equal to $n-1$. We may suppose that the e_i are primitive vectors of L. Then $f(e_i) = 1$ or 2.

Since $(v, e_i) = 0$, it follows that $e_i = \tau(e_i') + c_i v$, where $e_i' \in L_0$, $c_i \in \mathbf{Q}$ and $(e_i', e_j') = (e_i, e_j)$ for all i, j. In particular $f_0(e_i') = 1$ or 2. Furthermore the Gram matrix of $\{e_i', \dots, e_k'\}$ being of rank $n-1$, the e_i' generate a sublattice of finite index in L_0.

A DEFINITION. *Let f_0 be a unimodular positive definite integral quadratic form and L_0 be the corresponding lattice. We shall call f_0 a reflection form if the vectors $e \in L_0$, for which $f_0(e) = 1$ or 2, generate a sublattice of finite index in L_0.*

In these terms, we have proved

THEOREM 3.5. *If the group of units of an odd unimodular integral quadratic form of signature $(n, 1)$ contains a reflection subgroup of* *finite index, then every unimodular positive definite integral quadratic form on $n-1$ variables is a reflection form.*

COROLLARY. *Every unimodular positive definite integral quadratic form on $\leqslant 16$ variables is a reflection form.*

(Of course, this may be deduced from the classification of such forms due to M. Kneser [4].)

6. Theorem 3.5 has an analogue in the theory of even forms.

THEOREM 3.6. *If the group of units of an even unimodular integral quadratic form of signature $(n, 1)$ contains a reflection subgroup of finite index, then every even unimodular positive definite integral quadratic form on $n-1$ variables is a reflection form.*

The proof is the same as the one of Theorem 3.5, replacing the form (3.2) by $2y_0y_1$.

7. It is known [5] that there exists an even unimodular positive definite integral quadratic form on 24 variables, which does not represent 2. Therefore for $n \geqslant 25$ the group of units of any unimodular integral quadratic form of signature $(n, 1)$ contains no reflection subgroups of finite index. . q. e. d.

TABLE 1. Connected elliptic diagrams. (Lower index is equal to the rank)

$A_n,\ n \geqslant 1$

B_n or $C_n,\ n \geqslant 2$

$D_n,\ n \geqslant 4$

$E_n,\ n = 6, 7, 8$

F_4

$G_2^{(m)},\ m \geqslant 5$

H_3

H_4

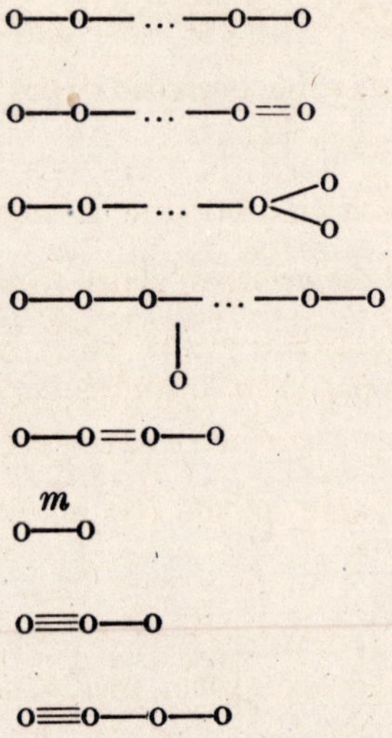

TABLE 2. Connected

(Lower index is

$\tilde{A}_1$

$\tilde{A}_n,\ n \geqslant 2$

$\tilde{B}_n,\ n \geqslant 3$

$\tilde{C}_n,\ n \geqslant 2$

$\tilde{D}_n,\ n \geqslant 4$

$\tilde{E}_6$

$\tilde{E}_7$

$\tilde{E}_8$

$\tilde{F}_4$

$\tilde{G}_2$

TABLE 4. The diagrams of the maximal reflection subgroups of signature $(n, 1), n \leqslant 17$.

A. ODD FORMS.

$2 \leqslant n \leqslant 17.$

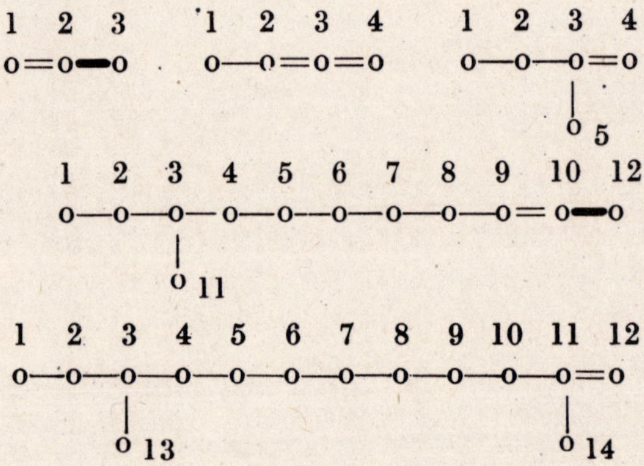

parabolic diagrams

equal to the rank)

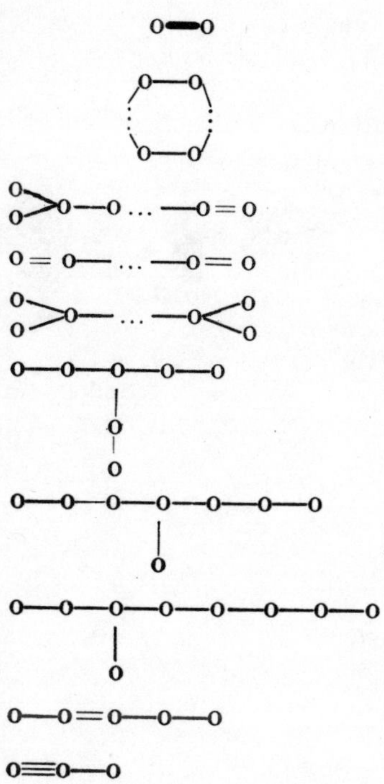

TABLE 3. Lanner's diagrams.
(n denotes the rank)

$n = 2$.

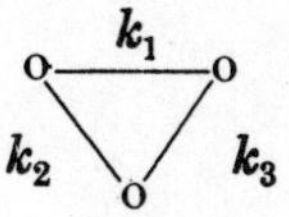

, $\frac{1}{k_1} + \frac{1}{k_2} + \frac{1}{k_3} < 1$

$n = 3$.

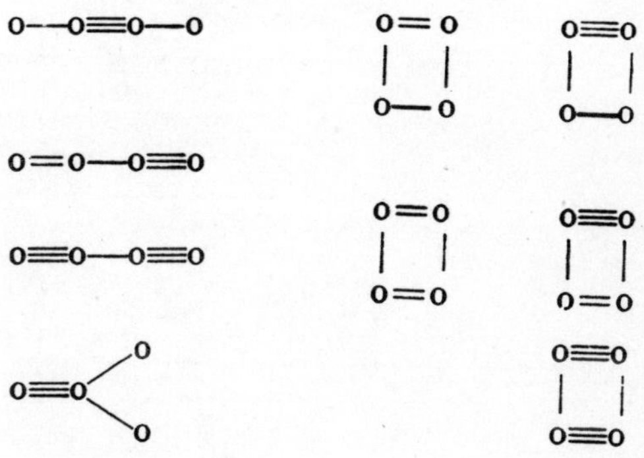

$n = 4$.

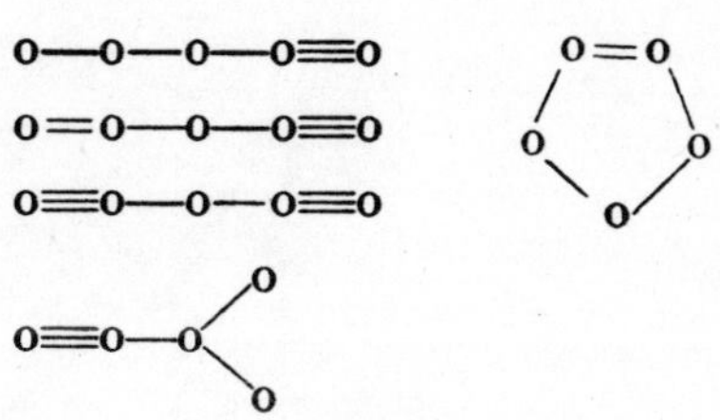

the groups of units of unimodular integral quadratic forms of
(The enumeration of vertices corresponds to the one in the text.)

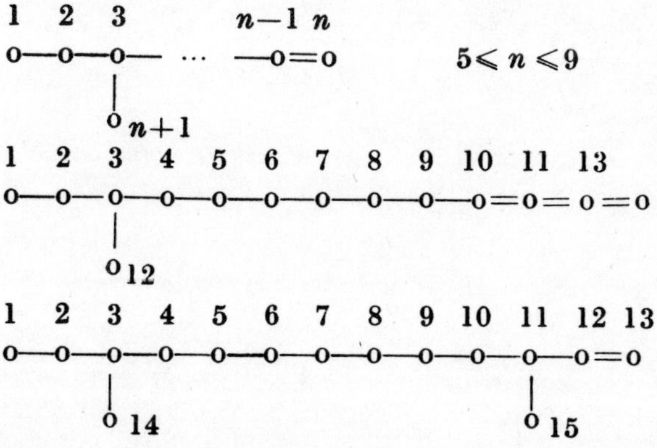

(contd.)

TABLE 4. (contd.)

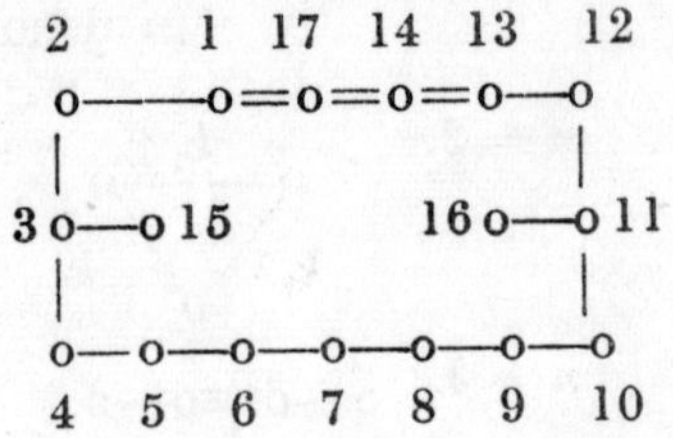

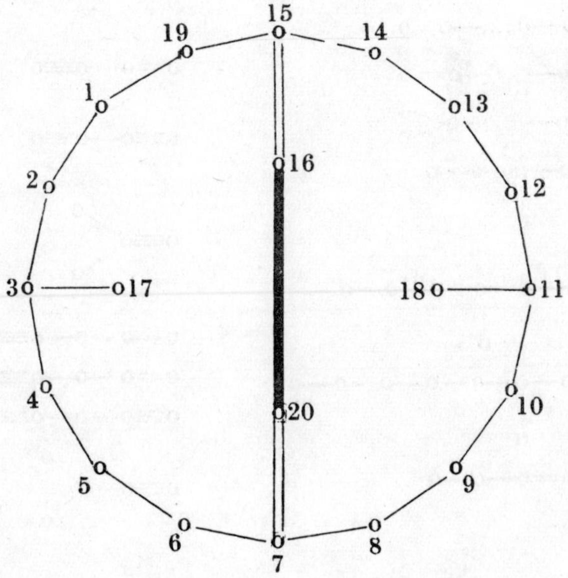

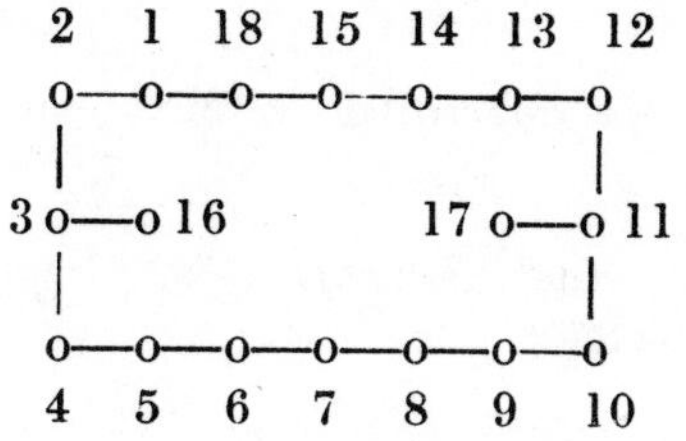

B. Even Forms. $n = 9, 17.$

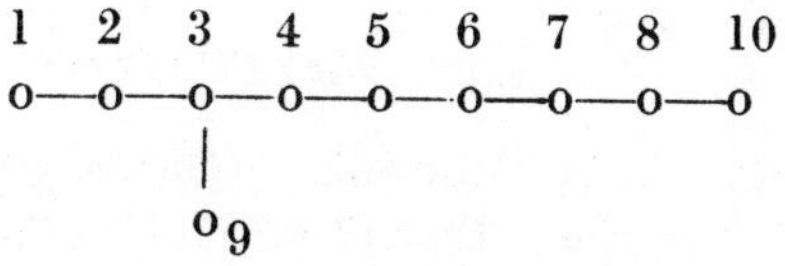

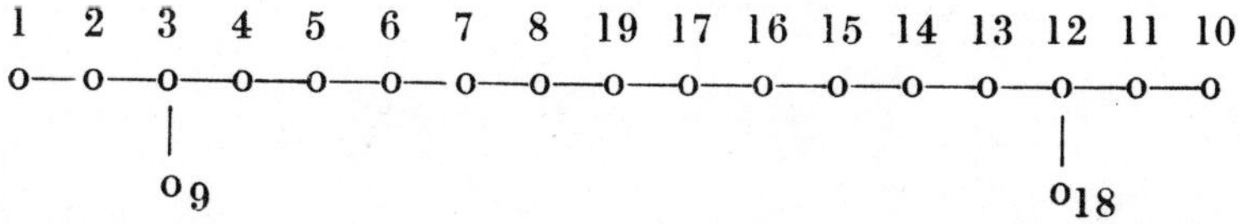

REFERENCES

1. H. S. M. Coxeter : Discrete groups generated by reflections, *Ann. Math.*, 35, (1934), 588-621.

2. C. L. Siegel : Einheiten quadratischer Formen, *Abh. Math. Sem. Univ. Hamburg*, 13, (1940), 209-239.

3. F. Lanner : On complexes with transitive groups of automorphisms, *Lunds Univ. Math. Sem.*, 11 (1950).

4. M. Kneser : Klassenzahlen definiter quadratischer Formen, *Arch. Math.*, 8, (1957), 241-250.

5. J. H. Conway : A characterization of Leech's lattice, *Invent. Math.*, 7, (1969), 137-142.

6. J.-P. Serre : *Cours d'arithmetique*, Presses Universitaires de France, Paris, (1970).

7. E. M. Andreev : On convex polyhedra in Lobaĉevskiĭ spaces, *Mat. Sbornik*, 81, (1970), 445-478.

8. E. M. Andreev : On convex polyhedra of finite volume in Lobaĉevskiĭ spaces, *Mat, Sbornik*, 83, (1970), 256-260.

9. E. M. Andreev : On the intersection of the planes of faces of polyhedra with acute angles, *Mat. Zametki*, 8, (1970), 521-527.

10. È. B. Vinberg : Discrete groups generated by reflections in Lobaĉevskiĭ spaces, *Mat. Sbornik*, 72, (1967), 471-488.

11. È. B. Vinberg : Some examples of crystallographic groups in Lobaĉevskiĭ spaces, *Mat. Sbornik*, 78, (1969), 633-639.

12. È. B. Vinberg : Discrete linear groups generated by reflections, *Izv. Akad. Nauk SSSR, ser. mat.*, 35, (1971), 1072-1112.

13. È. B. Vinberg : On the groups of units of some quadratic forms, *Mat. Sbornik*, 87 (1972), 18-36.

14. È. B. Vinberg : On unimodular integral quadratic forms, *Funkc. Analiz*, 6 (1972), 25-32.